Lecture Notes in Mathematics

Edited by A. Dold and B. Eckmann

810

Geometrical Approaches to Differential Equations

Proceedings of the Fourth Scheveningen Conference on Differential Equations, The Netherlands
August 26 – 31, 1979

Edited by R. Martini

Springer-Verlag
Berlin Heidelberg New York 1980

Editor

Rodolfo Martini
Twente University of Technology, Department of Applied Mathematics
Postbus 217
Enschede
The Netherlands

AMS Subject Classifications (1980): 35-XX

ISBN 3-540-10018-0 Springer-Verlag Berlin Heidelberg New York
ISBN 0-387-10018-0 Springer-Verlag New York Heidelberg Berlin

Library of Congress Cataloging in Publication Data. Scheveningen Conference on Differential Equations, 4th, 1979. Geometrical approaches to differential equations. (Lecture notes in mathematics; 810) Bibliography: p. Includes index. 1. Differential equations, Partial--Congresses. 2. Geometry, Differential--Congresses. I. Martini, Rodolfo, 1943- II. Title. III. Series: Lecture notes in mathematics (Berlin); 810. QA3.L28 no. 810 [QA374] 510s [515.3'5] 80-19204

Printed in Germany

Printing and binding: Beltz Offsetdruck, Hemsbach/Bergstr.
2141/3140-543210

P R E F A C E

This volume contains the text of the lectures delivered at the Fourth Scheveningen Conference on Differential Equations.

The conference was organised by B.L.J. Braaksma (University of Groningen), E.M. de Jager (University of Amsterdam), H. Lemei (Delft University of Technology) and R. Martini (Twente University of Technology) and financially supported by the Minister of Education and Sciences of the Netherlands.

Like the three preceding Scheveningen conferences (North-Holland Mathematics Studies, Vols. 13, 21 and 31) the aim of the conference was to bring together a number of research-workers active in a field closely related to differential equations and thus be informed about recent developments and the state of the art.

The emphasis of this conference was on geometrical aspects of differential equations. In connection with this the following topics may be mentioned: prolongation structures, Bäcklund transformations, solitons, Pfaffian systems, structural stability of phase portraits, Hamiltonian systems. However, there were also contributions of different nature.

Let all those who helped make this Conference a success find here an expression of our gratitude, contributors, participants and in particular Professor F.B. Estabrook and Professor J. Corones who have both accepted the invitation to deliver a series of lectures. Finally, the editor wishes to express his gratitude to Springer-Verlag for their most helpful service and effective production of the proceedings.

R. Martini
Enschede, The Netherlands, March, 1980

CONTENTS

Invited papers

Contributed papers

LIST OF PARTICIPANTS

Invited speakers

P.L. Christiansen	Technical University of Denmark, Denmark
J. Corones	Iowa State University, U.S.A.
F.B. Estabrook	California Institute of Technology, U.S.A.
R. Gérard	University of Strasbourg, France
A. Jeffrey	University of NewCastle upon Tyne, England
H.C. Morris	University of Dublin, Ireland
Y. Sibuya	University of Minnesota, U.S.A.

Other participants

B.L.J. Braaksma	University of Groningen
P.J.M. Bongaarts	University of Leiden
L.J.F. Broer	Eindhoven University of Technology
A.H.P. v.d. Burgh	Delft University of Technology
O. Diekman	Mathematical Centre, Amsterdam
A. Dijksma	University of Groningen
M.W. Dingemans	Delft University of Technology
H. ten Eikelder	Eindhoven University of Technology
L. Frank	University of Nijmegen
J.A. van Gelderen	Delft University of Technology
S.A. van Gils	University of Amsterdam
L. Gitter	Ramat-Gan, Israel
J. de Graaf	Eindhoven University of Technology
P. de Groen	University of Brussel, Belgium
Br. van Groesen	University of Nijmegen
R.J.P. Groothuizen	University of Amsterdam
G. Halvorsen	Institute of Mathematics, Trondheim, Norway
A. van Harten	University of Utrecht
D. Hilhorst	Mathematical Centre, Amsterdam
F.J. Jacobs	Kon. Shell, Rijswijk, Zuid-Holland
E.M. de Jager	University of Amsterdam
P. Jonker	Twente University of Technology
T.H. Koornwinder	Mathematical Centre, Amsterdam
H.A. Lauwerier	University of Amsterdam
H. Lemei	Delft University of Technology
W. Lentink	University of Utrecht
H. Majima	University of Strasbourg, France
R. Martini	Twente University of Technology
G.Y. Nieuwland	Free University, Amsterdam
J.P. Pauwelussen	Mathematical Centre, Amsterdam
H.G.J. Pijls	University of Amsterdam
J.W. Reyn	Delft University of Technology
H. Rijnks	Delft University of Technology
B. Sagraloff	University of Regensburg, Germany
R. Schäfke	University of Essen, Germany
D. Schmidt	University of Essen, Germany
P.C. Schuur	University of Utrecht
N.M. Temme	Mathematical Centre, Amsterdam
F. Twilt	Twente University of Technology

E.J.M. Veling	Mathematical Centre, Amsterdam
G.K. Verboom	Delft University of Technology
F. Verhulst	University of Utrecht
W. Wesselius	Twente University of Technology
J.H. Wevers	Twente University of Technology
P. Wilders	University of Amsterdam.

DIFFERENTIAL GEOMETRY AS A TOOL FOR APPLIED MATHEMATICIANS

Frank B. Estabrook
Jet Propulsion Laboratory
California Institute of Technology
Pasadena, CA 91103/USA

I. Introduction

The concepts of differential geometry in the style of É. Cartan can be intuitively understood without a deep knowledge of modern results, and then one has a tool for some systematic applied mathematics. The practical advantage of Cartan's notation is that it is manipulative--one learns to do, almost automatically, local algebraic differential (and integral) operations that lead to meaningful (covariant) results, and even better one is prevented from attempting essentially empty or fruitless games. Integrability conditions for sets of partial differential equations are, again almost automatically, included in one's analysis. With these really quite simple techniques a number of groups of researchers have recently obtained very interesting results for classes of nonlinear partial differential equations. I will in these lectures survey the manipulative techniques and attempt to justify, or at least make plausible (intuitive), the geometrical concepts behind the manipulative operations. But the primary concern of the student should be to do some manipulation himself, for only by doing exterior calculus does one become truly reconciled, or in some cases even addicted, to it.

The topics we will discuss are Cartan's local criteria of integrability of ideals of exterior forms, and the use of associated vectors and forms in Hamiltonian theory, variational calculus, invariance groups, Cauchy characteristics, prolongation and Bäcklund correspondence. The interested student will need to consult the references for many details and applications.

No doubt all sorts of other topics--singular solutions of partial differential equations (p.d.e.'s), bifurcations, boundary conditions, and so on--can be treated beautifully and deeply by differential geometry, but I have neither the experience, nor indeed in this brief course the time, even to mention them.

II. Vectors and Forms

It is especially appropriate that I acknowledge at this school in Scheveningen my great indebtedness to the Dutch mathematician, J. A. Schouten, whose splended book on the "Ricci-Calculus"[1] has guided me for years. From standard tensor texts on Riemannian geometry, I came to forms and p.d.e.'s through Schouten and Cartan.[2] Although I have more recently gotten used to omitting coordinate indices (which I suspect Schouten would not approve of!), this is only practical when (without a metric tensor, for example) one deals with limited classes of tensor objects. All the formulae we will use are to be found in Schouten, proved and justified in general, in coordinate language! Recourse to coordinates is always underneath our understanding of differential geometry, and seems sometimes absolutely necessary. It is the essence of applied mathematics that at some point we introduce coordinates, and solve for something!

The basic objects of differential geometry are vectors and 1-forms, in the older terminology called contravariant or covariant vectors, respectively. For vectors I will write V, W, etc., meaning objects which can be described by arrays of coordinate components V^i, W^i, etc., $i = 1 \dots n$, where n is the dimensionality of the manifold. For 1-forms I will write ω, σ, etc., meaning objects which can be described by arrays of components ω_i, σ_i, etc. Each of these will be smoothly varying fields, but at a point they are objects in associated linear vector spaces, and can be multiplied by scalars, and added to its own kind, component by component in any coordinate frame.

The linear vector space of vectors at a point can be intuitively imagined as displacements in a small tangent copy of the differentiable manifold. We see V + W as adding directed line segments

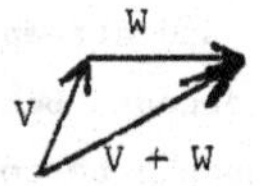

Clearly vectors have orientation and magnitude, relative to any coordinate frame.

The prototypical 1-form at a point is the set of coordinate components of the gradient of a function (scalar field), say $f = f(x^1, \dots x^n)$. $x^1 \dots x^n$ are coordinates and so $f_{,i} = \partial f/\partial x^i$ is the gradient. We write $f_{,i}$ as df. Schouten emphasizes how this object also has orientation and magnitude--his picture in the local coordinate frame uses two level surfaces of f (of dimension n-1): say $f = f_o$ and $f = f_o + 1$:

The vector addition law for 1-forms follows by considering the level surfaces of the sum of f + g:

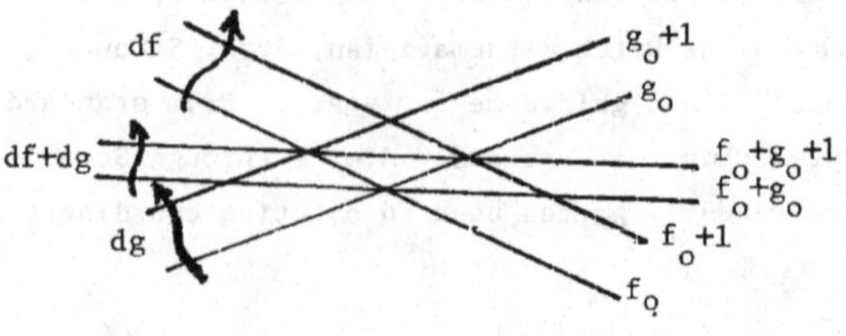

The operation + between two 1-forms is again commutative and associative. The notation df should not be thought to indicate that the 1-form is infinitesimal. dx^1, for example, simply expresses the orientation and magnitude (or spacing) of the level surfaces of the x^1 coordinate. But if one takes a small displacement vector whose coordinate components are infinitesimals also (but, from our present standpoint, confusingly) symbolized by $dx^1, \ldots dx^n$, the usual notation for total differential

$$df = f_{,i}\, dx^i \tag{1}$$

is now read as a finite statement that the (covariant) coordinate components of df are $f_{,i}$.

Having thus discussed a prototype gradient 1-form, one then simply writes a general--non-gradient--1-form, say ω, in terms of components as

$$\omega = \omega_i dx^i \tag{2}$$

If at a point we have level surfaces of dimension n-2, their orientation and magnitude of spacing is described by a "2-form" $\omega = \phi\wedge\sigma$ denoted the outer product of the 1-forms ϕ and σ. In coordinate language it is an antisymmetric covariant tensor field $\phi_i\sigma_j - \phi_j\sigma_i$. This particular 2-form could be seen geometrically as it is algebraically very special; for the general 2-form $\omega_{ij} = -\omega_{ji}$ (which would not satisfy $\varepsilon^{ijkl}\omega_{ij}\omega_{kl}$) the picture gets more complicated and isn't of much use. The general 2-form is locally an object in a $\frac{n(n-1)}{2}$ dimensional linear vector space. Similar definitions, as antisymmetric covariant tensors, are given for forms of higher rank. The collection of all p-forms at a point, where p = 0,1,...n, form a Grassmann algebra; the antisymmetric multiplication operation $\wedge$ and addition + satisfy the first three lines of Table 1.

A notation for a similar Grassmann collection of antisymmetric contravariant quantities could of course be set up, but this does not seem to be as useful because differential operations to compare such objects as fields, varying from point to point, are nonlinear--we see an example of this point later. There is such an operation for p-forms, which is the familiar exterior derivative, or curl.

A given vector (field), say V, and 1-form (field), say σ, can be contracted, yielding a scalar (field). In coordinate terms this is of course the invariant or scalar "inner product" $V^i\sigma_i$. Or it can be said that either serves at each point to map the other into the space of reals R (or perhaps C if one wants to use complex components and be more general--we won't). $\sigma(V)$ is one notation--I prefer $V \rfloor \sigma$, or even more simply (for a physicist) $V\cdot\sigma$. The contraction of a vector on a 2-form is a 1-form, etc. In coordinate language there is no ambiguity as to which indices are contracted, in modern index-free notations a convention--say, contraction from-the-left--must be understood: we adopt $V^\sigma\omega_{\sigma\nu} \doteqdot V\cdot\omega$. For 1-forms, $V\cdot(\phi\wedge\sigma) = V\cdot\phi\sigma - V\cdot\sigma\phi$. The various identities are summarized in Table 1.

Table 1. Summary of Vector and Form Manipulation

$$\omega\wedge\sigma = (-1)^{pq}\,\sigma\wedge\omega \qquad (p = \text{rank}\ \omega,\ q = \text{rank}\ \sigma)$$

$$\omega\wedge(\sigma\wedge\tau) = (\omega\wedge\sigma)\wedge\tau$$

$$(\omega+\sigma)\wedge\tau = \omega\wedge\tau + \sigma\wedge\tau$$

$$(V+W)\cdot\omega = V\cdot\omega + W\cdot\omega$$

$$(fV)\cdot\omega = fV\cdot\omega$$

$$V\cdot(\omega\wedge\sigma) = (V\cdot\omega)\wedge\sigma + (-1)^{p}\omega\wedge(V\cdot\sigma)$$

$$d(\omega\wedge\sigma) = d\omega\wedge\sigma + (-1)^{p}\omega\wedge d\sigma$$

$$dd\omega = 0$$

$$dc = 0$$

$$\text{coord. bases:}\quad dx^{i}\ ,\ \frac{\partial}{\partial x^{i}}$$

$$df = f_{,i}\,dx^{i}$$

$$\frac{\partial}{\partial x^{i}}\cdot dx^{j} = \delta^{j}_{i}$$

$$V = V^{i}\frac{\partial}{\partial x^{i}}$$

$$\mathcal{L}_{V} f = V\cdot df$$

$$\mathcal{L}_{V} x^{i} = V\cdot dx^{i} \equiv V^{i}$$

$$\mathcal{L}_{V}\omega = V\cdot d\omega + d(V\cdot\omega)$$

$$\mathcal{L}_{V}(\omega\wedge\sigma) = (\mathcal{L}_{V}\omega)\wedge\sigma + \omega\wedge(\mathcal{L}_{V}\sigma)$$

$$\mathcal{L}_{V}W \equiv [V,W] \equiv -\mathcal{L}_{W}V$$

$$\mathcal{L}_{V}W\cdot\sigma = [V,W]\cdot\sigma + W\cdot\mathcal{L}_{V}\sigma$$

$$\mathcal{L}_{fV}\omega = f\mathcal{L}_{V}\omega + df\wedge(V\cdot\omega)$$

$$[fU,V] = f[U,V] - V\cdot dfU$$

$$[U,[V,W]] + [V,[W,U]] + [W,[U,V]] = 0$$

$$\text{or}\ \mathcal{L}_{U}[V,W] = [\mathcal{L}_{U}V,W] + [V,\mathcal{L}_{U}W]$$

$$\mathcal{L}_{U}\mathcal{L}_{V} - \mathcal{L}_{V}\mathcal{L}_{U} = \mathcal{L}_{[U,V]}$$

$$\int_{v} d\omega = \int_{\partial v}\omega$$

$$\delta\int\omega = \varepsilon\int\mathcal{L}_{V}\omega$$

From a set of n linearly independent 1-forms ϕ^i one can derive a set of n dual vectors V_i by requiring $V_j \cdot \phi^i = \delta^i_j$. All vectors and forms can be expanded on these as bases. If natural 1-forms are used, say dx^i, determined from the level surfaces of scalar fields x^i, one has introduced a coordinate frame. The dual vectors are then conveniently written as $\frac{\partial}{\partial x^j}$. $\frac{\partial}{\partial x^1}$ can be visualized as a vector oriented along the intersection of the level surfaces of $x^2, = ..x^n$, and so on. This points up a caution: although dx^1 has good meaning whatever other scalars beside x^1 are, or are not, adopted as a complete set, $\frac{\partial}{\partial x^1}$ derives most of its meaning not from the choice of x^1 but rather from all the others!

Expanding a vector in a coordinate basis one has $V = V^i \frac{\partial}{\partial x^i}$ (incidentally, this illustrates one of the perennial problems of modern notation, which Schouten would abhore--one must be told what symbols are; above we had a set of vectors V_i, now we have a set of scalars V^i, components of a single vector!). Contracting with this on a 1-form $df = f_{,i} dx^i$ one gets $V \cdot df = V^i f_{,i}$. If V is "a small displacement vector, whose coordinate components are infinitesimals respectively denoted $dx^1, .., dx^n$", we recover the expression of Eq. (1) for the total differential of the scalar field f-- it is, of course, the directional derivative produced by the operator $V \cdot d$. We will avoid the older notations--so-called "total" derivatives are not a very useful concept. We concentrate instead on arrays of partials, or gradients, produced and symbolized by the operation of exterior derivation, d, and on the directional derivative, or Lie derivative, produced by a given vector field V.

III. Differential concomitants

A concomitant is a geometric object derived from the variation of other objects. In coordinate terms, a concomitant is formed from sets of partial derivatives of tensor fields, and is itself again a tensor field. Remarkable, and I suspect poorly understood, examples are to be found in the researches of Schouten and his collaborators.

The most well known and useful concomitants are the exterior derivative, and the Lie derivative, and it is with these two in modern ostensibly coordinate-free form that the exterior calculus deals. We will later on mention only briefly one other example.

The exterior derivative, or generalized curl, is of course the familiar operation denoted by d that, from a given p-form, obtains a p+1-form. Repeated twice it gives identically zero. Applied to a scalar, it yields a 1-form, so when applied to the coordinate fields x^i it yields the exact 1-forms already introduced as (natural) bases, dx^i. A form ω such that $d\omega = 0$ is denoted as closed--if further it can be written as $\omega = d\theta$, ω is called exact. This distinction is global; locally the two concepts are the same, and any closed form p-form ω can be written as $d\theta$, with the p-1-form θ known only up to the exterior derivative of a p-2-form, say, Ψ: $\theta' = \theta + d\Psi$; $\omega = d\theta' = d\theta$; $d\omega = 0$.

The Lie derivative of any tensor field with respect to a given vector field V is a concomitant which correctly generalizes the classical notion of directional (or material) derivative. Indeed a field V is the generator of a 1-parameter group of diffeomorphisms of the differentiable manifold. From any tensor field, a new field of the same tensorial character results. Lie derivative obeys Leibniz' rule. When the Lie derivative with respect to one vector field is applied to a second vector field the derived vector field is antisymmetric and bilinear in the first two, and is often known as a Lie bracket. Lie algebras can be represented in terms of these vector operators on manifolds, and the Jacobi identity appears when one applies a third operation of derivation to a Lie bracket.

The various results that follow from permuting the operations d and $\underset{V}{\mathcal{L}}$ are of course derived in many readily available texts[3][4], and need no further expounding here. They all are in Schouten[1], but scattered. I have attempted to give a more complete introduction in an earlier set of notes.[5] I believe all that needs to be known to pursue all the local operations of the exterior calculus can be written conveniently on one page--see Table 1! Everybody needs to come to personal terms with these manipulations--people may prefer different introductory texts, and build up different algebraic or geometric intuitions, and set-theoretic terminologies, but there is no substitute for actually applying some of these operations to some given fields of forms and vectors, expressed in a basis system, if one wants to be able to use all this as a tool for applied mathematics.

IV. Immersion and submersion

The intuitively easy concept of an m-dimensional "immersed" subspace or submanifold of an n-dimensional space is, in coordinate terms, seen as a map of a set of variables y^μ, $\mu = 1,\ldots m$, onto a larger set x^i, $i = 1,\ldots n$,

$$x^i = f^i(y^\mu) \qquad m \leq n \tag{3}$$

A vector field V lying in a submanifold can, again intuitively rather obviously, be visualized as existing also in the n-dimensional space. In coordinate language $V^\mu \to V^i = V^\mu \frac{\partial x^i}{\partial y^\mu}$. One could imagine n-m additional parameters z^a, $a = m+1,\ldots n$, so that the submanifold is only one of an n-space-filling family,

$$x^i = f^i(y^\mu, z^a). \tag{4}$$

Then, with this additional "rigging," one sees the process as traditional coordinate transformation $\{y^\mu, z^a\} \to \{x^i\}$, with $V^a = 0$. Rigging is not needed, however; the geometrical fact is that the point map ϕ: $y^\mu \to x^i$ also directly maps, or carries along, vector fields. A very important property of Lie commutators of vector fields is that they commute with mapping operations.

Fields of 1-forms, conversely, map with an inverse map, denoted ϕ^*. That is, for example, given a 1-form ω in the larger space of the x^i, with components, say ω_i (so that $\omega = \omega_i dx^i$), one can restrict or section it to the subspace by the unique prescription of components there

$$\omega_\mu = \omega_i \frac{\partial x^i}{\partial y^\mu}, \tag{5}$$

one writes

$$\phi^*: \quad \omega \rightarrow \tilde{\omega} = \omega_\mu dy^\mu \tag{6}$$

It is very important to understand that the relations of exterior algebra, and of exterior differentiation, of forms are preserved under restriction (inverse maps into submanifolds).

The other case of mapping is when $m \geq n$. Now ϕ: $y^\mu \rightarrow x^i$ can be said to be a projection operation; a modern terminology is that it is a submersion. The set of equations

$$x^i = f^i(y^\mu) \qquad m \geq n \tag{7}$$

implicitly describes submanifolds of the larger m-dimensional y^μ space, called fibers, each of which belongs to a point of the n-dimensional base space spanned by x^i. Rigging now would be accomplished by adding m-n equations for variables z^A, $A = 1...m-n$, that vary in the fiber. At any point, a vector V--components V^μ--can be directly mapped to V^i in the base space, as before, but this map must produce the same V^i starting from every point in the fiber over x^i, if a meaningful (unique) vector field is to result in the base space. The condition for this is that, for any vector Y lying in the fiber, [V,Y] must again be a vector lying in the fiber--this condition clearly suffices since any vector lying in the fiber is projected to zero in the base space.

A form in the base space can be lifted--mapped uniquely--into forms in the space of y^μ, by the inverse map, ϕ^*. With arbitrary rigging this can be seen as traditional coordinate transformation of a covariant field ω_i, $\omega_A = 0$ into ω_μ. Such a lifted form has the property that contraction with a Y gives zero. As before, Lie commutator relations between vectors comute with the direct map, when it is allowed, and the exterior algebraic and derivation operations on forms persist under the inverse map.

V. Cartan theory of partial differential equations

Given a set of forms, ω, $= \omega_i dx^i$, $\sigma = \sigma_{ij} dx^i \wedge dx^j$, etc., one may inquire as to those submanifolds on which the forms, when restricted, become identically zero.

Cartan noted that such a problem amounts to finding solutions of a coupled set of first-order partial differential equations--the dimensionality n of the submanifold sought means that n independent variables, say y^μ, can be introduced autonomously, and a set of linear homogeneous equations $\tilde{\omega} = \omega_\mu dy^\mu = 0$ arises from restriction of each given 1-form, a set of homogeneous quadratic equations from each given 2-form $\tilde{\sigma} = \sigma_{\mu\nu} dy^\mu \wedge dy^\nu = 0$, etc.

If there are, say, s_o independent 1-forms in the given set, $\omega^1, \omega^2, \ldots, \omega^{s_o}$ (note these superscripts are labels, not tensor indices!), any s_o linearly independent 1-forms formed from them (with arbitrary scalar functions as coefficients) would yield the same p.d.e.'s. If to a 2-form σ we had added any other, or combinations of terms such as $\psi \wedge \omega^i$ where ψ is an arbitrary 1-form, again no essential change would occur in the homogeneous equations. In sum, any set of generators of an <u>ideal</u> I of forms is sufficient. It is the <u>ideal</u> I that geometrizes the p.d.e.'s. This is the essence of Cartan's approach.

Since exterior derivation commutes with the (inverse) map, or sectioning (or restriction) into a so-called integral manifold, the exterior derivatives of all our generators can also be added in as generators (if they are not already in I). Sometimes it can happen that close inspection shows a few other forms can also be added, if one is only interested in integral manifolds of a given dimension. The ideal I is thus closed and completed. The partial differential problem, which may be quite nonlinear, is then expressed in a geometrical form which we believe expedites both the conceptualization, and systematic local analysis, of many otherwise ad hoc "games" of applied mathematics.

PROBLEM (after B. K. Harrison). Given a set of basis 1-forms $\xi_1, \ldots \xi_6$ whose exterior derivatives are $d\xi_1 = \xi_1 \wedge \xi_4$, $d\xi_2 = \xi_2 \wedge (\xi_5 - \xi_4)$, $d\xi_3 = \xi_3 \wedge \xi_5 - \xi_1 \wedge \xi_2$, $d\xi_4 = 0$, $d\xi_5 = 0$, $d\xi_6 = \xi_6 \wedge \xi_5$.

Show that the following set of 2-forms generates a closed ideal:

$$\xi_3 \wedge \xi_1 - \xi_2 \wedge \xi_4 \qquad \xi_3 \wedge \xi_2 - \xi_1 \wedge \xi_4$$
$$\xi_5 \wedge \xi_2 - \xi_1 \wedge \xi_6 \qquad \xi_5 \wedge \xi_1 - \xi_2 \wedge \xi_6$$
$$\xi_5 \wedge \xi_3 - \xi_4 \wedge \xi_6 \qquad \xi_4 \wedge \xi_5 - \xi_6 \wedge \xi_3$$

Cartan takes such a structure, a closed differential ideal I, and shows how, in the generic case, a large class of integral subspaces (on which $\tilde{I} = 0$) can be constructed in a stepwise procedure, using each as a boundary in turn for the next. Let us take a set of generators of I to be $\omega^A, \sigma^B, \ldots$. $A = 1 \ldots s_o$, etc. The 1-dimensional integral manifolds are generated by vector fields V_1 which must be such that $V_1 \cdot \omega^A = 0$. (All higher rank forms, when restricted to a 1-dimensional submanifold, vanish identically). Starting from an arbitrary given initial point, and setting $V_1^i = dx^i/dy^1$, this is a set of s_o automomous ordinary differential equations for n variables x^i. There will be at each step of the integration $\ell_1 = n - s_o$ arbitrary choices--so ℓ_1 arbitrary functions of the autonomous independent variable (y^1) enter the general solution.

Along a given 1-dimensional integral manifold one then finds a second vector, V_2, such that $V_2 \cdot \omega^A = 0$, and such that $V_1 \cdot V_2 \cdot \sigma^B = 0$ for all generating 2-forms. These linear homogeneous equations for V_2 obviously have rank $\geqslant s_o$, so Cartan denotes their rank as $s_o + s_1$. The second so-called Cartan "character"--the integer s_1--must thus be ≥ 0. These underdetermined coupled autonomous homogeneous linear ordinary differential equations for the components V_2^i, as functions of, say, y^2 (now with y^1 as a parameter), will have $\ell_2 = n-s_o-s_1$ arbitrary functions of y^1 and y^2. Since $V_2 = V_1$ solves them trivially, we must have $\ell_2 > 1$ to proceed to integrate to find a 2-dimensional integral manifold.

This second set of integrations for V_2 is not yet completely specified, however, as we only had determined V_1 initially. We also need the components of V_1, as parameters, at each point of the integrations for V_2. Cartan's deep insight was that since we are working with a closed ideal I, it is consistent to require $[V_1, V_2] = 0$. This determines V_1 as being "dragged along" the V_2 congruence. Now since

$$\pounds_{V_2}(V_1 \cdot \omega^A) = [V_2, V_1] \cdot \omega^A + V_1 \cdot (d(V_2 \cdot \omega^A) + V_2 \cdot d\omega^A) \qquad (8)$$

we see that all three right-hand terms vanish by our construction, and so the condition $V_1 \cdot \omega^A = 0$, true initially, is itself dragged along V_2, and preserved. V_1 thus is constructed throughout the 2-dimensional manifold as everywhere belonging to 1-dimensional integral manifolds, as indeed we initially took it. The result is that one constructs intersecting 1-dimensional integral manifolds from V_1 and V_2, that these are 2-forming, and the 2-manifolds are also integral manifolds. The autonomous variables y^1 and y^2 are introduced by writing the components of the vectors as $V_1{}^i = \frac{\partial x^i}{\partial y^1}$, $V_2{}^i = \frac{\partial x^i}{\partial y^2}$ and the construction guarantees that these are consistent.

The construction of 3-dimensional integral manifolds proceeds entirely analogously. This time we begin with a bounding 2-manifold, everywhere containing V_1 and V_2, and search for V_3 such that $V_3 \cdot \omega^A = 0$, $V_3 \cdot V_1 \cdot \sigma^A = 0$, $V_3 \cdot V_2 \cdot \sigma^A = 0$, and $V_1 \cdot V_2 \cdot V_3 \cdot \tau^A = 0$ (where τ^A are any 3-forms that may be in I). The rank is $s_o + s_1 + s_2$, so s_2 must be ≥ 0, $\ell_3 = n-(s_o + s_1 + s_2)$ degrees of freedom arise in the integration. We can proceed if $\ell_3 > 2$, and, as we go, drag the integral 2-manifold along by $[V_3, V_1] = 0$, $[V_3, V_2] = 0$, which preserves $[V_1, V_2] = 0$. $V_3{}^i = \frac{\partial x^i}{\partial y^3}$.

Integral manifolds constructed in this way, from nested integral manifolds of lower dimensionality, are called <u>regular</u>. Not all integral manifolds are regular. In any event, we recognize with Cartan that the positive integers s_o, s_1, s_2,...are <u>numerical concomitants</u> of the closed ideal I. A theory of canonical types, and representations and algebraic invariants, of ideals I (other than the simplest case, when I is generated by one 1-form and its exterior derivative) is much to be desired, and needed for the canonical classification of systems of partial differential equations. The Cartan characters surely will play a rôle.

Now at each integration we add more linear equations and can only become more constrained, $\ell_p \leq \ell_{p-1}$. But if we have a p-1-dimensional integral manifold, we also need $\ell_p > p-1$ to construct a p-dimensional one. The process must terminate, so the regular integral manifolds of I must have a maximum dimension, say g (Cartan's genus). If $\ell_g > g - 1$ but $\ell_{g+1} \leq g$ we cannot proceed past g-dimensions. This says in particular that if

$$\ell_g = n - (s_o + \ldots s_g) = g \tag{9}$$

there is no freedom in the final construction of a maximum dimensional integral manifold, and the relation $\ell_{g+1} \leq g$ then follows immediately without further calculation. Although there is no unique ideal I to represent a given set of partial differential equations, limiting the ideals considered to those that satisfy this criterion makes the choice of I as a practical matter quite limited. If autonomous variables are eliminated, g is of course still the number of independent variables, leaving n - g dependent.

PROBLEM. Find the Cartan characters and genus of the closed ideal in a space of 10 dimensions generated by the following two 3-forms:

$$\begin{aligned} \alpha &\equiv dx^4 \wedge dx^1 \wedge dx^0 + dx^5 \wedge dx^2 \wedge dx^0 + dx^6 \wedge dx^3 \wedge dx^0 \\ &\quad + dx^7 \wedge dx^2 \wedge dx^3 + dx^8 \wedge dx^3 \wedge dx^1 + dx^9 \wedge dx^1 \wedge dx^2 \\ \beta &\equiv dx^7 \wedge dx^1 \wedge dx^0 + dx^8 \wedge dx^2 \wedge dx^0 + dx^9 \wedge dx^3 \wedge dx^0 \\ &\quad - dx^4 \wedge dx^2 \wedge dx^3 - dx^5 \wedge dx^3 \wedge dx^1 - dx^6 \wedge dx^1 \wedge dx^2 \end{aligned} \tag{10}$$

VI. Associated vector fields: Cauchy characteristics and isovectors

Given a closed ideal I, one can search for and derive, in quite algorithmic fashion, various sorts of associated vector fields and/or forms. In the course of playing such games, one comes upon many concepts and techniques often already known to applied mathematicians as tools for local analysis--but one now sees them not in an ad hoc fashion but rather in an intuitively clear context where their mutual connections are revealed and natural generalizations suggested.

If a vector field can be found such that, when contracted on any form in I, a form (necessarily of lower rank) again in I results, it is denoted a Cauchy characteristic. We write $V \cdot I \subset I$. It is sufficient that if I is generated by a set of forms $\{\omega^A, \sigma^B, \ldots\}$, that the vector field satisfy $V \cdot \omega^A = 0$, $V \cdot \sigma^B = f^B_A\, \omega^A$ (f^B_A scalar functions), etc. These are homogeneous linear equations for coordinate components of V, and may well be overdetermined, with only a trivial solution. But for each independent solution that can be found, Cartan shows that by proper choice of all the coordinates it can be arranged that one coordinate will be omitted from the expression of a set of generators for I. An example of this would be for the coordinate x^1, and its basis 1-form dx^1, not to appear in the set $\{\omega^A, \sigma^B, \ldots\}$--the corresponding V would have only an x^1 component, i.e., $\frac{\partial}{\partial x^1}$. The ideal I could be regarded as having been generated by forms lifted from a base space of n-1 dimensions. Considering Cartan's construction of integral manifolds, it is clear that from any such an integral manifold of one higher dimension can be constructed by adjoining V at every point. If one has already a maximum dimension integral manifold (dimension g) then V must lie in it already. Since x^1 varies in the solution manifolds, it can be taken to be one of the independent variables in the set of partial differential equations describing these integral submanifolds--but since it does not appear in I it will not explicitly be seen! This means that when a Cauchy characteristic exists one actually solves partial differential equations in g-1 independent variables; when a Cauchy characteristic is imposed, as we next discuss, these are called similarity variables.

The most immediate generalization of Cauchy characteristics is to include all auxiliary vector fields V which are such that $\underset{V}{\pounds} I \subset I$. These have been called isovectors (Harrison and Estabrook,[6] and operating on themselves can immediately be shown to form a group. This is an invariance group, in the sense of Lie, of any of the various sets of partial differential equations for integral submanifolds. There is a nice way of understanding geometrically the meaning of sets of so-called similarity solutions, derived from knowledge of an invariance generating operator (isovector) V: if I is augmented by adjoining all forms $V \cdot I$, the new ideal $I' = \{I, V\cdot I\}$ is immediately shown to be again closed, and to have a Cauchy characteristic (viz. V!), and so the number of independent variables needed for its solution is reduced by one. The integral manifolds of I' are clearly a subset of those of I.

The equations for the components of V such that $\underset{V}{\pounds} I \subset I$ are linear, first-order, possibly overdetermined partial differential equations. If families of similarity solutions were the goal, it would be tempting to generalize the idea of isovector to vector fields such that $\underset{V}{\pounds} I \subset I'$, as I' would still be closed, and have a Cauchy characteristic, but the penalty paid is that the auxiliary equations to solve for the components of V now become nonlinear.

PROBLEM. Discuss the ideal I in 4 dimensions generated by the two 2-forms $dp \wedge dx + dq \wedge dy$ and $dq \wedge dx - dp \wedge dy$. What are the isovectors?

PROBLEM. Given an ideal I' generated by forms, say, ω^A. If I' has an isovector V, consider the ideal I generated by forms $\underset{V}{\pounds}\omega^A$. Show that I is a closed subideal of I'.

VII. Associated vector fields: The cotangent bundle

As another example of auxiliary vector field, which also is interesting because it illuminates the modern concept of fiber bundle, let us consider, in a space of $n = 2m$ dimensions, a 1-form ω whose exterior derivative is a closed 2-form

σ of maximum rank ($\sigma\wedge\sigma\wedge\ldots$taken m times does not vanish).

A remarkable theorem usually ascribed to Darboux tells us that locally coordinates p_i, q^i exist such that $\omega = p_i dq^i$, $i = 1\ldots m$ (summed on i). $\sigma = dp_i\wedge dq^i$. In a general m-dimensional subspace (or cross-section) in which none of the dq^i are restricted to zero $\tilde{\omega} = p_i(q)dq^i$ appears as a 1-form field. The m-space of variables q^i is itself the base space of a submersion, the fibers of which (q^i = const.) are coordinatized by the p^i. Can the functions $p_i(q)$ that live in the cross sections be regarded as components of a geometric object in the base space? The form ω, a single, given 1-form in the 2m-dimensional fibered space, is invariant under a group of homomorphisms generated by vector fields V such that $\underset{V}{\mathcal{L}}\omega = 0$. In the Darboux coordinates p_i, q^i these equations are readily solved for the components of V; the general result is

$$V = \phi^i \frac{\partial}{\partial q^i} - p_j \frac{\partial \phi^j}{\partial q^i} \frac{\partial}{\partial p_i} \tag{11}$$

where the ϕ^i are arbitrary functions of q^i only. The Lie product of a field V with a fiber vector $\frac{\partial}{\partial p_i}$ is $\phi^j_{,i} \frac{\partial}{\partial p_i}$, and so lies in the fiber; hence V <u>is projectible</u> into the base space. Fields V form a group. Corresponding to any V is an infinitesimal variation which can be written

$$\begin{aligned} q^i &\to \bar{q}^i = q^i + \varepsilon\phi^i \\ p_i &\to \bar{p}_i = p_i - \varepsilon p_j \frac{\partial \phi^j}{\partial q^i} = p_j \frac{\partial \bar{q}^j}{\partial q^i} \end{aligned} \tag{12}$$

This is usually regarded as saying that--seen in the base space--the functions p_i (as components of a covariant rank-1 tensor) transform contragradiently with an arbitrary coordinate transformation of the q^i. To say that the p_i are <u>components</u> of a geometric object is, we see, locally equivalent to specifying that our fibered space is a bundle <u>with structure group</u>, in particular the invariance group we have found to belong to the object ω. ω takes the rôle of a platonic ideal form, in 2m-space, with avatars on every m-dimensional cross section. Of course we assumed the concept of 1-form in first introducing ω. A modern definition of a 1-form field needs only the concept of cross section of the fibered space, and a specification of the group that maps fiber into fiber while it also acts to generate automorphisms of the base space.

PROBLEM. Consider, in a 2m-dimensional space spanned by scalars (or coordinates) V^i, q^i, $i = 1\ldots m$, an m-1-form $\tau = \varepsilon_{rs\ldots t} V^r dq^s\wedge\ldots\wedge dq^t$. (The ε-symbol simply sums over all permutations of m different integers 1, ..., m). Discuss the invariance group generated by vectors W such that $\underset{W}{\mathcal{L}}\tau = 0$.

VIII. Symplectic and co-symplectic spaces

If in the example of the previous section attention is directed at the existence of a closed two-form σ, the canonical variables p_i, q^i appear equally as coordinates in symplectic space. Again $i = 1,\ldots m$; and $n = 2m$. A vector field V such that $\underset{V}{\mathcal{L}}\sigma = 0$ is said to be the generator of a canonical transformation. The vectors derived in Section VII are a subset. Another subset consists of vectors which also transform the canonical coordinates linearly:

$$\begin{aligned} \underset{V}{\mathcal{L}} p_i &= a^j_i p_j + b_{ij} q^j \\ \underset{V}{\mathcal{L}} q^i &= c^{ij} p_j + e^i_j q^j \end{aligned} \tag{13}$$

(the symplectic group sp(m)).

To any scalar f may be associated a vector field F by setting

$$df = F\cdot\sigma . \tag{14}$$

Applying d, the integrability condition is that $\underset{F}{\mathcal{L}}\sigma = 0$, so F is canonical. Conversely, to every canonical V is associated a scalar, say v, defined up to an additive constant.

Calculating Cartan's characters, one finds that the genus is m, with a remaining degree of freedom. This is remedied in a Hamiltonian system, generated by a closed 1-form $d\mathcal{H}$ as well as the closed 2-form σ. $\mathcal{H}$ is a function of p_i, q^i--a Hamiltonian. The Cartan characters are now $s_o = 1$, $s_1 = 1$, $\ldots s_{m-1} = 1$. The maximal dimensional integral manifolds, of dimension m, are solutions of the Hamilton-Jacobi partial differential equation[7].

To the scalar $\mathcal{H}$ corresponds a vector H, the generator of a canonical transformation, and, even more strongly, a Cauchy characteristic of the ideal $\{d\mathcal{H},\sigma\}$:

$$\begin{aligned} H\cdot d\mathcal{H} &= 0 \\ H\cdot\sigma &= d\mathcal{H} \\ \underset{H}{\mathcal{L}}\sigma &= 0 . \end{aligned} \tag{15}$$

H of course describes geometrically the ordinary differential equations of the classical trajectories of Hamiltonian theory. It is important for that theory that H lies in the solution manifolds of the Hamilton-Jacobi p.d.e.

Modern developments put emphasis on the so-called co-symplectic structure dual to σ. If σ is of maximum rank, its determinant (in any coordinate frame) exists, and so does its dual: an antisymmetric second rank contravariant tensor. If we denote it as Φ^{ij}, a co-symplectic-form (poor terminology--to repeat, it's not a form, but rather contravariant!), we have

$$\Phi^{ij}\sigma_{kj} = \delta^i_k . \tag{16}$$

PROBLEM. Show that, from the exactness of σ, viz., $d\sigma = 0$ or $\sigma_{[jk,i]} = 0$, it follows that

$$\Phi^{i[j}\,\Phi^{k\ell]}{}_{,i} = 0 \; .$$

(This is a specialization of a tensor concommitant discovered by Nijenhuis and Schouten, cf. ref. (1), p. 67.)

Φ can be used to "raise indices," as in Riemannian geometry, so from any scalar f one can construct a vector F by setting

$$F^i = \Phi^{ij} f_{,j} \; . \tag{17}$$

This is equivalent to (14), and F is canonical, if σ is of maximum rank, but Hamiltonian theory can be generalized by simply postulating co-symplectic structures Φ without requiring them to be of maximum rank, and invertible, in an even dimensional space.

The calculus of Poisson brackets follows from considering the mutual Lie products of vectors F, G, etc. in terms of the associated scalars f, g, etc. A co-symplectic form thus leads naturally to an operator calculus for classical mechanics.

PROBLEM: If Γ^k_{ij} are structure constants of a Lie algebra, hence satisfying

$$\Gamma^\ell_{[ij}\Gamma^m_{k]\ell} = 0 \; , \tag{18}$$

and if we define an antisymmetric second rank contravariant field Φ by asserting that its components in a particular coordinate frame x^i are

$$\Phi^{ij} \overset{\text{def}}{=} \Gamma^k_{ij} x^k , \tag{19}$$

show that Φ is a co-symplectic form.

IX. Associated forms: Cartan forms, variational principles and Noether's theorem.

Differential geometric formulations of variational principles and Noether's theorem have been found by many mathematicians, often independently. I believe the first were LeSage and Dedecker, in Belgium; my own introduction was through the fine text of Robert Hermann.[(8)]

Again consider a given closed ideal I of 1-forms ω^A, 2-forms α^B (which include the $d\omega^A$), etc., in n-dimensional space; the maximum dimension of the regular integral manifolds of I is the "genus" g, and if some of the variables, say $x^1 \ldots . x^g$ are not constrained when restricted to such manifolds, they can be adopted as "independent" coordinates in writing an equivalent set of partial differential equations for the remaining "dependent" coordinates, say $y^1,\ldots y^{n-g}$.

Now consider any g-form θ, not necessarily in I. θ may be integrated over any g-dimensional subspace V to yield a scalar functional $\int_V \theta$. A decisive property of the operation of Lie derivation with respect to any vector field V, is that it can be used to calculate the change in such a scalar, when the region of integration V is infinitesimally displaced along the lines of V. In fact, if the displacement of each point of V (and in particular its boundary points) is εV, we have

$$\delta \int_V \theta = \varepsilon \int_V \underset{V}{\mathcal{L}} \theta \; . \tag{20}$$

Now we use the identity $\underset{V}{\mathcal{L}}\theta = V\cdot d\theta + d(V\cdot\theta)$, and follow this by use of the Stokes' theorem on the second term:

$$\delta \int_V \theta = \varepsilon \int_V V\cdot d\theta + \varepsilon \oint_{\partial V} V\cdot\theta \tag{21}$$

By ∂V we indicate the closed boundary of V, a g-1-space. As is quite customary in variational problems, we then ignore the last term in the above, as being a "boundary integral." Its value may be changed by adding a divergence term, say $\delta\phi$, to θ.

If the ideal I is complete --and those well formulated to express partial differential equations are--any form which vanishes when sectioned (or pulled back) into all g-dimensional integral manifolds of I must itself belong to I. We then read Eq. (21) as saying that for the arbitrary variation--the left side--to vanish when V is an integral manifold, it is necessary that $V\cdot d\theta$ belong to I. We write this as

$$V\cdot d\theta = 0 \bmod I \tag{22}$$

for arbitrary V. (22) is in any event sufficient for the variation (21) to vanish when V is a solution. With Hermann, we denote a form θ which satisfies Eq. (22) a Cartan form.

For each Cartan form we obtain a Lagrangian density L by writing $\int_V \theta$ in the space of independent variables $x^1,\ldots$ spanning any integral manifold V as $\int_V \tilde{\theta} = \int_V L\, dx^1\wedge\ldots dx^g$. And conversely, Hermann and others have shown how, given L, one may construct a θ satisfying (22). θ is only well determined up to a divergence, and is only non-trivial if it is not in I.

Consider now any vector W having the property that

$$\underset{W}{\mathcal{L}}\theta = 0 \bmod I \; . \tag{23}$$

If the right side of (23) is also a divergence (i.e., a conservation law of I), W may well in fact be an isovector of I, since operating on the left side of (22) with

$\underset{W}{\mathcal{L}}$ yields

$$[W,V]\cdot d\theta + V\cdot d(\underset{W}{\mathcal{L}}\theta) \; ; \tag{24}$$

the second term vanishes in this case, and the first is in I by (22). Such a W thus takes whatever forms of I occur on the right side of (22) again into I. (If a complete set of generators of I can be produced on the right side of (22), W is surely an isovector.) If W is an isovector, it is of a special kind, as we will now show, associated with a conservation law. For all solutions W of (23) to be isovectors is again a kind of completeness property.

To find a conservation law we have only to rewrite (23):

$$d(W\cdot\theta) = - W\cdot d\theta \text{ mod I} \tag{25}$$

so we have come upon a g-1-form

$$\Sigma = W\cdot\theta \tag{26}$$

which is such that

$$\oint_{\partial V}\Sigma = \int_V d\Sigma = -\int_V (W\cdot d\theta + \text{g-forms in I}) \tag{27}$$

and if V is an integral manifold--a solution--this last vanishes by (22). Hence the g-1-integral, which is taken over a compact bounding g-1-space, vanishes for any solution:

$$\oint_{\partial V}\tilde{\Sigma} = \oint_{\partial V}\widetilde{W\cdot\theta} = 0 \;\; (V \text{ an integral manifold}) \tag{28}$$

This is a <u>conservation law</u> for the set of partial integral equations, and our derivation of Σ from the variational form θ, and the invariance generator W, in (26), is essentially Noether's Theorem.

X. The discovery of Cartan forms

We achieve the possibility of systematic derivation of variational principles by specializing the form of dθ to be a superposition of terms, each of which is the outer product of two forms, each belonging to I. It is clear that if dθ is quadratic in I, (22) will be satisfied for arbitrary V. There will be arbitrary scalar coefficients of each term of dθ, say F, G, etc., functions of all the variables. Writing the integrability conditions d(dθ) = 0 gives a g+2-form, the separate vanishing of the coefficients of the independent basis g+2-forms in which gives a set of overdetermined coupled <u>linear</u> partial differential equations for F, G,.... Such sets of equations for auxiliary functions are also encountered in the derivation of isovectors. They may be integrated by repreated partial differentiation and back substitution. The solution consists of none, one or more sets F, G,..., for each of which a θ, and so an L results. For each θ, again an overdetermined linear problem can be solved to find the vectors W that leave it invariant, and from each of these, by the Noether Theorem, we immediately derive a conservation law for I.

Consider now as an example the differential ideal I in 5 dimensions (ϕ,u,v,x,t) generated by

$$\begin{aligned} \omega &= d\phi - udx - vdt \\ \sigma^1 &= d\omega = -du\wedge dx - dv\wedge dt \\ \sigma^2 &= (u^2+1)dv\wedge dx - 2\,uvdu\wedge dx \\ &\quad + (1-v^2)du\wedge dt \quad . \end{aligned} \tag{29}$$

The genus g = 2, and x and t can be adopted as independent variables in the 2-dimensional maximal integral manifolds. These manifolds V are the solutions of the 2-dimensional Born-Infeld equation

$$(1-\phi_t^2)\phi_{xx} + 2\,\phi_x\,\phi_t\,\phi_{xt} - (1+\phi_x^2)\phi_{tt} = 0\ . \tag{30}$$

We search for a closed 3-form $d\theta$ whose terms are quadratic in I, that is, which must be of the form

$$d\theta = F\sigma_1\wedge\omega + G\sigma_2\wedge\omega \tag{31}$$

where $F = F(\phi,u,v,x,t)$, $G = G(\phi,u,v,x,t)$.
Taking the exterior derivative of this, we find a 4-form required to be identically zero. The five coefficients of basis 4-forms such as $du\wedge dv\wedge dx\wedge dt$, etc., are

$$\begin{aligned} F_u + (1-v^2)G_v - 2\,vG &= 0 \\ F_v + 2\,uvG_v + (1+u^2)G_u + 4uG &= 0 \\ u(1-v^2)G_v + uF_u + 2\,F &= 0 \\ v(1+u^2)G_p + uF_p &= 0 \\ u(1+v^2)G_p + vF_p &= 0\ . \end{aligned} \tag{32}$$

This is the first linear overdetermined set to solve. We drop the possibility of dependence on x, t, find $F = -uvG$, $G_\phi = 0$, and by cross differentiation

$$G_v = \frac{-3vG}{v^2-u^2-1} \quad \text{and} \quad G_u = \frac{3uG}{v^2-u^2-1} \tag{33}$$

Integrating these, with a conventional normalization on G gives $G = -[v^2-u^2-1]^{-3/2}$, $F = uv[v^2-u^2-1]^{-3/2}$, so

$$d\theta = \frac{1}{[v^2-u^2-1]^{3/2}}\{-uvdv\wedge dt + uvdu\wedge dx \\ -(1-v^2)du\wedge dt - (1+u^2)dv\wedge dx\}\wedge\{d\phi - udx - vdt\} \tag{34}$$

which integrates immediately to

$$\theta = \frac{1}{[v^2-u^2-1]^{\frac{1}{2}}}\{-vd\wedge dx\phi - ud\phi\wedge dt - dx\wedge dt\} . \tag{35}$$

To find the usual Lagrangian for the Born-Infeld equation, we restrict (or section) θ by setting $d\phi = \phi_x dx + \phi_t dt$, $u = \phi_x$, $v = \phi_t$:

$$\tilde{\theta} = \sqrt{v^2-u^2-1}\, dx\wedge dt = L dx\wedge dt. \tag{36}$$

Next we write

$$\underset{W}{\mathcal{L}}\theta = 0$$

and solve that linear overdetermined set of p.d.e.'s for the components of W. There are six independent, superimposable solutions; listing the coordinate components in the order $[\phi,u,v,x,t]$, they are:

$$\begin{aligned} W_1 &= [0, -v,-u,t,x] \\ W_2 &= [-x,-1-u^2,-uv,\phi,o] \\ W_3 &= [t, -uv,1-v^2,0,\phi] \\ W_4 &= [0,0,0,1,0] \\ W_5 &= [0,0,0,0,1] \\ W_6 &= [1,0,0,0,0] \end{aligned} \tag{37}$$

In fact, these are all also isovectors of the ideal (29), as may be readily verified. The last three are generators of translations. The first three satisfy

$$[W_1, W_2] = - W_3,\ [W_3,W_1] = W_2,\ [W_2,W_3] = -W_1 \tag{39}$$

The resulting conservation laws are

$$
\begin{aligned}
\Sigma_1 &= (v^2-u^2-1)^{-\frac{1}{2}} (-(ux+vt)d\phi-xdx+tdt)\\
\Sigma_2 &= (v^2-u^2-1)^{-\frac{1}{2}} (-\phi vd\phi-vxdx+(\phi-ux)dt)\\
\Sigma_3 &= (v^2-u^2-1)^{-\frac{1}{2}} (-\phi ud\phi-(\phi-vt)dx+utdt)\\
\Sigma_4 &= (v^2-u^2-1)^{-\frac{1}{2}} (vd\phi-dt)\\
\Sigma_5 &= (v^2-u^2-1)^{-\frac{1}{2}} (ud\phi+dx)\\
\Sigma_6 &= (v^2-u^2-1)^{-\frac{1}{2}} (-vdx-udt)
\end{aligned}
\tag{40}
$$

XI. Associated forms: Conservation laws, potentials and pseudopotentials

Independently of any use of the Noether Theorem, conservation laws can be sought directly. To amplify the previous sections, a non-trivial conservation law belongs to each closed--or locally exact--form in I which is not merely the exterior derivative of a form (of one less rank) also in I. For if such a form is, say, $d\alpha$, of rank $p \le g$, then, applying Stokes' theorem to a p-manifold V contained in a g-dimensional integral manifold, (so $d\alpha$, restricted to V, vanishes there) one gets

$$0 = \int_V d\tilde{\alpha} = \int_V d\alpha = \oint_{\partial V} \alpha$$

So α, of rank $p-1 < g$, which does not vanish locally in a solution manifold, nevertheless vanishes when integrated over any closed surface ∂V immersed in a solution manifold. If the solution of interest happens to vanish asymptotically in certain directions, roughly speaking on the side walls of a cylinder, the result may be seen in traditional form as equality of integrals over suitably oriented end walls. The case when $p = g$ is of most interest.

To search systematically for conservation forms, one simply writes $d\alpha$ as an arbitrary superposition of p-forms in I; the vanishing of the exterior derivative then yields a set of coupled p.d.e.'s for the unknown scalar coefficients.

For each conservation form $d\alpha$ discovered in an ideal I there is the option of prolonging: (1) erecting fibers coordinatized by additional variables, and (2) in the fibered space considering an augmented ideal I' generated by the generators of I, lifted, also by the additional form α, and possibly by others for completeness. I' is thus again closed, and integral submanifolds, or solutions, of I' also are solutions of I.

The simplest example occurs when I is generated by 1-forms and 2-forms only (a "Pfaffian system"). If there exists a 2-form $d\alpha$ in I which is not the exterior derivative of a 1-form in I, then to generate I' we add in one additional 1-form, viz. $dy + \alpha$, where y is an additional variable. The total number of variables n needed to write I is thus increased to n+1 for I', s_0 increases by one, g stays the same. On any solution manifold one now gets in addition the dependence of y on the g independent variables. y is precisely what physicists call a potential field--

some of the original dependent variables, on the solution manifold, are seen to be expressed in terms of partial derivatives of y (this is because $dy + \alpha$ is now also required to vanish there.)

PROBLEM. Return to the PROBLEM at the end of Section VI, Eq. (10). Prolong the ideal I by adding in four variables y^0, y^1, y^2, y^3, and the two forms

$$d(y^0dx^0 + y^1dx^1 + y^2dx^2 + y^3dx^3) - x^4dx^2\wedge dx^3 - x^5dx^3\wedge dx^1$$

$$- x^6dx^1\wedge dx^2 + x^7dx^1\wedge dx^0 + x^8dx^2\wedge dx^0 + x^9dx^3\wedge dx^0,$$

and

$$dy^0\wedge dx^1\wedge dx^2\wedge dx^3 + dy^1\wedge dx^2\wedge dx^3\wedge dx^0 + dy^2\wedge dx^3\wedge dx^1\wedge dx^0 + dy^3\wedge dx^1\wedge dx^2\wedge dx^0$$

Calculate the Cartan characters and genus of I'.

The isogroup of the I of Eq. 10 was worked out by Harrison and Estabrook[6]--not surprisingly it consists of the 17 generators of time (x^0) and space (x^1, x^2, x^3) translations, rotations, Lorentz boosts, space-time scaling, conformal transformations, electromagnetic field (x^4-x^9) scaling and duality rotation. It would be interesting to know the isogroup of the I' above!

Hugo Wahlquist and I have introduced the concept of pseudo-potential as a natural generalization of these ideas of prolongation[9][10][11]. Where for potentials one searches for forms α such that $d\alpha\subset I$, now one searches for forms α such that $d\alpha\subset I' = \{I,\alpha\}$. Closure of I' still is automatic. The search still can be performed systematically; in the case of Pfaffian systems it is clear that the genus is still unchanged; the problem that arises is still the solution of overdetermined first-order p.d.e.'s for the coefficients that appear--the price that is paid is that these last are now nonlinear.

The good news, however, is that the nonlinearities can turn out to be only quadratic, of the commutator form of Lie products of vectors in the fiber space. The formalism that is uncovered, called by us the algebra of prolongation structures, seems to be the same as that used in the theory of connections on principle fiber bundles.[12] When found, the auxiliary variables--pseudo-potentials--satisfy linear equations that result from restriction of α (or sets of α's), and these are the linear equations used in the inverse scattering--or spectral transform--method for solving the boundary value problem for nonlinear equations such as the Korteweg-de Vries, sine-Gordon, etc.

The method of prolongation gives a systematic approach to the discovery of inverse spectral transforms, for equations for which such transforms exist, and it gives a geometric formulation for discussion of other remarkable properties that have been found for these equations, such as Bäcklund transformations. In the above, I is a closed subideal of I'. I also has Cauchy characteristics--the fibers. If there are several such, say I_1, $I_2 \ldots$, a solution (or integral) manifold of I' results in corresponding solutions of I_1, $I_2, \ldots$. A systematic method of discovery of closed subideals, especially those with Cauchy characteristics, should yield the most general Bäcklund correspondences. Auto-Bäcklund correspondences which generate isomorphic subideals are perhaps the basis of discrete soliton-creation operations.[12][13]

Acknowledgements

I have greatly benefited from all the personal and professional interactions at the 4th Scheveningen Conference. To Jim Corones, especially, my thanks for many useful conversations, and for his patience in insisting that I must understand group actions in fiber bundles! The hospitality provided by the Organizing Committee was outstanding. The research described in these notes was carried out at the Jet Propulsion Laboratory, California Institute of Technology, under NASA Contract NAS7-100.

References

(1) J. A. Schouten, Ricci-Calculus (Springer-Verlag, Berlin, 1954)

(2) É. Cartan, Les Systèmes différentiels extérieurs et leurs applications Géométriques (Hermann, Paris, 1945)

(3) W. Slebodzinski, Exterior Forms and their Applications (Polish Scientific Publishers, Warsaw, 1970)

(4) Y. Choquet-Bruhat, Géométrie différentielle et systèmes extérieurs (Dunod, Paris, 1968)

(5) F. B. Estabrook "Some Old and New Techniques for the Practical Use of Differential Forms" in R. Miura, Ed., Bäcklund Transformation, the Inverse Scattering Method, Solitons and their Application, Lecture Notes in Mathematics No. 515 (Springer-Verlag, Berlin, New York, 1976)

(6) B. K. Harrison and F. B. Estabrook, "Geometric Approach to Invariance Groups and Solution of Partial Differential Systems," J. Math. Phys. 12, 653-666 (1971)

(7) F. B. Estabrook and H. D. Wahlquist, "The Geometric Approach to Sets of Ordinary Differential Equations and Hamiltonian Mechanics, SIAM Review 17, 201-220 (1975)

(8) R. Hermann, Differential Geometry and the Calculus of Variations, 2nd Edition, Vol. XVII, Interdisciplinary Mathematics (Math Sci Press, Brookline, MA, 1977)

(9) H. D. Wahlquist and F. B. Estabrook, "Prolongation Structures of Nonlinear Evolution Equations" J. Math. Phys. 16, 1-7 (1975)

(10) F. B. Estabrook and H. D. Wahlquist, "Prolongation Structures of Non-linear Evolution Equations. II", J. Math. Phys. 17, 1293-7 (1976)

(11) F. B. Estabrook, H. D. Wahlquist and R. Hermann, "Differential-Geometric Prolongations and Bäcklund Transformations," in R. Hermann, Ed., The Ames Research Center (NASA) 1976 Conference on the Geometric Theory of Non-Linear Waves. Lie Groups: History Frontiers and Applications, Vol. VI, (Math Sci Press, Brookline, MA, 1977).

(12) R. Hermann, Geometric Theory of Non-Linear Differential Equations, Bäcklund Transformations and Solitons, Part A and Part B, Vols. XII and XIV, Interdisciplinary Mathematics (Math Sci Press, Brookline, MA, 1976 and 1977)

(13) F. B. Estabrook and H. D. Wahlquist, "Prolongation Structures, Connection Theory and Bäcklund Transformation" in F. Calogero, Ed., Nonlinear Evolution Equations Solvable by the Spectral Transform, Research Notes in Mathematics No. 26 (Pittman, London, San Francisco, Melbourne, 1978)

Some Heuristic Comments on Solitons, Integrability Conditions and Lie Groups

James Corones

Department of Mathematics and Ames Laboratory-USDOE
Iowa State University
Ames, Iowa 50011 U.S.A.

I. Introduction

What follows are some comments on solitons and associated mathematical structures that have arisen from investigations of integrability conditions that are satisfied "on" partial differential equations. Time and other constraints have prevented me from giving as complete a perspective on this approach as I would have liked. However, if the reader is familiar with some of the basic concepts and thrusts of soliton research, I think these notes are reasonably self contained. For those not yet familiar with soliton research the somewhat dated but still extremely valuable review by Scott, Chu and McLaughlin provides a useful starting point.

I, of course, make the standard disclaimer: there is a great deal of interesting and, in fact, beautiful work on solitons that is closely related to the contents of these lectures which is not referenced. I hope that the references given will be used as a starting point for people who wish to learn more about this rapidly growing and, it seems, continually surprising field.

I would like to thank my hosts and fellow participants for their interest in this work and for the opportunity to present it. Their patience in putting up with extended but unfortunately unavoidable delays in the preparation of this manuscript was most appreciated.

This work was supported by the United States Department of Energy under contract No. W-7405-eng-82 and by the National Science Foundation.

II. Solitons and Integrability Conditions

The central thesis of these lectures is that integrability conditions of a particular sort play an important role in what is currently understood about solitons. In particular they arise in connection with associated eigenvalue problems [EVP] and isospectral flows [ISF], and with Backlund transformations. I will give some illustrative examples of this and then discuss the general problem.

As a first example consider

$$L\psi = \lambda\psi \tag{2.1a}$$

$$i\psi_t = B\psi \tag{2.1b}$$

where $\psi_t \equiv \partial\psi/\partial t$ and

$$L = \begin{pmatrix} i\frac{d}{dx} & -iq(x,t) \\ ir(x,t) & i\frac{d}{dx} \end{pmatrix} \tag{2.2a}$$

$$B = \begin{pmatrix} a(x,t;\lambda) & b(x,t;\lambda) \\ c(x,t;\lambda) & -a(x,t;\lambda) \end{pmatrix} \quad . \tag{2.2b}$$

The function ψ is defined on a suitable function space, say Ψ. Clearly (2.1a) can be thought of as an eigenvalue problem for L and (2.1b), once B is fixed, defines a flow (one-parameter motion) on Ψ.

The following question can be asked, "What are the conditions on L and B such that the flow defined by B keeps the proper eigenvalues of L constant along the flow?" In short one can ask when the flow B is isospectral; $\lambda_t = 0$. It is not difficult to show that the answer is

$$iL_t = BL - LB \quad . \tag{2.3}$$

This is the Lax condition [2]. The operator L is the Zakhanov-Shabat/AKNS operator [3], [4], which has been extensively studied in the soliton context.

Certainly the condition (2.3) is interesting but it takes on real life if one requires that the B operator depends on the functions q and r that appear in L. For example consider the case when

$$q = \pm r = u \qquad (u \text{ Real}) \tag{2.4a}$$

$$a = 4\lambda^3 + 2\lambda q^2 \tag{2.4b}$$

$$b = i(4\lambda^2 + 2q^2)q + iq_{xx} + 2\lambda q_x \tag{2.4c}$$

$$c = -i(4\lambda^2 + 2q^2)q - iq_{xx} + 2\lambda q_x \quad . \tag{2.4d}$$

If this choice of q, r, a, b, c, is put into (2.3) it is found that the matrix equation is satisfied provided

$$u_t + 6u^2u_x + u_{xxx} = 0 \quad . \tag{2.5}$$

[In (2.4), (2.5) and in the sequel $q_x \equiv \partial q/\partial x$, etc.]. Equation (2.5) is the modified Korteweg-deVries (MKdV) equation.

Several remarks are in order. First, even with (2.4a) given there are many (in fact an infinite number of) B operators that are isospectral. This is true in general: one L operator has many associated ISF's. Second, each ISF yields a different equation when substituted into (2.3). Third, you might ask, how (2.4) were obtained? The answer is that in the early stages of soliton study EVP and ISF pairs were investigated and the resulting partial differential equations that were arrived at were derived from conditions placed on the B operators - say B is polynomial in λ [see [4] for example], etc. This type of result is still being very actively pursued [5], however, the problem of <u>starting</u> from a given p.d.e. and deriving an EVP and ISF for the equation is best approached by the methods discussed in these lectures, though as we will see the results are not completely satisfactory.

To obtain expressions that are more directly applicable to study via integrability conditions, it is necessary to rewrite (2.1) as

$$\psi_x = \tilde{\Gamma}_1\psi \tag{2.6a}$$

$$\psi_t = \tilde{\Gamma}_0\psi \tag{2.6b}$$

where

$$\tilde{\Gamma}_1 = \begin{pmatrix} -i\lambda & q \\ -r & i\lambda \end{pmatrix} \tag{2.7a}$$

$$\tilde{\Gamma}_0 = \begin{pmatrix} -ia & -ib \\ -ic & ia \end{pmatrix} \tag{2.7b}$$

with q, r, a, b, c as in (2.4) the MKdV equation is now equivalent to the condition

$$\psi_{xt} = \psi_{tx} \quad . \tag{2.8}$$

That is

$$\tilde{\Gamma}_{1,t} - \tilde{\Gamma}_{0,x} + \tilde{\Gamma}_1\tilde{\Gamma}_0 - \tilde{\Gamma}_0\tilde{\Gamma}_1 = 0 \quad . \tag{2.9}$$

Thus the MKdV, EVP and ISF can be rewritten as a system of first order equations for which MKdV is the compatibility condition. Clearly the same transcription can be made for any q, r, a, b, c. Indeed much of the recent work in solitons has as its starting point equations of the form (2.6) with the Γ's in general n x w matrices, for example [5,6]. Other work, for example the Gelfand-Dikii [7], theory starts with L operators of order ≥ 2 with associated B operators that also are differential operators. The simplest nontrivial example being

$$L = -\partial_x^{\,2} + u(x,t) \tag{2.10a}$$

$$B = -4i\partial_x^{\ 3} + 3i(u\partial_x + \partial_x u) \quad . \tag{2.10b}$$

In this case the Lax equation directly yields

$$u_t - 6uu_x + u_{xxx} = 0 \tag{2.11}$$

which is the Korteweg-deVries, KdV, equation. Clearly by introducing

$$\psi_1 = \psi \quad ; \quad \psi_2 = \psi_x \quad . \tag{2.12}$$

(2.1) can be rewritten as a first order system of the form (2.6) (with the Γ's, of course, containing no differential operators). Again KdV (in this case) would be equivalent to the compatibility conditions (2.8) for (2.6).

The conclusion then is that in the context of local equations and local EVP and ISF equations of the form (2.6) with the property that these compatibility conditions are a p.d.e. or in general a system of p.d.e.'s are prime candidates for study.

We next turn to Backlund transformations. The Backlund transformation, B.T., for many soliton equations has the same general form as EVP and ISF. To make this clear consider the B.T. for the sine-Gordon equation

$$\theta_{xt} = \frac{1}{8}\sin 2\theta \quad . \tag{2.13}$$

It is

$$(\hat{\theta} + \theta)_x = -2\eta \sin(\theta - \hat{\theta}) \tag{2.14a}$$

$$(\hat{\theta} - \theta)_t = \frac{1}{8}\sin(\theta + \hat{\theta}) \tag{2.14b}$$

where η is a real parameter. Observe that $\hat{\theta}$ is a solution of (2.13) provided θ is a solution of (2.13). By simply moving θ_x, θ_t to the right hand side of (2.14a,b) respectively these equations are immediately seen to be first order equations for $\hat{\theta}$ and indeed their compatibility condition is (2.13). So, again, first order systems (though in this case not linear systems) with p.d.e.'s as compatibility conditions arise in soliton theory. The fact that $\hat{\theta}$ also satisfies (2.13) is an extra feature.

Another example is provided by the B.T. for KdV (see Chen in [8])

$$(w + w')_x = k^2 - (w' - w)^2 \tag{2.15a}$$

$$\begin{aligned}(w - w')_t = {} & 2(4k^3 + 4kw_x - 2w_{xx})(w - w' + k) - \frac{1}{2}(8k^2 + 8w_x)(w - w' + k)^2 \\ & + 2(-4k^2 w_x + 2kw_{xx} - w_{xxx} - 4w_x^{\ 2})\end{aligned} \tag{2.15b}$$

where $w_x = u$ and u_x satisfies (2.11) (with coefficient +12 rather than -6 - this is a simple scale change). I do not wish to dwell on the explicit form of B.T.'s, but rather wish only to notice that by moving w_x and w_t to the right hand side of (2.15) a first order system for w' is manifest. The compatibility conditions are, of course, the potential KdV equation. Here potential in the sense that the equation is that satisfied by w, $w_x = u$.

I wish to add a cautionary note here. I do not claim that all B.T.'s are of the form of the integrability conditions discussed. I simply do not know a general way of looking at B.T.'s. I am only arguing that integrability conditions of the form considered embrace a sufficiently large class of B.T.'s to shed some light on these rather imperfectly understood objects.

I hope these examples have been sufficient to convince you that integrability conditions of a particular sort often arise in soliton theory. It is natural to pursue a study of the integrability conditions themselves and to attempt to relate their general properties to the soliton problem. Indeed, with the aid of historical perspective, one of the main contributions of Wahlquist and Estabrook [9] was to focus attention on these integrability conditions and to illustrate the rich structure that they possess.

To begin this investigation and to place the objects of interest in the context of better known mathematical objects consider a system of partial differential equations of the form

$$\frac{\partial q^a}{\partial x^\lambda} = \Gamma^a_\lambda(x,q) \tag{2.16}$$

where $\lambda = 0, \ldots, n-1$, $a = 1, \ldots, d$. Such systems are sometimes called Mayer or Mayer-Lie systems [10,11]. In the classical terminology (2.16) is completely integrable provided that to each choice of initial conditions $q_0 = (q^1_0, \ldots, q^d_0)$ is a solution of (2.16)

$$q^a = S^a(x,q_0) \tag{2.17}$$

that satisfies these initial conditions. It turns out that a necessary and sufficient condition for local existence and uniqueness is given by

$$\frac{\partial q^a}{\partial x^\mu \partial x^\lambda} = \frac{\partial q^a}{\partial x^\lambda \partial x^\mu} \quad . \tag{2.18}$$

Now, it is clear that the examples discussed above all have the general shape of (2.16). In all of them n = 2, i.e., $x^0 = t$, $x^1 = x$. In (2.6) and (2.7) d = 2 and the Γ^a_λ's are linear functions of the q's (ψ's). In (2.14) and (2.16) d = 1 and if $\hat{\theta}$ or w' is renamed the Γ^a_λ in these cases are nonlinear functions of the q's. What of the integrability conditions?

Returning to (2.18) the left hand side is

$$\frac{\partial q^a}{\partial x^\mu \partial x^\lambda} = \frac{\partial \Gamma^a_\lambda}{\partial x^\mu} + \frac{\partial \Gamma^a_\lambda}{\partial q^b}\frac{\partial q^b}{\partial x^\mu}$$

$$= \frac{\partial \Gamma^a_\lambda}{\partial x^\mu} + \frac{\partial \Gamma^a_\lambda}{\partial q^b} \Gamma^b_\mu \quad . \tag{2.19}$$

(Here and throughout the summation convention will be used.) Thus (2.18) can be rewritten as

$$R^a_{\mu\nu} \equiv \frac{\partial q^a}{\partial x^\mu \partial x^\lambda} - \frac{\partial q^a}{\partial x^\lambda \partial x^\mu}$$

$$= \frac{\partial \Gamma^a_\lambda}{\partial x^\mu} - \frac{\partial \Gamma^a_\mu}{\partial x^\lambda} + \frac{\partial \Gamma^a_\lambda}{\partial q^b} \Gamma^b_\mu - \frac{\partial \Gamma^b_\mu}{\partial q^b} \Gamma^b_\lambda = 0 \quad . \tag{2.20}$$

The quantities $R^a_{\mu\nu}$ have been called curvature quantities [10]. The integrability conditions for (2.16) are thus equivalent to the vanishing of these curvature quantities.

In passing, note [11] that if the 1-forms

$$w^a = dq^a - \Gamma^a_\lambda dx^\lambda \tag{2.21}$$

are introduced, the fact that the $R^a_{\mu\nu}$ vanish implies

$$dw^a = \frac{\partial \Gamma^a_\lambda}{\partial q^b} dx^\lambda \wedge w^b \quad . \tag{2.22}$$

That is, the w^a forms a differentially closed ideal of one-forms. Conversely if the Mayer system is completely integrable then (2.22) follows as does (2.20).

Now the Mayer system (2.16) is a very general object that becomes extremely interesting if a certain type of restriction is placed on the independent (x^λ) dependence on the right hand side. The following situation is prototypical and gives the essential and essentially new emphasis supplied by Wahlquist and Estabrook [9]. I will comment on the general situation at the end of this lecture.

Suppose

$$u_t = k(u, u_1, \ldots) \tag{2.23}$$

where $u_1 = u_x = \partial u/\partial x$, $u_2 = u_{xx}$, etc. and k is in general a nonlinear function of the u_i. That is, suppose u satisfies a nonlinear evolution equation in two variables ($\lambda = 0, 1$; $x^0 = t$, $x^1 = x$). For definiteness further suppose that i is at most m + 1. Thus u could satisfy

$$u_t + u^2 u_1 + u_3 = 0 \tag{2.24}$$

(the MKdV equation).

Now, let $z = \{u, u_1, \ldots, u_m\}$. It is required that

$$\frac{\partial q^a}{\partial x^\lambda} = \Gamma^a_\lambda(q,z) \tag{2.25}$$

and that (2.20) holds, subject to the condition that (2.23) is satisfied. Notice we are not requiring that Γ^a_λ be of any particular form nor is the number of q's fixed. We are asking for the most general form of the Mayer system that has (2.23) as its integrability condition. The one assumption that has been made is how the Γ^a_λ depend on u and its derivatives, i.e., only via elements of the set z.

The requirements placed on (2.25) are minimal yet the fact is that they place a extraordinarily tight restriction on the Γ^a_λ. The easiest way to see why this might be so is to plow ahead and do a sample computation. I will begin it here, relegate the remainder to an appendix, and finally quote the result. The example is (2.24), MKdV.

Since $\lambda = 0,1$ and $z = (u,u_1,u_2)$ (2.25) is a pair of equations

$$\frac{\partial q^a}{\partial x} = \Gamma^a_1(q,u,u_1,u_2)$$

$$\frac{\partial q^a}{\partial t} = \Gamma^a_0(q,u,u_1,u_2) \quad . \tag{2.26}$$

It is easy to see that

$$\frac{\partial q^a}{\partial t\partial x} = \frac{\partial\Gamma^a_1}{\partial u}u_t + \frac{\partial\Gamma^a_1}{\partial u_1}u_{1t} + \frac{\partial\Gamma^a_1}{\partial u_2}u_{2t} + \frac{\partial\Gamma^a_1}{\partial q^b}\frac{\partial q^b}{\partial t}$$

$$= \frac{\partial\Gamma^a_1}{\partial u}u_t + \frac{\partial\Gamma^a_1}{\partial u_1}u_{1t} + \frac{\partial\Gamma^a_1}{\partial u_2}u_{2t} + \frac{\partial\Gamma^a_1}{\partial q^b}\Gamma^b_0 \quad . \tag{2.27}$$

Likewise

$$\frac{\partial q^a}{\partial x\partial t} = \frac{\partial\Gamma^a_0}{\partial u}u_1 + \frac{\partial\Gamma^a_0}{\partial u_1}u_2 + \frac{\partial\Gamma^a_0}{\partial u_2}u_3 + \frac{\partial\Gamma^a_0}{\partial q^b}\Gamma^b_1 \quad . \tag{2.28}$$

At this point (2.27) is set equal to (2.28) and the relation between u_t, u_1, u_2 and u_3 is used, i.e., (2.24). This relation allows the replacement of u_t by spatial derivatives only. The resulting expression, i.e., the equality of (2.27) and (2.28) subject to (2.24) is treated as an identity in u and its derivatives. So, for example, the term u_{2t} introduces a term involving u_5 in (2.27). There is no term including u_5 in (2.28), thus the coefficient of u_5, $\partial\Gamma^a_1/\partial u_2$, must vanish. Likewise $\partial\Gamma^a_1/\partial u_1$ must also vanish. Using these two results we explicitly write the equality of (2.27) and (2.28) subject to (2.24).

$$\frac{\partial\Gamma^a_1}{\partial u}(-u^2u_1 - u_3) - \frac{\partial\Gamma^a_0}{\partial u}u_1 - \frac{\partial\Gamma^a_0}{\partial u_1}u_2 - \frac{\partial\Gamma^a_0}{\partial u_2}u_3 + [\Gamma_1,\Gamma_0]^a = 0 \tag{2.29}$$

where $\Gamma^a_1 = \Gamma^a_1(u,q)$. The notation

$$[\Gamma_1,\Gamma_0]^a \equiv \frac{\partial\Gamma_1^a}{\partial q^b}\Gamma_0^b - \Gamma_1^b\frac{\partial\Gamma_0^a}{\partial q^b} \tag{2.30}$$

has been used.

Since the only u_3 terms that appear appear explicitly it follows that

$$-\frac{\partial\Gamma_1^a}{\partial u} = \frac{\partial\Gamma_0^a}{\partial u_2} \quad . \tag{2.31}$$

Thus

$$\Gamma_0^a = u_2\frac{\partial\Gamma_1^a}{\partial u} + A^a(u,u_1,q) \tag{2.32}$$

where the function A^a depends on all variables not integrated over. If this result is substituted into the residue of (2.29) [that which is left after the u_3 dependence has been utilized] the result is

$$-u^2u_1\Gamma^a_{1,u} - u_1u_2\Gamma^a_{1,uu} - u_1A^a_u - u_2A^a{}_{u_1} + [\Gamma_1,A]^a + u_2[\Gamma_1,\Gamma_{1,u}]^a = 0 \quad . \tag{2.33}$$

Thus equating the u_2 terms

$$A^a{}_{u_1} = [\Gamma_1,\Gamma_{1,u}]^a - u_1\Gamma^a_{1,uu} \tag{2.34}$$

$$A^a = u_1[\Gamma_1,\Gamma_{1,u}]^a - \frac{1}{2}u_1{}^2\Gamma^a_{1,uu} + B^a(u,q) \tag{2.35}$$

and so it goes. The expression (2.35) is used in (2.33) B^a is determined as a function of u. As seen in the appendix the final result is

$$\Gamma_1^a(u,q) = \frac{1}{2}u^2X_1^a + uX_2^a + X_3^a \tag{2.36}$$

$$\begin{aligned}\Gamma_0^a(z,q) = {} & u_2\{uX_1^a + X_2^a\} + u_1\{[X_3,X_2]^a - \frac{1}{2}u_1{}^2X_1\} - \frac{1}{4}u^4X_1^a \\ & + \frac{1}{2}u^2\{-X_2^a + [X_2,[X_3,X_2]]^a\} + u[X_3,[X_3,X_2]]^a + X_4^a\end{aligned} \tag{2.37}$$

together with the equations in the appendix (A.7) and (A.9).

Equations (2.36), (2.37), (A.7), and (A.9) contain some good news and some bad news. The good news is that we have been able to compute explicitly the $z = (u, u_1, u_2)$ dependence of (2.25). The bad news is that we do not know the X_k^a explicitly. All we learned is that they must satisfy (A.2) and (A.9). Based on all the computations of this type known to me both of these results are generic.

Unfortunately there is no satisfacotry method now available to solve equations of the type (A.7) and (A.9). The original approach of Walquist and Estabrook was to observe that the bracket

$$[\Gamma_k,\Gamma_\ell]^a \equiv \frac{\partial\Gamma^a_k}{\partial q^b}\Gamma^b_\ell - \Gamma^b_k\frac{\partial\Gamma^a_\ell}{\partial q^b} \tag{2.38}$$

is a good Lie bracket (as can easily be checked) and to force all brackets that appear in equations such as (A.7) and (A.9) to be linear combinations of the X^a_k, thus forcing a Lie algebra structure on the system.

The present author [14,15] has exploited the fact that when a = 1, i.e., one q is present, the bracket equations possess some nice properties that allows their direct integration. This approach is rather limited since it only (eventually) leads to equations solvable by 2 x 2 matrix problems.

Kaup [12] has recently attempted to use the WE approach, as formulated here, to find EVP and ISF for a variety of nonlinear p.d.e. Aside from the explicit results that he obtains, the work is instructive in that it shows that even with our very imperfect understanding of solving equations of the form (A.7) and (A.9) the computational approach presented can still be a great utility. Work of Dodd and Gibbon [13] again shows that with a sufficiently strong arm it is possible to push through the computations to the end using some guess work and intuition along the way.

No general ideas have yet emerged that allow a systematic treatment of the structure equations. Put more precisely, there is no systematic procedure for finding a set of functions that a) do not all commute under the bracket operation and b) satisfy the structure equations. I think that it is important to observe that finding the general solution to the structure equations is not the relevant question. What is important in the context of the p.d.e. (which after all is the starting point of the computation) is finite sets of X's. These sets have in practice turned out to be closed under the bracket operations. Thus finding (finite-dimensional) Lie algebras that are solutions of the structure equations appear to be (from the evidence) the question of interests. The precise connection between these Lie algebras, the p.d.e., and Lie groups will be made in the next section.

It should also be noted that the method also gives sharp negative results on occasion. To understand this it must be remembered that a solution of equations of the (A.7) and (A.9) type are, by definition, nontrivial if at least one of the brackets, such as (2.38) does not vanish, i.e., the structure is not abelian (for a discussion of this see [14]). It can be shown that

$$u_t + f(u)u_1 + u_3 = 0 \tag{2.39}$$

has a nontrivial structure provided

$$f''' = 0 \quad . \tag{2.40}$$

Thus for $f'''(u) \neq 0$ equations of the form (2.39) cannot have (local) EVP and ISF.

However, it should not be thought that only soliton equations possess nontrivial structures of the type being discussed. The first example of this was provided in [14]. The Burger's equation, discussed in [13] is another example. It is clear that many other nonsoliton equations also possess nontrivial associated structures. The reason for this is still a mystery. There is however a great (perhaps perfect) overlap between equations that possess nontrivial structures of the type being discussed here and those that have infinite numbers of symmetries in the sense discussed in [16]

It is by no means clear at the moment why such apparently diverse calculations succeed on (probably) the same set of equations. Indeed the resolution of this question would add considerably to our knowledge of these equations.

I will close this lecture by commenting on the field variable dependence in (2.25). No _a priori reason_ was given to select the set $\{u,u_1,u_2\}$. I will give a rule of thumb. In the two variable, x,t, consider a p.d.e. or system of p.d.e. The variables occurring on the right hand side of the q^a_x equations should be Cauchy data for motion off the x-axis (u), and the variables occurring on the right hand side of the q^a_t equation should be "Cauchy" data off the t-axis, e.g., for (2.24) (third order in x!) u, u_1, and u_2. This rule also applied to equations that are not evolution equations, e.g., the sine-Gordon equation in characteristic coordinates or not. I emphasize that this is a rule of thumb _not_ a theorem. It probably can be justified.

As was pointed out in [ ] it is possible to attempt to include "higher" derivatives on the RHS of say (2.25). Formally this would call for treating not only the equation but its derivatives as well. So far there has been no practical or theoretical need to pursue this question. However due to the fact that conservation laws of arbitrary order would be recovered in the case when the X^a_k were trivial provided arbitrary derivatives (and derivatives of the equation) were included suggests that more attention might be given to the calculations.

Now that we have seen the role of first order systems that are integrable on given p.d.e. and the method of computing these objects the next task is to discuss the formal structure of the equations obtained.

III. Lie Groups and Solitons

In the previous section the importance of integrability conditions was stressed and a strategy for computing all integrability conditions satisfied on a partial differential equation or system of such equations was outlined. As we have seen the nature of the equations that the X_k^a satisfy together with the results of numerous calculations suggest that there is an intimate connection between Lie algebras and Lie groups and the type of integrability conditions studied. In this section it will be shown that such a connection does indeed exist and that there is a clear and unambiguous way of assigning a group theoretic content to all the auxillary functions that occur in soliton theory, as well as to the soliton equation itself.

To specify the problem more exactly recall that the out puts of the computational procedure sketched in the previous lecture were two fold. First a set of equations of the form

$$\frac{\partial q^a}{\partial x^\lambda} = f_\lambda^k(z) X_k^a(q) \tag{3.1}$$

where the $f_\lambda^k(z)$ are known (computed) functions of the set z. The solutions of this system, the q^a, exist and are unique, locally. The number of q's is not known. Second, by means that as we have seen are far from algorithmic, we have the fact that the $X_k^a(q)$ satisfy

$$[X_k, X_\ell]^a = C_{k\ell}^m X_m^a \tag{3.2}$$

I emphasize that (3.1) and (3.2) are the starting points of the present lecture. The reader is reminded that how to arrive at the $C_{k\ell}^m$ of (3.2) from structure equations such as (A.7,9) is by no means obvious and further work is required to make this passage smoother, and one would hope eventually algorithmic. However for the moment the problem is to account for (3.1) and (3.2) in group theoretic terms.

It should be noted here that if the X_m^a obey (3.2) and (3.1) is treated as a system of first order equations for the q^a then the system is integrable, i.e., solutions locally exist and are unique, if

$$\frac{\partial f_u^k}{\partial x^\lambda} - \frac{\partial f_\lambda^k}{\partial x^u} - C_{np}^k f_\lambda^n f_u^p = 0 \tag{3.3}$$

Alternatively, if (3.3) is satisfied then the q^a exist locally provided (3.2) is satisfied. Now as we have seen in the previous section in practice the f_λ^k are known functions of the field variable or variables. What is interesting about (3.3) and its relation to (3.1) is that (3.3) is equivalent to the p.d.e. and <u>does not</u> involve any auxillary variables, e.g., the eigenfunctions or pseudopotentials. As we will see (3.3) is a statement about a Lie group G and (3.1) is a statement about group actions of G.

Clearly elements of the group G must depend upon space-time in some way. The correct form of the space-time dependence can be motived in a variety of ways.

From the perspective of a physicist the left hand side of (3-3) is seen to be (formally) the Yang-Mills field strength F^k_{uv}. Thus (3-3) can be thought of as the vanishing of the Yang-Mills field strength. Recall that in general, that is when $F^k_{uv} \neq 0$, Yang [17] has shown that gauge fields can be associated with mappings from the set of all paths (curves) in space-time into a group G. And that in general the group elements depend upon the entire path. In general then a group element that depends upon a path that begins and ends at a point x^λ_o will not be the unit element of G. When this is the case for all closed paths the mapping does not depend upon the entire path but just upon the end points. This is the case when $F^k_{uv} = 0$, that is the case of interest here.

Another motivation is the following. Since the group elements must depend upon space-time in some fashion it is natural to first consider elements and a Lie group G that depend upon x^λ in a simple fashion. For example if $G = SU(2)$ suppose $g(x) \varepsilon SU(2)$

$$g(x) = \begin{pmatrix} a(x) & b(x) \\ -b^*(x) & a(x) \end{pmatrix} \tag{3.4a}$$

$$|a(x)|^2 + |b(x)|^2 = 1 \tag{3.4b}$$

where the notation $g(x^\lambda) = g(x)$ has been used. Clearly $g(x)$ varies as the space-time point x varies. It is also clear that the mapping $R^n \to SU(2)$ given explicitly by (3.4), once the functions a and b are fixed might or might not be onto G. This will certainly not be the case if the dimension of G is greater than the dimension of the space-time.

Now it happens that a slightly more complicated space-time dependence than that exemplified in (3.4) is necessary if any significant connection with p.d.e.'s is to be made. To motivate the subsequent definition I will present a rough heuristic argument.

Equation (3.1) very much would like to be a group action, after all the left hand side is a derivative and the right hand side is formally an element of the Lie algebra of G. Consider a Lie group that depends on x^λ that is a typical element is $g(x)$. Since everything we will do is a local theory, i.e., in the neighborhood of the identity of G suppose $g(x_o) = e$ further let $S^k(x)$ be the parameters (co-ordinates) of g. Expanding $g(x)$ about e means writing

$$\begin{aligned} g(x) &= g(x_o) + \frac{\partial g}{\partial s^k} \frac{\partial s^k}{\partial x^\lambda}\Big|_{x=x_o} dx^\lambda + \cdots \\ &= e + A_\lambda(x_o) dx^\lambda + \cdots \end{aligned} \tag{3.5}$$

Now the $A_\lambda(x_o)$ are related to the group actions of G. The difficulty is that they depend upon a fixed point x_o while what is of interest is a group action that is space-time dependent. As I said this is a rough argument (from which I will retreat if pushed). However, the difficulty suggested by it is real.

Taken together these motivations suggest the consideration of group elements that depend upon two space-time points (which can be thought of as the ends of Yang's paths if one is familiar with this approach). To state the relevant results in sufficient generality it is necessary to recall a few facts about Lie groups.

First recall that if a group element g_1, say of SU(2), has co-ordinate θ_1,ϕ_1,ψ_1 and a second element of SU(2) has co-ordinates θ_2,ϕ_2,ψ_2 then the group element $g_1g_2=g_3$ is the appropriate matrix product of matrices of the for (2.3) but the co-ordinates of g_3 are three functions of the θ_1,θ_2 etc. The general situation is that if t_1^k and t_2^k specify element of a group G, i.e., are their co-ordinates, then a group composition law $R^k(t_1,t_2)$ is an analytic (say) function of 2f variables such that

$$\begin{aligned} R^k(t,\bar{t}) &= R^k(\bar{t},t) = 0 \\ R^k(t,o) &= R^k(o,t) = 0 \\ R^k(t_1,R(t_2,t_3)) &= R^k(R(t_1,t_2),t_3) \end{aligned} \tag{3.6}$$

where the fact that $e\varepsilon G$ can always be taken as having the co-ordinates 0 is used and $\bar{t}^k$ denotes the co-ordinates of the group element inverse to g. These three facts are the co-ordinate statements of the properties

$$\begin{aligned} gog^{-1} &= g^{-1}og = e \\ goe &= eog = g \\ g_1o(g_2og_3) &= (g_1og_2)og_3 \end{aligned} \tag{3.7}$$

respectively.

Now, recalling the need to introduce space-time dependence that depends upon two points we consider group elements of the form $g(x)g^{-1}(y)$. In co-ordinates if $t^k(x)$ and $t^\ell(y)$ are the co-ordinates of $g(x)$ and $g(y)$ respectively then

$$r^k(x,y) = R^k(t(x),\bar{t}(y)) \tag{3.8}$$

are the co-ordinates of $g(x,y) = g(x)g^{-1}(y)$. Notice the g's of this form have the property that

$$\begin{aligned} g(x,y)g(y,z) &= g(x,z) \\ g(x,x) &= g(y,y) = e \\ g(x,y) &= g^{-1}(y,x) \end{aligned} \tag{3.9}$$

Now consider the functions

$$\tilde{f}_\lambda^k(y) = \left.\frac{\partial r^k(x,y)}{\partial y}\right|_{x=y} \tag{3.10}$$

It is not difficult to show that [18]

$$\frac{\partial \tilde{f}_u^k}{\partial x^\lambda} - \frac{\partial \tilde{f}_\lambda^k}{\partial x^u} - C_{np}^k \tilde{f}_\lambda^n \tilde{f}_m^1 = 0 \tag{3.11}$$

and conversely that a given set of functions $\tilde{f}_\lambda^k$ that satisfy (3.11) guarantee the local existence and uniqueness of the functions $r^k(x,y)$. To demonstrate this a slight bit of machinery from Lie theory is needed, machinery that would not be used again here. Thus

I have avoided the derivation. Full details can be found in [18].

Clearly from the perspective of nonlinear p.d.e. the connection between (3.11) and (3.8) is the point that is essential. Given functions f^k_λ that satisfy (3.11) the existence of group elements of the form (3.8) follows. However recall that in the case of interest (3.11) is equivalent to the nonlinear p.d.e. Thus the p.d.e. itself determines the connection with the group.

Having established a connection between Lie groups and nonlinear p.d.e. we are still left with the problem of accounting for the auxiliary variables that play such a critical role in, for example soliton theory. It is rather satisfactory that these variables do enter in a natural way when the Lie groups defined via (3.11) act as groups of transformations. To see how this works the basics of Lie groups of transformations must be recalled.

Suppose that $\tilde{Q}$ is a d-dimensional space with element $\tilde{g}\varepsilon\tilde{Q}$ having co-ordinates $\tilde{q}^a$, $a = 1,\cdots d$. And suppose that t^k_i are the co-ordinates of $g_i \varepsilon G$, $c = 1,2,3$. A set of functions $F^a(t,q)$ defines a group action of G on $\tilde{Q}$, $\tilde{q}_1 = F(t,\tilde{q})$ if

1) If $\tilde{q}_1 = F(t,\tilde{q})$

   then

   $$\tilde{q} = F(\bar{t},\tilde{q}_1)$$

2) $F(t_1,F(t_2,\tilde{q})) = F(t_3,\tilde{q})$

The generator functions of F are defined by

$$X^a_k(\tilde{q}) = -\frac{\partial F^a(t,\tilde{q})}{\partial t^k}\bigg|_{t^k=0} \tag{3.12}$$

It follows that [19] (with the bracket as in Section II).

$$[X_k,X_\ell]^a = C^m_{k\ell}\, X^a_m \tag{3.13}$$

This is all true independently of any space-time dependence in either G or $\tilde{Q}$.

The correct picture for our purposes is nearly obvious now. Suppose at each point of x there is a copy of $\tilde{Q}$, $\tilde{Q}(x)$ with $\tilde{q}\varepsilon\tilde{Q}$ having co-ordinates $q^a(x)$ and suppose that the group element of G are of the form $g(x)g^{-1}(y)$, i.e., have co-ordinates $r^k(x,y)$. The relevant, defining conditions for a group action are the same as above but, now with the added space-time dependence take the form

1) If $\tilde{q}(x) = F(r(x,y);\tilde{q}(y))$

   then

   $$\tilde{q}(y) = F(\overline{r(x,y)};\tilde{q}(x))$$

2) $F(r(y,x');F(r(x',x);q(x))) = F(r(y,x);q(x))$

The generator functions $X^a_k(q(x))$ are defined as above, (3.13) is satisfied and the differential equation that the $\tilde{q}(x)$ satisfy is easily seen to be

$$\frac{\partial\tilde{q}^a(x)}{\partial x^\lambda} = \tilde{f}^k_\lambda(x)\, \tilde{X}^a_k(q) \tag{3.14}$$

This is the result that was sought. From this point on we can remove the tilde.

Some examples are clearly in order. I will give two, both in the context of SU(2) for simplicity. One is completely generic the other is suggestive.

The generic example envolves linear group actions. In this case the $F^a(t,q)$ are linear functions of q as are the $X_k^a(q)$ and we write

$$\frac{\partial q^a(x)}{\partial x^\lambda} = f_\lambda^k(x) I_k{}^a{}_b\, q^b(x) \tag{3.15}$$

where the matrices I_k with matrix elements $I_k{}^a{}_b$ can easily be seen to satisfy

$$I_k I_\ell - I_\ell I_k = C_{k\ell}^n I_n \tag{3.16}$$

i.e., they form a matrix representation of the Lie algebra of G.

Now if G = SU(2) $C_{k\ell}^n = 2\,\varepsilon_{n\ell m}$ $n,\ell,m = 1,2,3$ $\varepsilon_{n\ell m}$ is the completely antisymmetric tensoiz and $\varepsilon_{123} = 1$. And if

$$\begin{aligned} f_1^1 &= \frac{i}{2}(u+u^*);\ f_1^2 = -\frac{1}{2}(u-u^*);\ f_1^3 = \lambda \\ f_0^1 &= \frac{1}{2}\{2\lambda(u-u^*) + i(u_x-u_x^*)\} \\ f_0^2 &= -\frac{1}{2}\{2\lambda(u-u^*) + i(u_x+u_x^*)\} \\ f_0^3 &= 2\lambda^2 + |u|^2 \end{aligned} \tag{3.17}$$

Direct substitutics of (3.17) into (3.3) shows

$$i\,u_t + u_{xx} - 2|u|^2 u = 0 \quad \text{(and c.c.)} \tag{3.18}$$

Thus the nonlinear Schroedinger equation is the integrability condition that guarantees a bilocal parameterization of SU(2). This illustrates (3.3).

Now suppose $I_k = c\,\sigma_k$ where the σ_k are the Pauli matrices. The linear group action associated with them for SU(2) as parameterized by (3.17) is by direct substitution into (3.14) (or 3.15)

$$\frac{\partial}{\partial x'}\begin{pmatrix} q^1 \\ q^2 \end{pmatrix} = \begin{pmatrix} -i\lambda & u \\ u^* & i\lambda \end{pmatrix}\begin{pmatrix} q^1 \\ q^2 \end{pmatrix} \tag{3.19a}$$

$$\frac{\partial}{\partial x^o}\begin{pmatrix} q^1 \\ q^2 \end{pmatrix} = \begin{pmatrix} -i(2\lambda^2+|u|^2) & 2\lambda u+iu_x \\ 2\,u^*-u_x^* & i(2\lambda^2+|u|^2) \end{pmatrix}\begin{pmatrix} q^1 \\ q^2 \end{pmatrix} \tag{3.19b}$$

It is easy to verify that (3.19) is the EVP and ISF for (3.18).

The example is clearly generic. All EVP and ISF are linear group actions for some G. Notice that the 2 x 2 matrix representations of su(2) and **was** used for simplicity. Any dimensional matrix representation could be used since, once the f_ℓ^k are known only the algebraic properties of the X_k^a (I_k) are needed. Thus it is easy to write down m dimensional EVP and ISF for say (3.18) one simply needs to look up linear representations of the Lie algebra of SU(2) and directly substitute them into (3.15). In a way this is an embarrassment of riches since it is not at all clear what, if anything, these more complicated objects tell us about the p.d.e. In fact this is an interesting open question. However one interesting thing is learned from this observation: asking for the maximum number of q's (in another language the maximum number of pseudopotentials) is a completely uninteresting question. In the example at hand, by the above construction, one can write down equations for <u>any</u> given number of pseudopotentials.

As a second example I will show how to arrive at Equation (2.14), the BT for sine-Gordon by group theoretic means. First observe that the functions

$$X_1(q) = \frac{i}{\sqrt{2}}\sin q \qquad X_2(q) = \frac{1}{\sqrt{2}} \qquad X_3(q) = \frac{1}{\sqrt{2}}\, i\cos q \tag{3.20}$$

satisfy

$$\frac{\partial X_i}{\partial q} X_j - X_i \frac{\partial X_j}{\partial q} = 2\,\varepsilon_{ijk} X_k \tag{3.21}$$

that is that satisfy (3.2) (for SU(2)) in the special case when one q is present). Said another way they are the generator functions for a <u>nonlinear</u> action of SU(2). It is likewise easy to see that

$$\begin{aligned} f_1^1 &= -2i\lambda\cos\theta;\ f_1^2 = -\theta_x;\ f_1^3 = 2i\lambda\sin\theta \\ f_0^1 &= \frac{-i}{8\lambda}\cos\theta;\ f_0^2 = \theta_t;\ \frac{-i}{8\lambda}\sin\theta \end{aligned} \tag{3.22}$$

where substituted into (3.3) yield the sine-Gordon equation. A brief calculation shows that when (3.20) and (3-21) are substituted into (3.14) the BT for sine-Gordon, i.e., (2.14) results.

It is, unfortunately far from clear that all BT's can be arrived at in this way. However, again the question deserves study particularly in light of BT's of the four discussed in the previous lecture. In any event, this example shows the important role that nonlinear actions of groups can play in soliton theory.

To summarize, in this lecture we have shown how integrability conditions that are satisfied on partial differential equations have a very natural interpretation as Lie group actions provided the Lie group elements are allowed to depend on space-time in the appropriate way, i.e., bilocally. In the next lecture we will show how to give some interesting geometric interpretations to these objects.

IV. A Brief Comment on Geometry and Solitons

In closing I will note some geometric aspects of soliton theory that are natural in the context of the last two lectures. The reader is referred to Frank Estabrook's lectures elsewhere in this volume for another facet of the geometry of soliton equations.

The first and most commonly repeated "geometric" statement about soliton equations is that they are closely related to a "curvature equals zero" condition. From the point of view developed above, this fact was signaled by defining the curvature quantities in Section II and by the interpretation of the soliton equations in Section III (and other equations as I noted) as equivalent to a Yang-Mills field strength vanishing. The intimate and beautiful connection between gauge fields and fiber bundles is too well documented in the literature to be repeated here. Hence, no discussion of the geometry of soliton equations via the route soliton equations → gauge fields → modern geometry. In fact, I will make an extremely conscientious statement: the above passage is interesting and true but in and of itself not very fertile. Having thus alienated a good number of my friends and colleagues, I better explain that statement.

To begin with, recall what it means to say that soliton equations define flat connections. Here is an example. Go back to (3.15) and define

$$\Gamma_\lambda(x) \equiv - f^k_\lambda I_k \tag{4.1}$$

and the covariant derivative of $q \equiv (q', \ldots, q^a)^T$ by

$$\nabla_\lambda q \equiv \frac{\partial q(x)}{\partial x^\lambda} + \Gamma_\lambda(x)\, q(x) \tag{4.2}$$

Since, by (3.3) the covariant derivative of q vanishes the curvature tensor

$$R^a_{b\lambda\mu} \equiv \frac{\partial\Gamma_{\mu b}^{\ a}}{\partial x^\lambda} - \frac{\partial\Gamma_{\lambda b}^{\ a}}{\partial x^\mu} + [\Gamma_\lambda,\Gamma_\mu]^a_b = 0 \quad . \tag{4.3}$$

So, clearly, soliton equations are equivalent to a vanishing curvature condition. The difficulty is that this statement is too general in and of itself to be relevant to soliton equations. In particular, there are two related but distinct points.

1) Equation (4.3) is a co-ordinate free staement yet, perversely, it is our ability to write (4.3) in very special co-ordinate systems, e.g., that associated with sine-Gordon or nonlinear Schroedinger that gives the condition content with respect to nonlinear p.d.e. Given a p.d.e. there is no clue in the geometrical content of the condition as to how to do this.

2) The distinctive and essential role of the "eigenvalue" is not apparent in this condition. Indeed, the failure, thus far to account for the eigenvalue in geometric terms is, it seems to me, the greatest failing of the various geometric approaches to soliton theory.

More detailed geometric information about particular soliton equations was, in the recent literature, initiated by Lamb [20] and has subsequently been pursued by many others [21-24]. Much of this work has been done in the context of two-surface theory with special attention to surfaces of constant curvature. At the moment it is difficult to see what the outcome of this line of attack will be. It is always interesting to look at complex phenomena, such as solitons, from various points of view, however, there is a danger of spending a great deal of effort to gain simply a literal translation of what is known from one mathematical language to another without gaining additional insight.

I do not wish to end these discussions on a negative note. In fact, I am convinced that a great deal more can be learned about solitons and their (still illusive) multi-dimensional generalizations by using geometric and group theoretic methods. The most intriguing possibility is that true multi-dimensional equations can be found that fit into the integrability scheme. That is, is it possible to find functions $f^k_\lambda(\hat{z})$ where a set of field variables over say, four-dimensional space-time, and any number of derivatives such that Eq. (3.3) is satisfied? The fact that the number of integrability conditions goes from 2 to 3 to 6 as the number of space-time dimensions goes from 2 to 3 to 4 suggests that systems of equations probably will be models, however, a search for such systems must be pursued and the properties of both the equations and their solutions (we hope!) should be explored.

Appendix A

From the text we arrive at

$$-u^2 u_1 \Gamma^a_{1,u} - u_1\{u_1[\Gamma_1,\Gamma_{1,uu}]^a - \frac{1}{2} u_1^2 \Gamma^a_{1,uuu} + B^a_u\} + u_1[\Gamma_1,[\Gamma_1,\Gamma_{1,u}]]^a$$

$$- \frac{1}{2} u_1^2 [\Gamma_1,\Gamma_{1,uu}]^a + [\Gamma_1,B]^a = 0 \tag{A.1}$$

when (2.35) is substituted into (2.33). Equating powers of u_1

$$u_1^3: \quad \Gamma^a_{1,uuu} = 0 \tag{A.2}$$

$$u_1^2: \quad [\Gamma_1,\Gamma_{1,uu}]^a - \frac{1}{2} [\Gamma_1,\Gamma_{1,uu}]^a = 0 \tag{A.3}$$

$$u_1: \quad -u^2 \Gamma^a_{1,u} - B^a_u + [\Gamma_1, [\Gamma_1,\Gamma_{1,u}]] = 0 \tag{A.4}$$

$$u_1^0: \quad [\Gamma_1,B]^a = 0 \quad . \tag{A.5}$$

Clearly from (A.2)

$$\Gamma^a_1 = \frac{1}{2} u^2 X^a_1(q) + u X^a_2(q) + X^a_3(q) \quad . \tag{A.6}$$

While from (A.3)

$$[X_1,X_3]^a = [X_1,X_3]^a = 0 \qquad \text{(A.7)}$$

Using these results in (A.4) and integrating yields

$$B = -\frac{1}{4}u^4 X_1 + \frac{1}{2}u^2\{-X_2 + [X_2,[X_3,X_2]]\} + u[X_3,[X_3,X_2]] + X_4 \quad . \qquad \text{(A.8)}$$

Finally, substituting (A.8) in (A.5) and equating powers of u to zero separately yields

$$u^3:\ [X_2,[X_2,[X_2,X_3]]] = 0$$

$$u^2:\ \frac{1}{2}[X_1,X_4] - \frac{1}{2}[X_3,X_2] + \frac{1}{2}[X_2,[X_2,[X_3,X_2]]] + \frac{1}{2}[X_3,[X_2,[X_3,X_2]]] = 0$$

$$u:\ [X_2,X_4] + [X_3,[X_3,[X_3,X_2]]] = 0$$

$$u^0:\ [X_3,X_4] = 0 \qquad \text{(A.9)}$$

References

1. A. C. Scott, F. Y. F. Chu and D. W. McLaughlin, Proc. IEEE 61, 1443 (1973).
2. P. D. Lax, Comm. Pure Appl. Math 21, 467 (1968).
3. V. E. Zakharov and A. B. Shabat, Soviet Phys. JETP, 34, 62 (1972).
4. M. J. Ablowitz, D. J. Kaup, A. C. Newell and H. Segur, Stud. Appl. Math., 53, 249 (1974).
5. V. E. Zakharov and A. V. Mikhailov, Soviet Phys. JETP, 47, 1017 (1978).
6. D. J. Kaup, A. Rieman, and A. Bers, Rev. Mod. Phys. 51, 275 (1979).
7. I. M. Gelfand and L. A. Dikii, Funct. Anal. and its Appl., 10, 259 (1976).
8. H. Chen in Backlund Transformations, Lecture Notes in Mathematics #515, edited by R. Miura (Springer, Berlin 1976).
9. H. D. Wahlquist and F. B. Estabrook, J. Math. Phys. (N.Y.), 16, 1 (1975).
10. C. Loewner, Theory of Continuous Groups Notes by H. Flanders and M. H. Protter, (MIT Press, Cambridge, Mass, 1971).
11. H. Flanders, Differential Forms (Academic Press, New York, 1963).
12. D. J. Kaup, Lectures on the Estabrook-Wahlquist Method, with Examples of Application, MIT Plasma Research Report, PRR 79/6 (1979).
13. R. K. Dodd and J. D. Gibbon, Proc. R. Soc. Lond. A. 358, 287 (1977) and 359, 411 (1978).
14. J. P. Corones in Reference 8.
15. J. P. Corones, J. Math. Phys. 17, 756 (1976).
16. For example, N. H. Ibragimov and R. L. Anderson, J. Math. Anal. Applic. 59, 145 (1977), R. J. Oliver, J. Math. Phys. 18, 1212 (1977), S. Kumei, J. Math. Phys, 18, 256 (1977), A. S. Fokas, R. L. Anderson, Lett. Math. Phys., 3, 117 (1979).
17. C. N. Yang, Phys. Rev. Lett. 33, 445 (1974).
18. J. P. Corones, B. L. Markovski and V. A. Rizov, J. Math. Phys. 18 2207 (1977).
19. L. P. Eisenhart Continuous Groups of Transformations (Princeton U.P., Princeton N. J., 1933).
20. G. L. Lamb, Phys. Rev. Lett. 37, 235 (1976) and J. Math. Phys. (N.Y.) 18, 1658 (1977).
21. F. Lund, Phys. Rev. Lett. 38, 1175 (1977) and Phys. Rev. D15, 1540 (1977).
22. A. Sym and J. Corones, Phys. Rev. Lett. 42, 1099 (1979).
23. M. Lakshmanan, Phys. Lett. 61A, 53 (1977), and 64A, 353 (1978).
24. M. Crampin, Phys. Lett. 66A, 243 (1978).
25. M. Crampin, F. Pirani and D. Robinson, Lett. Math. Phys. 2, 15 (1977).

ON BÄCKLUND TRANSFORMATIONS AND SOLUTIONS TO THE 2 + 1 AND 3 + 1 - DIMENSIONAL SINE - GORDON EQUATION

P L Christiansen

Laboratory of Applied Mathematical Physics
The Technical University of Denmark
DK-2800 Lyngby, Denmark

Abstract. A Bäcklund transformation for the 3 + 1 - dimensional sine-Gordon equation is applied successively with different Bäcklund parameters two, three, and four times. The resulting matrix Bianchi relations are useful for generation of scalar Bianchi relations from which solutions to the sine-Gordon equation can be obtained. The Bianchi-Lamb parallelogram is generalized to a new Bianchi-Lamb parallelepiped for three successive Bäcklund transformations, and to a hyperparallelepiped for four successive Bäcklund transformations. Constraints on the Bäcklund parameters relevant for soliton wave solutions are interpreted geometrically in connection with these generalized Bianchi-Lamb diagrams. It is also shown that the constraints lead to conservation in time of the area between three line solitons moving in the XY-plane and of the volume between four plane solitons moving in the XYZ-space. The latter result which is a necessary condition for plane solitons is believed to be new.

Table of contents.

1. Introduction.

The sine-Gordon equation (SGE) governs a number of non-linear wave phenomena in superconduction (Josephson junctions), dislocation theory, and field theory [1]. Exact solutions to the SGE can be obtained by methods like inverse scattering and the Bäcklund transformation (BT), [2] and [3]. The present contribution deals with the use of the latter method for the derivation of soliton wave solutions and their properties.

The BT for the SGE in 1 + 1 dimensions is a classical result [2]. Leibbrandt [4] found a BT for 2 + 1 dimensions and Christiansen [5], Leibbrandt [4], and Christiansen and Olsen [6] found a BT for 3 + 1 dimensions. Soliton solutions to the SGE in 1 + 1 and 2 + 1 dimensions were obtained by a different method by Hirota [7] and [8] and in higher dimensions by Kobayashi and Izutsu [9]. The properties of the soliton solutions were analyzed by Gibbon and Zambotti [10].

2. A Bäcklund transformation.

For the 3 + 1 - dimensional SGE

$$(\partial_x^2 + \partial_y^2 + \partial_z^2 - \partial_t^2)v = \sin v \tag{2.1}$$

Christiansen [5] and Leibbrandt [4] have found the BT

$$\underline{\underline{O}}^{\pm}(\partial_x,\partial_y,\partial_z,\partial_t)\ \frac{v \mp iw}{2} = \underline{\underline{N}}^{\pm}(\alpha,\beta,\gamma)\ \sin\frac{v \pm iw}{2} \tag{2.2}$$

where

$$\underline{\underline{O}}^{\pm}(\partial_x,\partial_y,\partial_z,\partial_t) = \underline{\underline{I}}\partial_x \pm i\underline{\underline{J}}\partial_y \pm i\underline{\underline{K}}\partial_z \pm i\underline{\underline{L}}\partial_t \tag{2.3}$$

and

$$\underline{\underline{N}}^{\pm}(\alpha,\beta,\gamma) = \underline{\underline{I}}\cos\alpha \pm i\underline{\underline{M}}(\beta)\sin\alpha\cos\gamma \pm i\underline{\underline{K}}\sin\alpha\sin\gamma \tag{2.4}$$

with

$$\underline{\underline{I}} = \begin{Bmatrix} 1 & 0 \\ 0 & 1 \end{Bmatrix}, \quad \underline{\underline{J}} = \begin{Bmatrix} 0 & 1 \\ 1 & 0 \end{Bmatrix} = \sigma_x, \quad \underline{\underline{K}} = \begin{Bmatrix} 1 & 0 \\ 0 & -1 \end{Bmatrix} = \sigma_z$$

$$\underline{\underline{L}} = \begin{Bmatrix} 0 & -1 \\ 1 & 0 \end{Bmatrix} = -i\,\sigma_y, \quad \text{and} \quad \underline{\underline{M}}(\beta) = \begin{Bmatrix} 0 & e^{-\beta} \\ e^{\beta} & 0 \end{Bmatrix}. \tag{2.5}$$

Here the relationship to the Pauli spin matrices σ_x, σ_y, and σ_z [11] has been indicated. Thus

$$\underline{\underline{N}}^{\pm}(\alpha,\beta,\gamma) = \begin{Bmatrix} \cos\alpha \pm i\sin\alpha\sin\gamma & \pm ie^{-\beta}\sin\alpha\cos\gamma \\ \pm ie^{\beta}\sin\alpha\cos\gamma & \cos\alpha \mp i\sin\alpha\sin\gamma \end{Bmatrix}. \tag{2.6}$$

We shall denote the elements of this matrix $n_{rs}^{\pm}$ with r = 1,2 and s = 1,2. In (2.2) α, β, and γ are real Bäcklund parameters and v and w are real functions. If v satisfies the SGE (2.1) and v and w satisfy the BT (2.2) then w is a solution to the hyperbolic SGE[1)]

$$(\partial_x^2 + \partial_y^2 + \partial_z^2 - \partial_t^2)w = \sinh w\,. \tag{2.7}$$

Vice versa when w is a solution to the hyperbolic SGE (2.7) and w and v satisfy the inverted BT

$$\underline{\underline{O}}^{\pm}(\partial_x,\partial_y,\partial_z,\partial_t)\ \frac{iw \mp v}{2} = \underline{\underline{N}}^{\pm}(\alpha',\beta',\gamma')\ \sin\frac{iw \pm v}{2} \tag{2.8}$$

---

1) Throughout the paper v denotes a solution to the SGE (2.1) while w denotes a solution to the hyperbolic SGE (2.7).

with Bäcklund parameters α', β', and γ' then v is a solution to the SGE (2.1). It is seen that (2.8) reduces to (2.2) for $(\alpha',\beta',\gamma') = (\alpha+\pi,\beta,\gamma)$. Note that upper and lower subscripts in the BT (2.2) yield equations which are complex conjugates of each other.

For the 2+1-dimensional SGE

$$(\partial_x^2 + \partial_y^2 - \partial_t^2)v = \sin v \tag{2.9}$$

we get the BT by letting $\partial_z = 0$ and $\gamma = 0$ in (2.2)

$$\underline{\underline{0}}^{\pm}(\partial_x,\partial_y,0,\partial_t)\,\frac{v \mp iw}{2} = \underline{\underline{N}}^{\pm}(\alpha,\beta,0)\,\sin\frac{v \pm iw}{2}\,. \tag{2.10}$$

For the 1+1-dimensional SGE

$$(\partial_x^2 - \partial_t^2)v = \sin v \tag{2.11}$$

the BT can be obtained by letting $\partial_y = 0$ and $\beta = i\,\frac{\pi}{2}$ and $\alpha = ic$, where c is real in (2.10). The result

$$\underline{\underline{0}}^{\pm}(\partial_x,0,0,\partial_t)\,\frac{v \mp iw}{2} = \underline{\underline{N}}^{\pm}(ic,i\tfrac{\pi}{2},0)\,\sin\frac{v \pm iw}{2} \tag{2.12}$$

is equivalent to the classical BT

$$(\partial_x \pm \partial_t)\,\frac{v \mp w}{2} = e^{\mp c}\,\sin\frac{v \pm w}{2}\,. \tag{2.13}$$

Separating real and imaginary parts in the BT (2.2) we obtain [5] the eight equations for the eight derivatives of v and w with respect to x, y, z, and t

$$\left.\begin{aligned}
\partial_x v &= 2\cos\alpha\,\sin\frac{v}{2}\cosh\frac{w}{2}\,,\\
\partial_y v &= 2\sin\alpha\,\cosh\beta\,\cos\gamma\,\sin\frac{v}{2}\cosh\frac{w}{2}\,,\\
\partial_z v &= 2\sin\alpha\,\sin\gamma\,\sin\frac{v}{2}\cosh\frac{w}{2}\,,\\
\partial_t v &= 2\sin\alpha\,\sinh\beta\,\cos\gamma\,\sin\frac{v}{2}\cosh\frac{w}{2}\,,\\
\partial_x w &= -2\cos\alpha\,\cos\frac{v}{2}\sinh\frac{w}{2}\,,\\
\partial_y w &= -2\sin\alpha\,\cosh\beta\,\cos\gamma\,\cos\frac{v}{2}\sinh\frac{w}{2}\,,\\
\partial_z w &= -2\sin\alpha\,\sin\gamma\,\cos\frac{v}{2}\sinh\frac{w}{2}\,,\\
\partial_t w &= -2\sin\alpha\,\sinh\beta\,\cos\gamma\,\cos\frac{v}{2}\sinh\frac{w}{2}\,.
\end{aligned}\right\} \tag{2.14}$$

From these equations it is seen that the solutions to the BT (2.2) are restricted to the functions of S, V and W,

$$v = V(S(\alpha,\beta,\gamma,\delta)) \quad \text{and} \quad w = W(S(\alpha,\beta,\gamma,\delta)) \tag{2.15}$$

with

$$S(\alpha,\beta,\gamma,\delta) = x\cos\alpha + y\sin\alpha\cosh\beta\cos\gamma + z\sin\alpha\sin\gamma + \\ + t\sin\alpha\sinh\beta\cos\gamma + \delta \tag{2.16}$$

where δ is a phase constant.

For $iw \equiv 0$ ("the vacuum solution") the BT (2.2) yields the solution to the SGE (2.1)

$$v = 4\tan^{-1}\exp S \tag{2.17}$$

where S is given by (2.16). Similarly, $v \equiv 0$ in (2.2) yields the solution to the hyperbolic SGE (2.7)

$$w = 4\tanh^{-1}\exp S\ . \tag{2.18}$$

In the 3 + 1 - dimensional case the solution is a plane soliton[1)] (2.17) moving in XYZ-space in the direction characterized by the unit vector

$$\hat{\underline{k}}(\alpha,\beta,\gamma) = \frac{(\cos\alpha,\ \sin\alpha\cosh\beta\cos\gamma,\ \sin\alpha\sin\gamma)}{\sqrt{1+\sin^2\alpha\sinh^2\beta\cos^2\gamma}} \tag{2.19}$$

with velocity

$$u(\alpha,\beta,\gamma) = \frac{-\sin\alpha\sinh\beta\cos\gamma}{\sqrt{1+\sin^2\alpha\sinh^2\beta\cos^2\gamma}}\ . \tag{2.20}$$

In the 2 + 1 - dimensional case where

$$S = x\cos\alpha + y\sin\alpha\cosh\beta + t\sin\alpha\sinh\beta + \delta \tag{2.21}$$

the solution (2.17) becomes a line soliton[1)] moving in the XY-plane. In the 1 + 1 - dimensional case

$$S = x\cosh c - t\sinh c \tag{2.22}$$

and the solution becomes identical to the classical soliton.

In the following section we shall develop the application of two, three, and four successive BT's.

---

1) This denotation will be justified in Section 4.

3. Two, three, and four successive Bäcklund transformations.

We first consider two BT's (2.2) with two different sets of Bäcklund parameters $(\alpha_1,\beta_1,\gamma_1) \neq (\alpha_2,\beta_2,\gamma_2)$

$$\underline{\underline{0}}^{\pm} \frac{v_a \mp iw_a}{2} = \underline{\underline{N}}^{\pm}(1) \sin \frac{v_a \pm iw_a}{2} \tag{3.1}$$

$$\underline{\underline{0}}^{\pm} \frac{v_a \mp iw_b}{2} = \underline{\underline{N}}^{\pm}(2) \sin \frac{v_a \pm iw_b}{2} . \tag{3.2}$$

Here we have introduced the abbreviations $\underline{\underline{0}}^{\pm}$ for $\underline{\underline{0}}^{\pm}(\partial_x,\partial_y,\partial_z,\partial_t)$ and $\underline{\underline{N}}^{\pm}(j)$ for $\underline{\underline{N}}^{\pm}(\alpha_j,\beta_j,\gamma_j)$ with $j = 1,2$. In (3.1) and (3.2) v_a is a known solution to the SGE (2.1). Then w_a and w_b are two different solutions to the hyperbolic SGE (2.7). If we apply the inverted BT (2.8) with $(\alpha',\beta',\gamma') = (\alpha_2,\beta_2,\gamma_2)$ to w_a in (3.1) we get

$$\underline{\underline{0}}^{\pm} \frac{iw_a \mp v_b}{2} = \underline{\underline{N}}^{\pm}(2) \sin \frac{iw_a \pm v_b}{2} . \tag{3.3}$$

Similarly, application of the inverted BT (2.8) with $(\alpha',\beta',\gamma') = (\alpha_1,\beta_1,\gamma_1)$ to w_b in (3.2) yields

$$\underline{\underline{0}}^{\pm} \frac{iw_b \mp v_b}{2} = \underline{\underline{N}}^{\pm}(1) \sin \frac{iw_b \pm v_b}{2} . \tag{3.4}$$

In both cases we get the solution v_b to the SGE (2.1) provided the BT's in (3.1) - (3.4) commute. The situation is illustrated in the Bianchi-Lamb diagram shown in Fig 1.a. The diagram is a parallelogram. The four vertices of the parallelogram are seen to correspond to two solutions to the SGE, v_a and v_b, and to two solutions to the hyperbolic SGE, w_a and w_b. Each of the four sides corresponds to a BT or an inverted BT. Addition and subtraction of (3.1) - (3.4) yield the algebraic equation

$$\underline{\underline{N}}^{\pm}(1)\left(\sin \frac{v_a \pm iw_a}{2} - \sin \frac{v_b \pm iw_b}{2}\right) = \underline{\underline{N}}^{\pm}(2)\left(\sin \frac{v_a \pm iw_b}{2} - \sin \frac{v_b \pm iw_a}{2}\right) \tag{3.5}$$

which we shall denote a Bianchi relation. Again upper and lower signs yield equations which are complex conjugates of each other.

Instead of starting with v_a we may begin with the known solution to the hyperbolic SGE (2.7), w_a, and arrive at v_a and v_b and w_b as illustrated in the inverted Bianchi-Lamb diagram shown in Fig 1.b. In this case (3.5) is replaced by the Bianchi relation

$$\underline{\underline{N}}^{\pm}(1)\left(\sin\frac{v_a \pm iw_a}{2} - \sin\frac{v_b \pm iw_b}{2}\right) = \underline{\underline{N}}^{\pm}(2)\left(\sin\frac{v_b \pm iw_a}{2} - \sin\frac{v_a \pm iw_b}{2}\right). \tag{3.6}$$

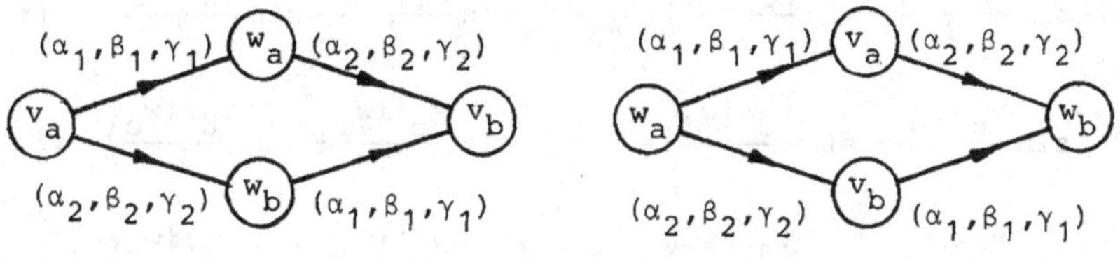

Figure 1. Bianchi-Lamb diagrams illustrating the commutative property of the BT's. The transformations from the v's to the w's are given by the BT (2.2) with two different sets of Bäcklund parameters. The transformations from the w's to the v's are given by the corresponding inverted BT's (2.8). In Fig 1.a the solution to the SGE, v_a, is the starting point, while in Fig 1.b the solution to the hyperbolic SGE, w_a, is the starting point.

We next consider three successive BT's with three different sets of Bäcklund parameters $(\alpha_1,\beta_1,\gamma_1)$, $(\alpha_2,\beta_2,\gamma_2)$, and $(\alpha_3,\beta_3,\gamma_3)$. Starting with the known solution to the hyperbolic SGE (2.7), w_a, we arrive at three different solutions to the SGE (2.1), v_a, v_b, and v_c, through the inverted BT's (2.8) with the three sets of Bäcklund parameters. The procedure can be followed in Fig 2 which illustrates the Bianchi-Lamb parallelepiped[1]) for three BT's. Next v_a is the starting point for two BT's (2.2) with Bäcklund parameters $(\alpha_2,\beta_2,\gamma_2)$ and $(\alpha_3,\beta_3,\gamma_3)$ leading to w_b and w_c. Similarly, we get from v_b to w_b and w_d by means of BT's with parameters $(\alpha_1,\beta_1,\gamma_1)$ and $(\alpha_3,\beta_3,\gamma_3)$ and from v_c to w_c and w_d by means of BT's with parameters $(\alpha_1,\beta_1,\gamma_1)$ and $(\alpha_2,\beta_2,\gamma_2)$. Finally, we get from w_b, w_c, and w_d to v_d by means of inverted BT's (2.8) with parameters $(\alpha_3,\beta_3,\gamma_3)$, $(\alpha_2,\beta_2,\gamma_2)$, and $(\alpha_1,\beta_1,\gamma_1)$ respectively. Commutativity is assumed throughout the procedure. The eight vertices of the parallelepided are seen to correspond to four solutions to the SGE, v_a, v_b, v_c, and v_d, and to four solutions to the hyperbolic SGE, w_a, w_b, w_c, and w_d. Each of the twelve edges corresponds to an inverted BT or a BT. The six faces of the parallelepiped correspond to three inverted Bianchi-Lamb diagrams and three Bianchi-Lamb diagrams. The corresponding Bianchi relations become

$$\underline{\underline{N}}^{\pm}(1)\left(\sin\frac{v_a \pm iw_a}{2} - \sin\frac{v_b \pm iw_b}{2}\right) = \underline{\underline{N}}^{\pm}(2)\left(\sin\frac{v_b \pm iw_a}{2} - \sin\frac{v_a \pm iw_b}{2}\right) \tag{3.7}$$

1) drawn as a cube like in [12].

$$\underline{\underline{N}}^{\pm}(1)\left(\sin\frac{v_a \pm iw_a}{2} - \sin\frac{v_c \pm iw_c}{2}\right) = \underline{\underline{N}}^{\pm}(3)\left(\sin\frac{v_c \pm iw_a}{2} - \sin\frac{v_a \pm iw_c}{2}\right) \quad (3.8)$$

$$\underline{\underline{N}}^{\pm}(2)\left(\sin\frac{v_b \pm iw_a}{2} - \sin\frac{v_c \pm iw_d}{2}\right) = \underline{\underline{N}}^{\pm}(3)\left(\sin\frac{v_c \pm iw_a}{2} - \sin\frac{v_b \pm iw_d}{2}\right) \quad (3.9)$$

$$\underline{\underline{N}}^{\pm}(1)\left(\sin\frac{v_c \pm iw_c}{2} - \sin\frac{v_d \pm iw_d}{2}\right) = \underline{\underline{N}}^{\pm}(2)\left(\sin\frac{v_c \pm iw_d}{2} - \sin\frac{v_d \pm iw_c}{2}\right) \quad (3.10)$$

$$\underline{\underline{N}}^{\pm}(1)\left(\sin\frac{v_b \pm iw_b}{2} - \sin\frac{v_d \pm iw_d}{2}\right) = \underline{\underline{N}}^{\pm}(3)\left(\sin\frac{v_b \pm iw_d}{2} - \sin\frac{v_d \pm iw_b}{2}\right) \quad (3.11)$$

$$\underline{\underline{N}}^{\pm}(2)\left(\sin\frac{v_a \pm iw_b}{2} - \sin\frac{v_d \pm iw_c}{2}\right) = \underline{\underline{N}}^{\pm}(3)\left(\sin\frac{v_a \pm iw_c}{2} - \sin\frac{v_d \pm iw_b}{2}\right) . \quad (3.12)$$

In this system one of the equations in redundant since it can be derived from the five others. This fact agrees well with the geometrical representation in the Bianchi parallelepiped. Thus we are left with five Bianchi relations in the case of three successive BT's.

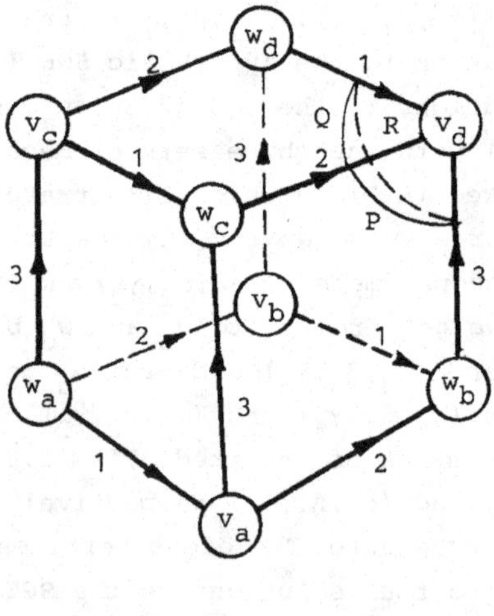

Figure 2. The Bianchi-Lamb parallelepiped drawn as a cube illustrating the commutative property of the BT's. Each vertex corresponds to a solution to either the SGE or the hyperbolic SGE. Each edge corresponds to an inverted BT or a BT. The Bäcklund parameter set is indicated by numbers 1, 2, and 3. Each face corresponds to either a Bianchi-Lamb diagram (Fig 1.a) or an inverted Bianchi-Lamb diagram (Fig 1.b). The angles between sides 2 and 3, 1 and 3, and 1 and 2 are denoted P, Q, and R respectively.

Alternatively we may start with v_a instead of w_a and end up with w_d instead of v_d after three successive BT's. The resulting inverted Bianchi-Lamb parallelepiped can be obtained from Fig 2 by replacing the w's by v's and vice versa. Also in this case five Bianchi relations result.

Finally we consider four successive BT's with four different sets of Bäcklund parameters $(\alpha_1,\beta_1,\gamma_1)$, $(\alpha_2,\beta_2,\gamma_2)$, $(\alpha_3,\beta_3,\gamma_3)$, and $(\alpha_4,\beta_4,\gamma_4)$. Starting with v_a we get w_a, w_b, w_c, and w_d after one different BT in each case. The procedure can be followed in Fig 3 which illustrates a new Bianchi-Lamb hyperparallelepiped for four BT's. In the next step v_b, v_c, v_d, v_e, v_f, and v_g are produced by the inverted BT's. Then follow w_e, w_f, w_g, and w_h after the BT's again and finally v_h results after the inverted BT's again. The sixteen vertices of the hyperparallelepiped correspond to the solutions $v_a, \cdots, v_h, w_a, \cdots, w_h$. Each of the 32 edges corresponds to a BT or an inverted BT. The 24 faces of the hyperparallelepiped correspond to twelve Bianchi-Lamb diagrams and twelve inverted Bianchi-Lamb diagrams. For space reasons we shall omit the corresponding 24 Bianchi relations! Each face is shared by two parallelepipeds in the hyperparallelepiped which can be shown to consist of eight such parallelepipeds. The redundancy of one equation per parallelepiped carries over from the three-dimensional case to the four-dimensional case. This means that we are left with <u>sixteen Bianchi relations in the case of four successive BT's</u>. So far we have been unable to prove further redundancy in the system of equations. Also in this case an inverted Bianchi-Lamb hyperparallelepiped can be constructed.

We note that all the Bianchi relations obtained in this section are matrix equations of the form

$$\underline{\underline{N}}^{\pm}(j)\; f^{\pm} = \underline{\underline{N}}^{\pm}(k)\; g^{\pm} \tag{3.13}$$

where $f^{\pm}$ and $g^{\pm}$ are complex scalar functions. This simply means that the elements of the matrices are proportional

$$n^{\pm}_{rs}(j)\; f^{\pm} = n^{\pm}_{rs}(k)\; g^{\pm}\;. \tag{3.14}$$

Here $n^{\pm}_{rs}(j)$ are the elements of $\underline{\underline{N}}^{\pm}(j)$. Using (2.6) we arrive at the conditions in the particular case of (3.5)

$$\frac{\cos\alpha_1}{\cos\alpha_2} = \frac{\sin\alpha_1 \cosh\beta_1 \cos\gamma_1}{\sin\alpha_2 \cosh\beta_2 \cos\gamma_2} = \frac{\sin\alpha_1 \sin\gamma_1}{\sin\alpha_2 \sin\gamma_2} =$$

$$\frac{\cos\frac{v_a - v_b}{2}\cosh\frac{w_a - w_b}{2} - 1 \pm i\sin\frac{v_a - v_b}{2}\sinh\frac{w_a - w_b}{2}}{\cos\frac{v_a - v_b}{2} - \cosh\frac{w_a - w_b}{2}} . \tag{3.15}$$

Similar conditions can be obtained from the other matrix equations. If the solutions w_a and w_b ($w_a \neq w_b$) are of the form (2.18) then Eqs (3.15) and (2.19) show that the corresponding waves must move in parallel directions. Furthermore $\sin (v_a - v_b)/2$ must vanish such that no essentially new solution v_b to the SGE (2.1) is obtained from the matrix Bianchi relations. Despite this severe restriction the matrix Bianchi relations are useful for generation of scalar Bianchi relations as we shall see in the following section.

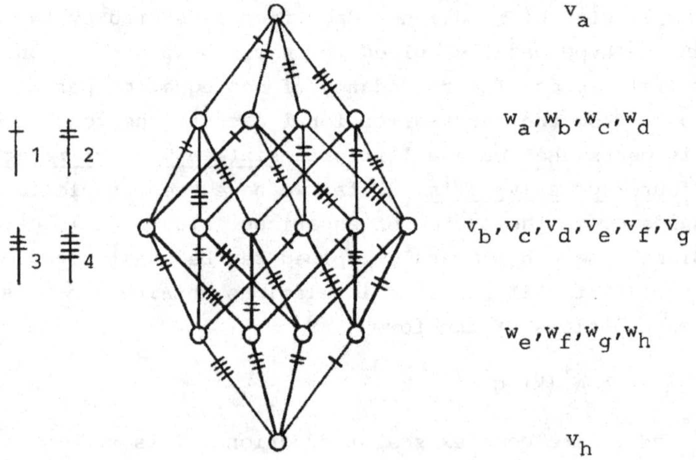

Figure 3. Projection of the Bianchi-Lamb hyperparallelepiped illustrating the commutative property of four successive BT's and inverted BT's. Each point (vertex) corresponds to a solution to either the SGE or the hyperbolic SGE. Each line (edge) corresponds to a BT or an inverted BT. The Bäcklund parameter set is indicated by bars corresponding to numbers 1, 2, 3, and 4.

4. Generation of a scalar Bianchi relation.

In order to get beyond the restrictions resulting from the matrix Bianchi relations we shall generate scalar Bianchi relations. We note that Eq (2.6) yields

$$\underline{\underline{N}}^{+}(j)\ \underline{\underline{N}}^{-}(j) = \underline{\underline{I}} \tag{4.1}$$

where $\underline{\underline{I}}$ is given by (2.5). Furthermore

$$\underline{\underline{N}}^{+}(j)\ \underline{\underline{N}}^{-}(k) + \underline{\underline{N}}^{+}(k)\ \underline{\underline{N}}^{-}(j) = 2\, L_{jk}\ \underline{\underline{I}} \tag{4.2}$$

with

$$L_{jk} = \cos\alpha_j \cos\alpha_k +$$

$$+ \sin\alpha_j \sin\alpha_k (\sin\gamma_j \sin\gamma_k + \cosh(\beta_j - \beta_k) \cos\gamma_j \cos\gamma_k) . \tag{4.3}$$

In order to generate a scalar relation we first use (3.5) as a typical example in the following manner

$$\underline{\underline{N}}^{+}(1)\left(\sin \frac{v_a + iw_a}{2} - \sin \frac{v_b + iw_b}{2}\right) \underline{\underline{N}}^{-}(1)\left(\sin \frac{v_a - iw_a}{2} - \sin \frac{v_b - iw_b}{2}\right)$$

$$= \underline{\underline{N}}^{+}(2)\left(\sin \frac{v_a + iw_b}{2} - \sin \frac{v_b + iw_a}{2}\right) \underline{\underline{N}}^{-}(2)\left(\sin \frac{v_a - iw_b}{2} - \sin \frac{v_b - iw_a}{2}\right). \tag{4.4}$$

By virtue of (4.1) this equation yields a single scalar equation which, however, turns out to be trivially fulfilled. Instead we multiply both sides of (3.5) with upper signs by $\underline{\underline{N}}^{-}(2)$ from the right and both sides of (3.5) with lower signs by $\underline{\underline{N}}^{+}(2)$ from the left. As a result we get

$$\underline{\underline{N}}^{+}(1)\ \underline{\underline{N}}^{-}(2)\ \left(\sin \frac{v_a + iw_a}{2} - \sin \frac{v_b + iw_b}{2}\right) =$$

$$\underline{\underline{N}}^{+}(2)\ \underline{\underline{N}}^{-}(2)\ \left(\sin \frac{v_a + iw_b}{2} - \sin \frac{v_b + iw_a}{2}\right) \tag{4.5}$$

and

$$\underline{\underline{N}}^{+}(2)\ \underline{\underline{N}}^{-}(1)\ \left(\sin \frac{v_a - iw_a}{2} - \sin \frac{v_b - iw_b}{2}\right) =$$

$$\underline{\underline{N}}^{+}(2)\ \underline{\underline{N}}^{-}(2)\ \left(\sin \frac{v_a - iw_b}{2} - \sin \frac{v_b - iw_a}{2}\right) \tag{4.6}$$

respectively. Addition of (4.5) and (4.6) now yields the single scalar Bianchi relation

$$L_{12}\left(\tan^2 \tfrac{1}{4}(v_a - v_b) + \tanh^2 \tfrac{1}{4}(w_a - w_b)\right)$$

$$= \tan^2 \tfrac{1}{4}(v_a - v_b) - \tanh^2 \tfrac{1}{4}(w_a - w_b) \tag{4.7}$$

or

$$\tan \tfrac{1}{4}\ (v_a - v_b) = \pm \sqrt{\frac{1 + L_{12}}{1 - L_{12}}}\ \tanh \tfrac{1}{4}\ (w_a - w_b) \tag{4.8}$$

where L_{12} is given by (4.3). Here (4.1), (4.2), and trigonometric addition formulae have been used. The same result was obtained in [4] and [6]. The scalar Bianchi relation corresponding to the matrix equation for the inverted Bianchi-Lamb diagram (3.6) is found to be

$$L_{12}\ \left(\tan^2 \tfrac{1}{4}(v_a - v_b) + \tanh^2 \tfrac{1}{4}(w_a - w_b)\right)$$

$$= -\ \tan^2 \tfrac{1}{4}(v_a - v_b) + \tan^2 \tfrac{1}{4}(w_a - w_b) \tag{4.9}$$

or

$$\tan \tfrac{1}{4}\ (v_a - v_b) = \pm \sqrt{\frac{1 - L_{12}}{1 + L_{12}}}\ \tanh \tfrac{1}{4}\ (w_a - w_b)\ . \tag{4.10}$$

Similar results can be derived from all the other matrix Bianchi relations in Section 3.

For $v_a \equiv 0$ ("the vacuum solution") the BT's (3.1) and (3.2) yield in accordance with (2.18)

$$w_a = 4\ \tanh^{-1}\ \exp\ S(1) \tag{4.11}$$

and

$$w_b = 4\ \tanh^{-1}\ \exp\ S(2) \tag{4.12}$$

respectively. Here we have introduced the abbreviation $S(j)$ for $S(\alpha_j, \beta_j, \gamma_j, \delta_j)$ with δ_j being the phase constant in $S(j)$. Insertion of the functions v_a, w_a, and w_b into (4.8) yields

$$v_b = 4\ \tan^{-1}\left(\mp\sqrt{\frac{1 + L_{12}}{1 - L_{12}}}\ \frac{\exp\ S(1) - \exp\ S(2)}{1 - \exp(S(1) + S(2))}\right)\ . \tag{4.13}$$

Now it can be shown by insertion [4] that v_b is a solution to the SGE (2.1). Analogously, Eq (4.10) yields the solution

$$w_b = 4\ \tanh^{-1}\ \left(\mp\sqrt{\frac{1 + L_{12}}{1 - L_{12}}}\ \frac{\exp\ S(1) - \exp\ S(2)}{1 + \exp(S(1) + S(2))}\right) \tag{4.14}$$

to the hyperbolic SGE (2.7).

The solution v_b (4.13) represents the non-linear superposition of two non-parallel plane solitons in 3 + 1 dimensions, of two non-parallel line solitons in 2 + 1 dimensions, and two solitons moving at different velocities in 1 + 1 dimensions. In all three cases the two soliton waves retain their identity after the non-linear intereaction between the two waves. Thus it is justified to denote the waves solitons.

5. Scalar Bianchi relations for three successive Bäcklund transformations.

In the case of three successive BT's we have obtained the six matrix Bianchi relations (3.7) - (3.12). The corresponding scalar Bianchi relations of the form (4.10) and (4.8) become

$$\tan\frac{1}{4}(v_a - v_b) = \pm\sqrt{\frac{1-L_{12}}{1+L_{12}}}\tanh\frac{1}{4}(w_a - w_b) \tag{5.1}$$

$$\tan\frac{1}{4}(v_a - v_c) = \pm\sqrt{\frac{1-L_{13}}{1+L_{13}}}\tanh\frac{1}{4}(w_a - w_c) \tag{5.2}$$

$$\tan\frac{1}{4}(v_b - v_c) = \pm\sqrt{\frac{1-L_{23}}{1+L_{23}}}\tanh\frac{1}{4}(w_a - w_d) \tag{5.3}$$

$$\tan\frac{1}{4}(v_c - v_d) = \pm\sqrt{\frac{1+L_{12}}{1-L_{12}}}\tanh\frac{1}{4}(w_c - w_d) \tag{5.4}$$

$$\tan\frac{1}{4}(v_b - v_d) = \pm\sqrt{\frac{1+L_{13}}{1-L_{13}}}\tanh\frac{1}{4}(w_b - w_d) \tag{5.5}$$

$$\tan\frac{1}{4}(v_a - v_d) = \pm\sqrt{\frac{1+L_{23}}{1-L_{23}}}\tanh\frac{1}{4}(w_b - w_c)\ . \tag{5.6}$$

One of these relations, (5.6) say, is redundant.

Starting with $w_a \equiv 0$ ("the vacuum solution") we get the solutions to the SGE (2.1)

$$v_a = 4\tan^{-1}\exp S(1) \tag{5.7}$$

$$v_b = 4\tan^{-1}\exp S(2) \tag{5.8}$$

$$v_c = 4\tan^{-1}\exp S(3) \tag{5.9}$$

in accordance with (2.17). From (5.1) - (5.3) we then get

$$w_b = 4\tanh^{-1}\left(\mp\sqrt{\frac{1+L_{12}}{1-L_{12}}}\ \frac{\exp S(1) - \exp S(2)}{1+\exp(S(1)+S(2))}\right) \tag{5.10}$$

$$w_c = 4\tanh^{-1}\left(\mp\sqrt{\frac{1+L_{13}}{1-L_{13}}}\ \frac{\exp S(1) - \exp S(3)}{1+\exp(S(1)+S(3))}\right) \tag{5.11}$$

$$w_d = 4\tanh^{-1}\left(\mp\sqrt{\frac{1+L_{23}}{1-L_{23}}}\ \frac{\exp S(2) - \exp S(3)}{1+\exp(S(2)+S(3))}\right) \tag{5.12}$$

in accordance with (4.14). Now insertion of v_c, w_c, and w_d into (5.4) yields one determination of v_d while insertion of v_b, w_b, and w_d into (5.5) yields another determination of v_d. If the BT's commute these two results might be expected to agree. The enormous task of checking this has not been carried out. Furthermore it is doubtful whether v_d is a solution to the SGE (2.1). In any case it is easily seen that v_d cannot be an ordinary superposition of soliton waves.

In order to study superpositions of soliton waves we consider the special case where $\underline{\underline{N}}^{\pm}(1)$, $\underline{\underline{N}}^{\pm}(2)$, and $\underline{\underline{N}}^{\pm}(3)$ are linearly dependent

$$p\,\underline{\underline{N}}^{\pm}(1) + q\,\underline{\underline{N}}^{\pm}(2) + r\,\underline{\underline{N}}^{\pm}(3) = 0 \ . \qquad (5.13)$$

Here p, q, and r are complex scalar functions. By virtue of (4.1) it can be shown that p, q, r must be real (apart from a common factor). Furthermore Eq (4.2) yields

$$\left.\begin{aligned} p\,L_{13} + q\,L_{23} + r &= 0 \\ p\,L_{12} + q + r\,L_{23} &= 0 \\ p + q\,L_{12} + r\,L_{13} &= 0 \end{aligned}\right\} \qquad (5.14)$$

and

$$\left.\begin{aligned} p^2 - q^2 - r^2 - 2qr\,L_{23} &= 0 \\ q^2 - r^2 - p^2 - 2rp\,L_{13} &= 0 \\ r^2 - p^2 - q^2 - 2pq\,L_{12} &= 0 \end{aligned}\right\} \qquad (5.15)$$

with L_{ij} given by (4.3). Figure 4 shows a trigonometrical interpretation of these results with

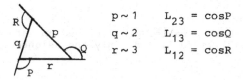

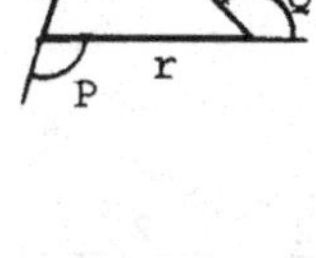

$p \sim 1 \qquad L_{23} = \cos P$

$q \sim 2 \qquad L_{13} = \cos Q$

$r \sim 3 \qquad L_{12} = \cos R$

Figure 4. Triangle illustrating the consequences, (5.14) and (5.15), of the linear dependence of $\underline{\underline{N}}^{\pm}(1)$, $\underline{\underline{N}}^{\pm}(2)$, and $\underline{\underline{N}}^{\pm}(3)$. The sides p, q, and r can be identified with the edges 1, 2, and 3 in Fig 2.

$$L_{23} = \cos P, \quad L_{13} = \cos Q, \quad \text{and} \quad L_{12} = \cos R, \tag{5.16}$$

Here p, q, and r are the sides of a triangle and P, Q, and R the corresponding opposite exterior angles.

From

$$P + Q + R = 2\pi \tag{5.17}$$

follows the result [10]

$$L_{23}^2 + L_{13}^2 + L_{12}^2 = 1 + 2\, L_{23}\, L_{13}\, L_{12} \,. \tag{5.18}$$

Now Eq (4.2) permits that the angles P, Q, and R can be interpreted as the angles in the Bianchi-Lamb parallelepiped (in Fig 2) between the edges 2 and 3, 1 and 3, and 1 and 2 at any of the vertices v_a, v_b, v_c, and v_d. Thus the sides p, q, and r correspond to the edges 1, 2, and 3. The result (5.17) then means that the three faces at each of the vertices v_a, v_b, v_c, and v_d in the Bianchi-Lamb parallelepiped lie in a plane. (A similar result for the three faces at w_a, w_b, w_c, and w_d can be found). Thus the Bianchi-Lamb parallelepiped is ripped up into a plane diagram of which we have shown the part around the vertex v_d in Fig 5. This part consists of the three parallelograms known from the Bianchi-Lamb parallelepiped. The corresponding Bianchi relations are (5.4) - (5.6). Elimination of w_b, w_c, and w_d from these equations should yield the Hirota result [8]

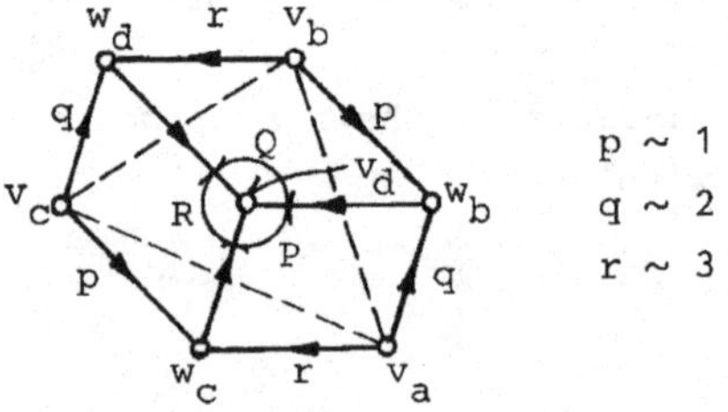

Figure 5. Bianchi-Lamb parallelepiped ripped up into a plane diagram when (5.13) or (5.17) is fulfilled. Only the neighbourhood of v_d is shown.

$$\tan\frac{v_d}{4} = \frac{\tan\frac{v_a}{4} + \tan\frac{v_b}{4} + \tan\frac{v_c}{4} + \tan^2\frac{P}{2}\tan^2\frac{Q}{2}\tan^2\frac{R}{2}\tan\frac{v_a}{4}\tan\frac{v_b}{4}\tan\frac{v_c}{4}}{1 + \tan^2\frac{R}{2}\tan\frac{v_a}{4}\tan\frac{v_b}{4} + \tan^2\frac{Q}{2}\tan\frac{v_a}{4}\tan\frac{v_c}{4} + \tan^2\frac{P}{2}\tan\frac{v_b}{4}\tan\frac{v_c}{4}} \,. \tag{5.19}$$

Note that

$$\tan^2 \frac{P}{2} = \frac{1 - L_{23}}{1 + L_{23}}\,, \quad \tan^2 \frac{Q}{2} = \frac{1 - L_{13}}{1 + L_{13}}\,, \text{ and } \quad \tan^2 \frac{R}{2} = \frac{1 - L_{12}}{1 + L_{12}} \tag{5.20}$$

and

$$\tan \frac{P}{2} \tan \frac{Q}{2} \tan \frac{R}{2} = \tan \frac{P}{2} + \tan \frac{Q}{2} + \tan \frac{R}{2} \tag{5.21}$$

due to (5.17). We may say that v_a, v_b, v_c, and v_d form a Bianchi-Lamb triple diagram which is described by the Bianchi relation (5.19). Insertion of v_a, v_b, and v_c, given by (5.7) - (5.9), into (5.19) gives a solution to the SGE (2.1) [8] and shows that the resulting v_d is a superposition of three soliton waves. By means of (2.6) the linear dependence of $\underline{\underline{N}}^{\pm}(1)$, $\underline{\underline{N}}^{\pm}(2)$, and $\underline{\underline{N}}^{\pm}(3)$, (5.13), yields the determinant condition

$$\begin{vmatrix} \cos\alpha_1 & \sin\alpha_1 \cosh\beta_1 & \sin\alpha_1 \sinh\beta_1 \\ \cos\alpha_2 & \sin\alpha_2 \cosh\beta_2 & \sin\alpha_2 \sinh\beta_2 \\ \cos\alpha_3 & \sin\alpha_3 \cosh\beta_3 & \sin\alpha_3 \sinh\beta_3 \end{vmatrix} = 0 \tag{5.22}$$

in the 2 + 1 - dimensional case where the Bäcklund parameters $\gamma_1 = \gamma_2 = \gamma_3 = 0$. The area between the three line solitons in (5.19) is defined as

$$A = \frac{1}{2} \frac{\begin{vmatrix} a_1 & b_1 & c_1 \\ a_2 & b_2 & c_2 \\ a_3 & b_3 & c_3 \end{vmatrix}^2}{\begin{vmatrix} a_1 & b_1 \\ a_2 & b_2 \end{vmatrix} \begin{vmatrix} a_2 & b_2 \\ a_3 & b_3 \end{vmatrix} \begin{vmatrix} a_3 & b_3 \\ a_1 & b_1 \end{vmatrix}} \tag{5.23}$$

when

$$S(j) = a_j x + b_j y + c_j \qquad j = 1,2,3 \,. \tag{5.24}$$

By comparison with (2.21) we see that

$$\begin{aligned} a_j &= \cos\alpha_j \,, \quad b_j = \sin\alpha_j \cosh\beta_j \\ c_j &= t \sin\alpha_j \sinh\beta_j + \delta_j \,. \end{aligned} \tag{5.25}$$

Insertion of (5.25) into (5.23) and use of (5.22) show the Gibbon-Zambotti theorem [10] that <u>A does not depend on t</u>. The area A is a function only of the Bäcklund parameters α_j and β_j and the phase constants δ_j.

6. Four successive Bäcklund transformations and plane solitons.

In the case of four successive BT's the 16 non-redundant matrix Bianchi relations can be scalarized in the forms (4.8) and (4.10). Of particular interest for the study of the interaction between four plane solitons in 3 + 1 dimensions is the case of linear dependence between $\underline{\underline{N}}^{\pm}(1)$, $\underline{\underline{N}}^{\pm}(2)$, $\underline{\underline{N}}^{\pm}(3)$, and $\underline{\underline{N}}^{\pm}(4)$

$$p\,\underline{\underline{N}}^{\pm}(1) + q\,\underline{\underline{N}}^{\pm}(2) + r\,\underline{\underline{N}}^{\pm}(3) + s\,\underline{\underline{N}}^{\pm}(4) = 0 \tag{6.1}$$

where p, q, r, and s are real functions (apart from a common factor). The condition (6.1) means that the four faces at each vertex of the Bianchi-Lamb hyperparallelepiped lie in a plane. The author believes to have found a Bianchi relation for the corresponding Bianchi-Lamb quadruple diagram (similar to (5.19) for the triple diagram). This result will be published elsewhere. In any case (6.1) seems to be a necessary condition for the existence of a superposition of four soliton waves. By means of (2.6) we derive the determinant condition

$$\begin{vmatrix} \cos\alpha_1 & \sin\alpha_1\cosh\beta_1\cos\gamma_1 & \sin\alpha_1\sin\gamma_1 & \sin\alpha_1\sinh\beta_1\cos\gamma_1 \\ \cos\alpha_2 & \sin\alpha_2\cosh\beta_2\cos\gamma_2 & \sin\alpha_2\sin\gamma_2 & \sin\alpha_2\sinh\beta_2\cos\gamma_2 \\ \cos\alpha_3 & \sin\alpha_3\cosh\beta_3\cos\gamma_3 & \sin\alpha_3\sin\gamma_3 & \sin\alpha_3\sinh\beta_3\cos\gamma_3 \\ \cos\alpha_4 & \sin\alpha_4\cosh\beta_4\cos\gamma_4 & \sin\alpha_4\sin\gamma_4 & \sin\alpha_4\sinh\beta_4\cos\gamma_4 \end{vmatrix} = 0 \tag{6.2}$$

from (6.1) in the 3 + 1 - dimensional case. The volume between four plane solitons is defined as

$$V = \frac{1}{6}\,\frac{\begin{vmatrix} a_1 & b_1 & c_1 & d_1 \\ a_2 & b_2 & c_2 & d_2 \\ a_3 & b_3 & c_3 & d_3 \\ a_4 & b_4 & c_4 & d_4 \end{vmatrix}^3}{\begin{vmatrix} a_1 & b_1 & c_1 \\ a_2 & b_2 & c_2 \\ a_3 & b_3 & c_3 \end{vmatrix}\begin{vmatrix} a_1 & b_1 & c_1 \\ a_2 & b_2 & c_2 \\ a_4 & b_4 & c_4 \end{vmatrix}\begin{vmatrix} a_1 & b_1 & c_1 \\ a_3 & b_3 & c_3 \\ a_4 & b_4 & c_4 \end{vmatrix}\begin{vmatrix} a_2 & b_2 & c_2 \\ a_3 & b_3 & c_3 \\ a_4 & b_4 & c_4 \end{vmatrix}} \tag{6.3}$$

when

$$S(j) = a_j\,x + b_j\,y + c_j\,z + d_j \qquad j = 1,2,3,4\ . \tag{6.4}$$

Comparison with (2.16) yields

$$a_j = \cos\alpha_j \quad b_j = \sin\alpha_j \cosh\beta_j \cos\gamma_j$$

$$c_j = \sin\alpha_j \sin\gamma_j \quad d_j = t \sin\alpha_j \sinh\beta_j \cos\gamma_j + \delta_j \ . \tag{6.5}$$

By insertion of (6.5) into (6.3) and use of (6.2) we find the new theorem [12] that the volume V does not depend on t. This seems to be a necessary condition for four plane solitons moving in XYZ-space.

Acknowledgements.

The author wishes to thank Professor R K Bullough, Department of Mathematics, The University of Manchester Institute of Science and Technology, who generously stimulated this investigation during his visit at Nordita, Copenhagen. Dr C Rogers, Department of Applied Mathematics, University of Waterloo, Dr. O H Olsen, Laboratory of Applied Mathematical Physics, The Technical University of Denmark, and Dr F D Pedersen, Department of Mathematics, The Technical University of Denmark, are acknowledged for helpful discussions.

References.

[1] Barone, A, Esposito, F, Magee, C J, Scott, A C, Theory and applications of the sine-Gordon equation, Riv Nuovo Cimento 1, 227-267 (1971).

[2] Scott, A C, Propagation of magnetic flux on a long Josephson tunnel junction, Nuovo Cimento, 69B, 241-261 (1970).

[3] Dodd, R K, Bullough, R K, Bäcklund transformations for the sine-Gordon equations, Proc Roy Soc London, A 351, 499-523 (1976).

[4] Leibbrandt, G, New exact solutions of the classical sine-Gordon equation in 2+1 and 3+1 dimensions, Phys Rev Letters 41, 435-438 (1978).

[5] Christiansen, P L, A Bäcklund transformation for the 3+1-dimensional sine-Gordon equation, Proc from the 8th International Conference on Non-Linear Oscillations held in Prague, September 11-15, 1978 (to appear).

[6] Christiansen, P L, Olsen, O H, Ring-shaped quasi-soliton solutions to the two- and three-dimensional sine-Gordon equation, Physica Scripta 20, 531-538 (1979).

[7] Hirota, R, Exact solution of the sine-Gordon equation for multiple collisions of solitons, J Phys Soc Japan 33, 1459-1463 (1972).

[8] Hirota, R, Exact three-soliton solution of the two-dimensional sine-Gordon equation, J Phys Soc Japan 35, 1566 (1973).

[9] Kobayashi, K K, Izutsu, M, Exact solution of the n-dimensional sine-Gordon equation, J Phys Soc Japan 41, 1091-1092 (1976).

[10] Gibbon, J D, Zambotti, G, The interaction of n-dimensional soliton wave fronts, Nuovo Cimento, 28B, 1-16 (1975).

[11] Dirac, P A M, The Principles of Quantum Mechanics, 4th ed, Oxford, 150 (1958).

[12] Christiansen, P L, Application of new Bäcklund transformations for the 2+1 and 3+1-dimensional sine-Gordon equations, Z Angew Math Mech (to appear).

Bäcklund Transformations

R.K. Dodd and H.C. Morris
Trinity College, Dublin 2, Ireland .

It is nearly a hundred years since Bäcklunds' memoirs appeared which introduced the transformation which bears his name [1],[2]. It is therefore a suitable occassion upon which to present a review of the classical and contemporary work upon this subject. As discussed by Lamb [3] interest in the transformations continued from their inception until about 1925 when reference to them disappeared from the literature. Recent interest can really be said to have begun at the beginning of the seventies, although there are a few earlier important physical applications of them [4], [5], [6], with their reappearance as a characteristic feature of equations which can be associated with an inverse problem. There is also a marked difference in approach between the two periods; the earlier authors took as their starting point the definition of a Bäcklund transformation and then proceeded to classify the different types of equations in complete generality. Modern work however is more interested in the converse question: *given an equation when is it possible to associate a Bäcklund transformation with it ?* As we shall see this is just another way of asking when an equation has an inverse problem associated with it because in fact the inverse problem is a Bäcklund transformation of a special type.

§1 Classical Theory

Much of the classical work on transformation theory was motivated by geometric ideas. In particular the theory of first order p.d.e.'s had been considerably advanced by Lie's *contact transformations*.

In order to make this precise we now introduce the necessary machinery. We shall only deal with equations in one dependent variable. All mappings are assumed smooth (C^∞) and since the theory is local all manifolds are submanifolds of R^n for some n . Denote the space of all homogeneous j-forms on a manifold M by $\Lambda^j(M)$ ($\Lambda^0(M)$ is $C^\infty(M,R)$). Introduce the *rth jet bundle* $J^r(n,1) \equiv J^r(R^n,R)$ of maps $f:R^n \to R$ given by $z = f(u)$, ((u), (z) local coordinates on R^n, R respectively) which has local coordinates $(uz_r) = (u,z,z_\alpha : |\alpha|<r)$. We have used a multi-index notation here, $\alpha = (\alpha_1,\ldots\alpha_n)$, $|\alpha| = \sum_{i=1}^{n} \alpha_i$ and $\alpha_i \varepsilon Z_+$. We also distinguish between points and maps by using Roman and *Script* lower case type respectively. Thus if $p \varepsilon R^n$ then $u(p) = u$ and we adopt the convention of denoting the general point p by u in (u) etc. The map f prolongs to a map $j^r f:R^n \to J^r(n,1)$ called the *rth jet extension* of f which defines

a n-dimensional submanifold of $J^r(n,1)$

$$u \mapsto (u, z=f(u), z_1=f_{,1}(u), \ldots\ldots, z_r=f_{,r}(u))$$

$$\text{where } f_{,l}(u) = \left\{ \frac{\partial^{|\alpha|} f(u)}{\partial u_1^{\alpha_1} \ldots \partial u_n^{\alpha_n}} \;:\; |\alpha| = \ell \right\}$$

Example 1

The case of principal interest is when $n = 2$, $(u) = (x,y)$. It is also convenient to have a notation for the coordinates of $J^r(2,1)$ and the partial derivatives of functions in this case. We write $f_{,mxny} \equiv \frac{\partial^{m+n} f}{\partial_x^m \partial_y^n}$ and the rth jet extension of f is then given by

$$(x,y) \mapsto (x,y,z=f(x,y),\ z_x=f_{,x}(x,y),\ldots,z_{mx(r-m)y}=f_{mx(r-m)}(x,y),$$
$$\ldots, z_{ry}=f_{,ry}(x,y))$$

$r=1$ $(xz_1) \equiv (x,y,z,z_x,z_y)$ since $(z_\alpha : |\alpha|<1) \equiv (z_{(1,0)}, z_{(0,1)})$

$r=2$ $(xz_2) \equiv (x,y,z,z_x,z_y,z_{2x},z_{xy},z_{2y})$ since

$$(z_\alpha : |\alpha|<2) \equiv (z_{(1,0)}, z_{(0,1)}, z_{(2,0)}, z_{(1,1)}, z_{(0,2)})$$

On $J^r(n,1)$ there is defined a submodule of the module of one forms over the zero forms called the *contact module* $\Omega^r(n,1)$. It is spanned in (uz_r) by the canonical one forms,

$$\theta_\alpha = dz_\alpha - z_{\alpha(i)} du^i \qquad |\alpha| < r-1$$

where $\alpha(i) = (\alpha_1,\ldots,\alpha_{i-1},\alpha_i+1,\ldots,\alpha_n)$ and we have adopted the dummy index notation; the same index which occurs in a raised and a lowered position is a summed index. A *system of equations* of order r is a submanifold of $J^r(n,1)$. This is determiend by a differentiable manifold S_r and an embedding $i : S_r \hookrightarrow J^r(n,1)$. Normally the system of equations is given in terms of a vector valued function $e_r \in C^\infty(J^r(n,1), R^m)$ and $i(S_r)$ is realised implicitly as the kernel of e_r. If $f: R^n \to R^m$ then the map $f^* : \Lambda(R^m) \to \Lambda(R^n)$ is defined by its action on 0 and 1 forms,

$$f^* g = g \circ f \;, \quad f^* dv = \frac{\partial v}{\partial u^i} du^i \qquad g \in \Lambda^0(J^r(n,1))$$

and the properties

$$f^* d = d f^* \;, \quad f^*(\lambda \wedge \eta) = f^*(\lambda) \wedge f^*(\eta) \;, \; \lambda, \eta \in \Lambda(R^m).$$

In the generic case with which we shall principally be dealing *a solution* $f \in C(R^n, R)$ of the equations satisfies $j^r f(R^n) \subset i(S_r)$. A solution annuls the contact module since locally,

$$j^r f^* \theta_\beta \;=\; \frac{\partial f_{,\beta}}{\partial u^i} du^i \;-\; f_{,\beta(i)} du \;\equiv\; 0 \qquad |\beta|<r-1$$

Similarly if $g:R^n \to S_r$ then there exists a $h \in C(R^n,R)$ such that $i o g = j^r h$ provided $(iog)^* \Omega^r(n,1) = 0$ and iog is *transversal* to the fibers of $\alpha: J^r(n,1) \to R^n$. Locally the transversality condition means that on $(iog)R^n$, $du^1 \wedge \ldots \wedge du^n \neq 0$.

Definition

A contact transformation f *on* $J^1(n,1)$ *is a local diffeomorphism which preserves the contact module,*

$$f^* \Omega^1(n,1) \subset \Omega^1(n,1)$$

Example 2

Let n=2, r=1, the contact module is spanned in (uz_1) by the canonical one form,

$$\theta_o \;=\; dz - z_x dx - z_y dy$$

If $(uz_1') = (x',y',z',z_x',z'_y)$ and $(uz_1') = f(uz_1)$ is a local diffeomorphism of $J^1(n,1)$ which preserves the contact module then there exists a $g \in (J^1(n,1))$ such that

$$f^* \theta_o \;=\; g\theta_o \text{ or } f^*(dz'-z_x'dx' - z_y'dy') \;=\; g.(dz-z_x dx-z_y dy) \qquad (1.1)$$

The *contact transformation*, that is the coordinate functions of f can be written in this coordinate system as

$$x' = X(uz_1),\; y'=Y(uz_1),\; z' = Z(uz_1),\; z_x' = Z_X(uz_1),\; z_y' = Z_Y(uz_1) \qquad (1.2)$$

Contact transformations were introduced by Lie in his work on transformations of first order p.d.e.'s [7]. Since

$$\begin{array}{ccc} S_1 & \xrightarrow{\;i_1\;} & J^1(n,1) \\ & \searrow_{i_2} & \downarrow f \\ & & J^1(n,1) \end{array}$$

we see that contact transformations have the property that jet extensions of solutions of a given system of equations are transformed into the jet extensions of solutions of another system provided the transversality condition is met. Thus solutions of the original system can be obtained from solving the transformed system. In the case of a first order differential equation defined locally by, $e_1(uz_1) = 0$, e_1 can always be made one of the constituent transformations. This fact together with Lie's theory of "*function groups*" [8], provides an effective method for investigating the *complete*, *general*, and *singular* solutions of first order systems.

Example 3

$$z(x,y) - xz,_x(x,y) - yz,_y(x,y) = 0$$

In $J^1(n,1)$ let $z' = Z(uz_1)$ where $Z \equiv z - x.z_x - y.z_y$. On the functions $\Lambda^o(J^1(2,1))$ we can introduce a *Poisson bracket* defined by

$$\{f,g\} = -X_f(g) \qquad f,g \in \Lambda^o(J^1(2,1))$$

where $X_f = (f,_{z_x} D_x^1 + f,_{z_y} D_y^1) - (D_x^1 f)\frac{\partial}{\partial z_x} - (D_y^1 f)\frac{\partial}{\partial z_y}$

is the *Cauchy characteristic vector field* associated with f, and $D_x^1 = \frac{\partial}{\partial x} + z_x\frac{\partial}{\partial z}$ is the *total derivative operator*. Then $\phi = (X, Y, Z, Z_X, Z_Y)$ defines a contact transformation provided,

$$\{X,Y\} = 0, \ \{Y,Z\} = 0, \ \{X,Z\} = 0$$

A solution is $X = z_x$, $Y = z_y$, $Z = z - x.z_x - y.z_y$. The functions Z_X, Z_Y are then determined by

$$Z_X = \frac{\partial Z}{\partial X} = \frac{\partial Z}{\partial z_x} = -x, \ Z_Y = \frac{\partial Z}{\partial Y} = \frac{\partial Z}{\partial z_y} = -y$$

and we find that the contact transformation ensures that

$$\phi^* (dz' - z'_x dx' - z'_y dy') = dz - z_x dx - z_y dy.$$

Clearly $z' = 0$, $x' = a$, $y' = b$, a, b constants satisfies $dz' - z'_x dx' - z'_y dy' = 0$. This corresponds to the *complete integral (solution)*, $z = ax + by$ of the original equation. This relationship can also be satisfied by putting $x' = f(y')$, f arbitrary, which requires that $z'_x . f,_{y'} + z'_y = 0$ or equivalently that $x.f \circ Y,_{z_y} + Y = 0$. It follows that $z_y = h(y/z)$ and $z_x = g(y/x)$ so that the *general solution* is $z = x(z_x + y/xz_y) = x(h(y/x) + (y/x)g(y/x))$ that is $z = x\ell(y/x)$ where ℓ is an arbitrary function of one variable. There is no *singular integral* (solution).

In general the contact transformations on $J^1(n,1)$ form a *psuedo group* G. Because of this one usually deals with the *infinitesimal transformations* of G. Each such transformation V is a local vector field on whose domain of definition there is a one parameter family of contact transformations $f_t \in G$, $-\varepsilon < t < \varepsilon$, f_0 = *identity*, such that $V = \frac{\partial f}{\partial t}t|_{t=0}$. The set of all such V obtained from G is called the *Lie algebra* of G. If f_t is the one parameter family of V then the definition of contact transformation in terms of the infinitesimal transformation is

$$\lim_{t \to 0} \frac{(f_t^* - f_0^*)}{t} \Omega^1(n,1) \subset \Omega^1(n,1)$$

or $\mathcal{L}_v \Omega^1(n,1) \subset \Omega^1(n,1)$ where $\mathcal{L}_v$ denotes the *Lie derivative*.

If one attempts to extend the definition of a contact transformation to one of higher order, then one finds the following theorem to be valid.

Theorem 1.1

An rth order contact transformation when the target manifold has dimension one, is the prolongation of a contact transformation [9], [10], [11], [12].

Proof: Define the map δ as the composition of exterior differentiation with the natural projection

$$\Omega^r \longleftarrow d \longrightarrow \Lambda^2(J^r(n,1)) \longrightarrow \Lambda^2(J^r(n,1)) \mod \Omega^r$$

where we have written $\Omega^r \equiv \Omega^r(n,1)$. Then in particular we have the short exact sequence,

$$0 \rightarrow \Omega^{r(1)} \rightarrow \Omega^r \xrightarrow{\delta} d\Omega^r \mod \Omega^r \rightarrow 0$$ where $\Omega^{r(1)}$ is the kernel of δ which is called the *first derived system* of Ω^r. Inductively we have that the *jth derived system* is defined by

$$0 \rightarrow \Omega^{r(j)} \rightarrow \Lambda^{r(j-1)} \xrightarrow{\delta} d\Omega^{r(j-1)} \mod \Omega^{r(j-1)} \rightarrow 0$$ and $\Omega^{r(r)} = \{0\}$.

Now if f is an rth order contact transformation then

$$f^*\Omega^r \subset \Omega^r$$

and since $f^* d = d f^*$, it follows that

$$f^*\Omega^{r(j)} \subset \Omega^{r(j)}$$ and that consequently

$$f^*\Omega^{r(r-1)} \subset \Omega^{r(r-1)}$$

But $\Omega^{r(r-1)}$ is locally generated in $(u z_r)$ by $\theta_0 = dz - z_{(i)} du^i$ so that f is the prolongation of a first order contact transformation.

By the *jth total prolongation* $f^{(j)}$ of a map $f: R^m \rightarrow R^n$ we mean the map formed from f and the set of distinct derivatives of the function up to the jth order.

Example 4

$$e_2 = z_t + z.z_x + z_{2x}$$
$$e_2^{(1)} = (e_2, z_{xt} + (z_x)^2 + z.z_{2x} + z_{3x}, z_{2t} + z_t.z_x + z.z_{xt} + z_{2xt})$$
$$e_2^{(2)} = (e_2^{(1)}, z_{2xt} + 3z_x.z_{2x} + z.z_{3x} + z_{4x}, z_{x2t} + 2z_x.z_{xt} + z_t.z_{2x}$$
$$+ z.z_{2xt} + z_{3xt}, z_{3t} + z_{2t}.z_x + 2z_t.z_{xt} + z.z_{x2t} + z_{2x2t})$$

Theorem I.1 implies that the *finite* higher order contact transformations are only applicable to equations which are the total prolongations of a first order system.

A *Bäcklund transformation* represents a generalisation of a contact transformation which is applicable to equations of any order greater than one. Posed for the two independent variable case the formulation of the "*Bäcklund Problem*" for first order as given by Clairin [13] is as follows.

Given four equations $F_i\ (uz_1, uz_1') = 0\ (i=1,\ldots,4)$ *between the coordinates of two systems of surface elements* (E), (E') *to determine the surfaces of the system* (E) *which correspond to the surfaces of* (E').

A surface, integral or contact element in the system (E) is determined by a subspace of the tangent space at (u,z) which is annihilated by $\theta_0 = dz - z_x dx - z_y dy$.

<u>*Example 5*</u>

Consider the Bäcklund transformation

$$z'_x + z_x - a\exp\tfrac{1}{2}(z - z') = 0 \qquad x' = x$$

$$z'_y - z_y - 2a^{-1}\exp\tfrac{1}{2}(z + z') = 0 \qquad y' = y$$

A solution surface in the primed variables $z' = f'(x,y)$ annihilates the contact form, θ_0. Using the transformation this pulls back to

$$dz' - (z_x - a\exp\tfrac{1}{2}(z-z'))dx - (z_y + 2a^{-1}\exp\tfrac{1}{2}(z+z'))dy$$

For each solution $z = f(x,y)$ we require there to be corresponding solutions $z' = f'(x,y)$. This imposes the condition that the relationship should be *completely integrable* when restricted to a solution submanifold,

$$d\,(f,_x - a\exp\tfrac{1}{2}(f - z'))dx - (f,_y + 2a^{-1}\exp\tfrac{1}{2}(f + z')\ dy = 0,$$

$dx\wedge dy \neq 0$ which is satisfied provided f satisfies

$$f,_{xy} + \exp f = 0 \quad (\textit{Liouville equation } [14]).$$

Conversely if we are given a solution $z' = f'(x,y)$ then the complete integrability of the pulled back form θ_0 when restricted to a solution submanifold requires that

$$f',_{xy}(x,y) = 0$$

Thus this Bäcklund transformation establishes a *correspondence* between the solutions of the equations,

$$z_{xy} + e^z = 0 \;\xleftarrow{B.T.}\!\!\rightarrow\; z'_{xy} = 0$$

The general solution of $z'_{xy} = 0$ is $z' = h(x) + g(y)$ and consequently by inverting the Bäcklund transformation we are able to obtain the general solution of the Liouville equation,

$$z = \log \frac{-2G(y)H(x)}{(G(y) + H(x))^2}$$

where $G(y) = a^{-1}\int^{y} e^{g} dy$ and $H(x) = -a/2\int^{x} e^{-h} dx$ are unique up to an arbitrary constant.

In example 5 we see that a given solution of one equation corresponds to a single infinity of solutions of the other and vice-versa. For under the transformation the pull back of the form θ is to be completely integrable when restricted to a solution submanifold and so the transformed solution involves an arbitrary constant. The Bäcklund transformation of first order is therefore not a transformation in the ordinary sense between the solution spaces of the transformable equations; it is though of course a map $J^1(2,1)\times R \to J^1(2,1)$.

The Bäcklund Problem stated earlier arose from Bäcklund's investigations into systems of first order equations and transformations between surfaces in Euclidean 3-space. Before developing the theory we present the transformation he discovered relating 2-surface of *constant negative curvature*. The following theorems are classical [15].

Theorem 1.2

The metric of a surface of constant negative curvature $K = -1$, *when referred to its lines of curvature has the form,*

$$ds^2 = \cos^2\frac{z}{2}dx^2 + \sin^2\frac{z}{2}dy^2$$

where the function z *is a solution of the equation,*

$$z_{,2x} - z_{,2y} - \sin z = 0$$

Theorem 1.3

From a given solution z *of the sine-Gordon equation a double infinity of further solutions may be obtained by solving the first order completely integrable system,*

$$(\sin a)(z'_x + z_y) = 2(\sin\tfrac{1}{2}z'\cos\tfrac{1}{2}z - \cos a\cos\tfrac{1}{2}z\sin\tfrac{1}{2}z')$$
$$(\sin a)(z'_y + z_x) = 2(-\cos\tfrac{1}{2}z'\sin\tfrac{1}{2}z + \cos a\sin\tfrac{1}{2}z'\cos\tfrac{1}{2}z)$$

The double infinity of solutions obtained from the Bäcklund transformation , which we denote by $B(a)$ arises from an arbitrary constant of integration and from the arbitrariness of the parameter a.

In Bäcklund's original work the transformations were between surfaces in 3 dimensions. Thus locally the solutions of the associated differential equations could be written as $z = f(x,y)$, $z' = g(x',y')$. Of course this is the case of principal interest and lends itself well to the jet bundle formation introduced earlier. For two contact forms $\omega_1 = dz - z_x dx - z_y dy$, $\omega_2 = dz' - z'_x dx' - z'_y dy'$ connected by four relations $F_i(uz_1, uz'_1) = 0$ the *general Bäcklund Problem* would be: *to obtain all the two dimensional solution manifolds of the associated Pfaffian system.* This generalisation is due to Goursat [16]. Let M be a six dimensional manifold with local coordinates (v) and let i: $M \to N$, $N = J^1(2,1) \otimes J^1(2,1)$ be realised as the submanifold defined by the kernel of $F = (F_1, \dots, F_4)$. If p_1, p_2 are the projections onto the first and second factors of N define

$$I(N) \quad = \quad p_1{}^* \, \Omega^1(2,1) \oplus p_2{}^* \, \Omega^1(2,1)$$

I(N) is spanned by $\{\omega_1, \omega_2\}$ so the associated Pfaffian system $i^* I(N)$ is spanned by $\{i^*\omega_1, i^*\omega_2\}$. In studying the solution manifolds of such a system one is automatically led to consider the exterior system generated by $\{i^*\omega_1, i^*\omega_2\}$.

Definition

Let I(M) be a Pfaffian system on M spanned by $\{\theta_i\}$. Then the exterior system generated by $\{\theta_i\}$ and denoted by E(M), $\overline{\{\theta_i\}}$ or $\overline{I}(M)$ is the differential ideal of $\Lambda(M)$ defined by

$$E(M) \quad = \quad \{\omega : \ \omega = \lambda^i \wedge \theta_i \ + \ \mu^i d\theta_i, \lambda^i, \mu^i \in \Lambda(M)\}$$

The reason for this is that Cartan (Les Systemes Differentielles Exterieures et Leurs Applications Geometrique, Hermann, Paris, 1946), showed in the analytic case that given such a system one can locally construct integral (solution) manifolds of the system up to a maximum dimension called the *genus* of the system. In the case under consideration the genus is 2. A solution or integral manifold of E(M) = I(M), I(M) = $i^*I(N)$, is a submanifold W, dim W = 2, $f: W \to M$ such that $f^*E(M) = 0$. This definition of a solution allows for the possibility that besides giving a correspondence between surfaces, Bäcklund transformations also transform between curves and surfaces and points and surfaces. Thus for example let M = X x Y, dim Y = 1 and let $i = (i_1, i_2)$ and suppose that $f: W \to M$ is a solution such that

$$(i \circ f)^* \, I(N) \ = (p_1 \circ i_1 \circ f)^* \ \Omega^1(2,1) \oplus (p_2 \circ i_2 \circ f)^* \Omega^1(2,1) = 0$$

so that locally $(i_1 \circ f)^* \ (dz - z_x dx - z_y dy) \ = \ 0$ and $(i_2 \circ f)^* \ (dz' - z'_x dx' -$

$z'_y dy')$ = 0. Thus although W is a 2-dimensional solution submanifold of M, the Bäcklund transformation is effected between a surface and a curve.

For simplicity we restrict ourselves to the case when the transformation is between surfaces, and when the coordinates which parametrise the surfaces are the same. It follows that M is fibred by W and that solutions are cross sections. In this case Cartan [17] refers to E(M) as being *in involution* with respects to x and y, (x,y) a local coordinate system on W. This corresponds to the *Restricted Bäcklund Problem* (Goursat [16]) considered by Bäcklund [1]: *to determine the 2 dimensional solution submanifolds of* E(M) *which are cross sections of* $\Pi: M \to W$).

Notice that the transformations F are only determined up to a contact transformation. For let $g \in \Lambda^0(J^1(2,1))$ be a contact transformation then $g^* \Omega^1(2,1) \subset \Omega^1(2,1)$ so that $j = io(g,g)$, $j: M \to N$ defines another Bäcklund. The method used by Goursat was to obtain *canonical forms* for the generators of E(M) through an analysis of its *singular integral elements*. Let V(M) denote the *module of vector fields* on M over $\Lambda^0(M)$, and $\lrcorner$ the *contraction operator*. The contraction operator is defined in local coordinates (x) on M for $X \in V(M)$ $\omega \in \Lambda^r(M)$ by

$$X \lrcorner \omega = X^a \omega_{abc\ldots}\, dx^b \wedge dx^c \wedge \ldots \qquad X = X^a \frac{\partial}{\partial x^a}, \quad \omega = \omega_{abc\ldots}\, dx^a \wedge dx^b \wedge \ldots$$

Definition

Let I(M) be a Pfaffian system on M then the completely integrable vector field system (involutive distribution)

$\mathrm{Char}(I) = \{X \in V(M): X \lrcorner I(M) \subset I(M)\}$ is called the *characteristic system* of I(M).

The characteristic system of I(M) defines a foliation of M by submanifolds, the *Cauchy characteristic foliation*.

Definition

The codimension of the Cauchy characteristic foliation is called the *class* of I(M).

Locally it is equal to the smallest number of variables necessary to write down the local generators of I(M). The class of a form is defined in an analogous fashion.

Example 6

Let i: $S_1 \rightarrow J^1(n,1)$ be the equation implicitly defined by the hypersurface $f(uz_1) = 0$. Then on S_1 we have the Pfaffian system $I(S_1)$ generated by $\{i^*\theta_0\}$ and $X \in$ Char I is determined by $X \lrcorner I(S_1) \subset I(S_1)$. Solutions of the equation are determined by jet extensions of maps f: $R^n \to J^1(n,1)$ which annihilate the exterior system $E(J^1(n,1)) = \{f, df, \theta_0, d\theta_0\}$. Char I has a generator X such $i_*X = Y$, $Y \in V(J^1(n,1))$ where Y $I(J^1(n,1)$ $I(J^1(n,1)$, and $I(J^1(n,1))$ is the exterior system generated by the Pfaffian system spanned by $\{df, \theta_0\}$. In $(uz_2) = (u^a, z, p^a)$ let $Y = Y^a_x \frac{\partial}{\partial u^a} + Y_z \frac{\partial}{\partial z} + Y^a_p \frac{\partial}{\partial p^a}$

Then since (a) $V \lrcorner df = 0$, (b) $Y \lrcorner \theta_0 = 0$, (c) $Y \lrcorner d\,\theta_0 = h\theta_0 + g.df$ $h, f \in \Lambda^0(J^1(n,1)$ we find that

$$i_*X = f,p^a \frac{\partial}{\partial x^a} + (f,_z .\ p^a .\ f,_{p}a) \frac{\partial}{\partial z} - (f,_z p^a + f,_x a) \frac{\partial}{\partial p^a}$$

class $\Omega^1(2,1) = 5$

class $(I(N)) =$ class $(p_1^* \Omega^1(2,1) \oplus p_2^* \Omega^1(2,1)) = 10$

From the outset we will assume that E(M) has class 6, that is that there are no Cauchy characteristic vector fields. However besides Cauchy characteristics more general characteristics defined by *singular vector fields* may exist. Hermann [18] has called these *Cartan characteristics*. Associated with the generators of E(M) is, as we show below, an equation of the second order called the *resolvent* of E(M). If one of the generators corresponds to a *singular form* of E(M) then the resolvent is of the *first type*, otherwise it is of the *second type*.

Resolvents of the first type

Let dM be a volume element on M and define the symmetric conformal $\Lambda^0(M)$ bilinear form $\langle , \rangle$ on I(M) by

$$d\lambda \wedge d\mu \wedge \phi_1 \wedge \phi_2 = \langle \lambda, \mu \rangle\ dM$$

where $\lambda, \mu, \phi = i^* w$, $\phi_2 = i^* \omega_2 \in I(M)$. Form $\Theta = h\phi, + \ell\phi_2$, $h, \ell \in \Lambda^0(M)$.
Then $\langle \Theta, \Theta \rangle = h^2 \langle \phi_1, \phi_1 \rangle + 2h\ell \langle \phi_1, \phi_1 \rangle + \ell^2 \langle \phi_2, \phi_2 \rangle \equiv Q'(h, \ell)$

The roots of Q, (h_1, ℓ_1), (h_2, ℓ_2) determine two one forms Θ, Θ', the singular forms of E(M) having the property that

$$(d\theta)^2 \wedge \phi_1 \wedge \phi_2 = 0$$

which is contact invariant. Consequently we are able in general to generate E(M) in precisely two ways from either

$\{\theta,\pi\}$ on $\{\theta\,|\,\pi'\}$ where π,π' are linear combinations of ϕ_1,ϕ_2 which complete the basis. It follows that up to a contact transformation, *there are precisely two resolvants e, e' of the first type associated with* $E(M)$. Let us consider canonical forms for $\{\theta,\pi\}$. Assuming that θ is of class 5 then we can choose $(\nu) = (x,y,x_1,x_2,x_3,x_4)$ so that

$$\theta = dx_1 - x_2dx - x_3dy$$

and

$$\pi = Xdx + Ydy + X_2dx_2 + X_3dx_3 + X_4dx_4$$

$$(X,Y,X_1,X_2,X_3,X_4)\in\Lambda^0(M)$$

Let Car I be the set of Cartan characteristic vector fields on M defined by I(M). Then if $V\varepsilon$ CarI it is necessary that $V\lrcorner I(M) = 0$ and $\dim\,(\,(V\lrcorner dI(M))\bmod I(M)) < \dim \Lambda^1(M) - \dim I(M) = 6-2 = 4$. These conditions are needed to ensure that V lies in a solution element of I(M), that is locally it is a singular element. The condition that θ is a singular form requires that there exists a basis of $\Lambda^1(M)$, $(\theta,\pi,\omega_3,\omega_4,\omega_5,\omega_6)$ such that

$$dV\lrcorner\theta = 0 \bmod I(M)$$

in other words $\quad dV\lrcorner\theta = f\theta + g\pi \quad f,g\ \varepsilon\ \Lambda^0(M)$

With $V = V^x\frac{\partial}{\partial x} + V^y\frac{\partial}{\partial y} + V^{x^i}\frac{\partial}{\partial x^i}$ we find that it is necessary that

$X_4 = 0$ and that the orbits of V are partially determined by

$$\frac{dx}{x_2} = \frac{dy}{x_3} = \frac{dx^1}{(x^2X_2+x^3X_3)} = \frac{-dx^2}{X} = \frac{-dx^3}{Y} \qquad (1.3)$$

Notice however that there is no equation for dx^4. Goursat also considers the cases when class $\theta < 5$. Thus we have shown that when there are two distinct singular forms θ,θ', E(M) can be generated by either of the sets $\{\theta,\pi\}$, $\{\theta',\pi\}$ which can be put into the canonical forms

$$\theta = dz - z_xdx - z_ydy$$

$$\pi = Xdx + Ydy + Z_xdz_x + Z_ydzy \qquad (1.4)$$

in the coordinate systems $(\nu) = (x,y,z,z_x,z_y,u)$, (ν) on M. Let ϕ: $R^2 \to M$, $(x,y) \to (x,y,z = f(x,y),\ z_x = f_{,x}(x,y),\ z_y = f_{,y}(x,y),\ u = g(x,y))$ be a solution of E(M). Then $\phi^*\pi = 0$ becomes

$$(\phi^*X + \phi^*Z_xf_{,xx} + \phi^*Z\,f_{,xy})(x,y) = 0 \qquad (1.5)$$

$$(\phi^*Y + \phi^* Z_x f_{xy} + \phi^* Z_y f_{,yy})(x,y) = 0$$

and the elimination of g between these equations leads to a p.d.e. of second order

$$e\,(x,y,z,z_x,z_y,z_{xx},z_{xy},z_{yy}) = 0 \qquad (1.6)$$

defined on $J^2(2,1)$, $(uz_2) = (x,y,z,z_x,z_y,z_{xx},z_{xy},z_{yy})$. This equation is a *resolvent of the first type*. Eliminating x^4 between equations (1.3) gives four one forms which together with θ generate a Pfaffian system which determines the characteristics of the first order for (1.6)(i.e. the defining equations don't involve the coordinates z_{xx},z_{xy},z_{yy}). *Thus solving the restricted Bäcklund problem can be brought in two different ways into an equivalent problem, that of solving a second order p.d.e. which admits a family of characteristics of the first order.*

Resolvents of the second type

Suppose $\{\phi_1,\phi_2\}$ is not singular. Then there exists a coordinate system (x,y,z,z_x,z_y,w) on M such that

$$\begin{aligned}\phi_1 &= dz - z_x dx - z_y dy\\ \phi_2 &= dw - X dx - Y dy - Z_x dz_x - Z_y dz_y\end{aligned} \qquad (1.7)$$

The sixth differential dw is necessary here for without it ϕ_1 is singular and we are back to the previous case. Since we are looking for solutions $z = f(x,y)$, $w = g(x,y)$ it is natural to take $M = J^1(2,1) \times W$ so that solutions h: $R^2 \to R^2$, $h = (f,g)$, when prolonged, $h^{(2)} = (j^2 f, g)$, annihilate $E(M^{(1)}) = \overline{\{\Pi^*\phi_2, p^*{}_1\Omega^2(2,1)\}}$ where $M^{(1)} = J^2(2,1) \times W$ and $\Pi : M^{(1)} \to M$, $(uz_2,w) \to (uz_1,w)$. For simplicity we shall only consider the simplest case when $Z_x \equiv 0$, $Z_y \equiv 0$ which corresponds to two of the transformations being point transformations $(x' = x, y' = y)$. In this case the generators have the canonical form

$$\begin{aligned}\phi_1 &= dz - z_x dx - z_y dy\\ \phi_2 &= dw - X dx - Y dy\end{aligned} \qquad (1.8)$$

and we find that

$$d\phi_2 = (D_y^2 X - D_x^2 Y + [X,Y])\,dx\,dy + (Y_{,w} dy - X_{,w} dx)\wedge\phi_2 \quad \text{mod } \Omega^2(2,1) \qquad (1.9\text{ a})$$

where

$$D_a^k = \frac{\partial}{\partial x} + z_x\frac{\partial}{\partial z} + \dots + z_{(m+1)xny}\frac{\partial}{\partial z_{mxny}} \dots + z_{kx}\frac{\partial}{\partial z_{(k-1)x}} \qquad (1.9\text{ b})$$

and we have written X abusively for Π^*X etc. The relationships (1.9 a) are deduced from the formula

$$d(\Pi^{k^*}_{k-1}G) \equiv \mathcal{D}^k_x(\Pi^{k^*}_{k-1}G)dx + \mathcal{D}^k_y(\Pi^{k^*}_{k-1}G)dy \bmod \Omega^k(2,1),\ G\epsilon\Lambda^o(J^{k-1}(2,1)$$

where π^k_{k-1} is the submersion $\pi^k_{k-1}J^k(2,1) \to J^{k-1}(2,1)(uz_k) \to (uz_{k-1})$

It follows that ϕ_2 is *completely integrable* provided

$$e \equiv \mathcal{D}^2_y X - \mathcal{D}^2 Y + [X,Y] = 0,\ [X,Y] = XY,_w - YX,_w \qquad (1.10)$$

If (1.10) does not depend upon w then it is called *a resolvent of the second type*. Upon substituting in (1.9 b) it is seen to be a special type of *Monge-Ampere equation*. In this case the Cartan characteristics of E(M) correspond to the *Monge characteristics* of e. Thus as before we obtain the singular forms θ,θ' (assumed distinct for simplicity) and determine a basis of $\Lambda^1(M)$, $(\theta,\theta',w^3,w^4,w^{3'},w'^4)$ such that there exists $V\ V' \in V(M)$ satisfying

$$\theta(V) = 0, \quad \theta'(V) = 0, \quad d\theta = \omega_3 \wedge \omega_4$$

and $\quad V \lrcorner d\theta = 0,\ V \lrcorner d\theta' = w_3'(V)\omega_4' - w_4'(V)w_3'$

with an analogous set of equations for V'. It follows that $V, V' \epsilon$ Car I. On the other hand the annihilators of the Pfaffian systems generated by $\{\phi,\omega_3\omega_4\}$ are tangent to the *first order characteristics* for e which belong to the characteristic vector held system for this equation.

Example 7

Given the system

$$\phi_1 = dz - z_x dx - z_y dy$$

$$\phi_2 = dw - Zdx - Ydy - Z_x dz_x - Z_x dz$$

and the singular forms θ, θ' to compute the basis $\{\theta,\theta',\omega_3,\omega_4,\omega'_3\ \omega'_4\}$ for $\Lambda(M)$.

Let $\theta = \mu\phi_1 + \phi_2$ then we find that

$$d\theta = Adx\wedge dy + Bdx\wedge dz_x + Cdy\wedge dz_y + Ddy\wedge dz_x + Edx\wedge dz_y + Fdz_x\wedge dz_y \quad \bmod \{\phi_1,\phi_2\}$$

where $A = \mathcal{D}'_x Y - \mathcal{D}'_y X + \{X,Y\}$ $\quad B = \mathcal{D}'_x Z_x - X,_{z_x} + \{X,Z_x\} - \mu;$
$C = \mathcal{D}'_y Z - Y,_{z_y} + \{Y,Z_y\} - \mu;$ $\quad \mathcal{D} = \mathcal{D}'_y Z_x - Y,_{z_x} + \{Y,Z_x\}$
$E = \mathcal{D}'_x Z_y - X,_{z_y} + \{X,Z_y\};$ $\quad F = Zy,_{z_x} - Z_x,_{z_y} + \{Z_x,Z_y\}$ and $\{X,Y\} = XY,_w - YX,_w$.

Taking $w_3 = \alpha dz_x + \beta dx + \gamma dy$, $w_4 = \delta dz_y + \rho dx + \zeta dy$ and equating $w_3\wedge w_4$ to the expression for $d\theta$ we find that $w_3 = Fdz_x + Edx + (C-\mu_1)dy$, $w_4 = F^{-1}(Fdx_y - (B-\mu_1)dx - \mathcal{D}dy)$ and w_3' and w_4' are given by similar expressions to those for w_3 and w_4 but with μ_2 replacing μ_1, where μ_1,

μ_2 are the roots of

$$\mu^2 - (B+C)\mu - (AF + DE) = 0$$

Bäcklund transformations

It is straightforward to obtain the following classification of Backlund transformations which is originally due to Clairin [13].

Corresponding Resolvents	Type of B.T.	Correspondence between Solutions
I $\longleftrightarrow$ I	B_1	$1 \longrightarrow 1$, $1 \longleftarrow 1$
I $\longleftrightarrow$ II	B_2	$1 \longrightarrow 1$, $\infty \longleftarrow 1$
II $\longleftrightarrow$ II	B_3	$1 \longrightarrow \infty$ $\infty \longleftarrow 1$

B_1 transformations

Determine a singular form for

$$\theta = dz - z_x dx - z_y dy$$
$$\pi = X dx + Y dy + Z_x dz_x + Z_y dz_y$$

and then by our earlier work after a possible contact transformation this must be θ' written in canonical form

$$\langle\theta,\theta\rangle = 0, \quad \langle\theta,\pi\rangle = \{Z_x, X\}_w + \{Z_y, Y\}_w$$

$$\langle\pi,\pi\rangle = 2[X(Z_x,Y,Z_y) + Y(X,Z_x,Z_y) + Z_x(X,Z_y,Y) + Z_y(X,Y,Z_x)]$$
$$- 2z_x\{Y,Z_y,Z_x\} - 2z_y\{X,Z_x,Z_y\}$$

where $(A,B,C) = \{A,B\}_{w,c} + \{C,A\}_{w,b} + \{B,C\}_{w,a}$

$\{A,B,C\} = A\{B,C\}_{w,z} + C\{A,B\}_{w,z} + B\{C,A\}_{w,z}$

and $\{A,B\}_w = AB,_w - BA,_w$, $\{A,B\}_{c,d} = A_c B,_d - A_d B,_c$

It follows that the singular form is given by

$$\theta' = \lambda_1 \theta + \lambda_2 \pi$$

where $2\lambda_1\langle\pi,\theta\rangle + \lambda_2\langle\pi,\pi\rangle = 0$

Example 8

The equation $z,_{xy} = \sin z$ (Sine-Gordon equation) is a resolvant of the first type for the system

$$\phi_1 = dz - z_x dx - z_y dy, \quad \phi_2 = dz_x - w dx - \sin z\, dy$$

Consequently $X = -w$ $Y, = \sin z$, $Z_x = 1$, $Z_y = 0$ and $<\phi,\pi> = -1$ $<\pi,\pi> = 0$ and $\theta' = dz' - z'_x dx - z'_y dy = g^*(dz_x - wdx - \sin z dy)$ for some contact transformation g. It follows that $z' = z_x$ $z'_x = w$ $z'_y = \sin z$ and the resolvent for $\{\theta',\Pi'\}$ is $z'_{,xy} = z'_{,x}(1 - z'^2_{,y})^{\frac{1}{2}}$.

The equation $z_{,xy} - A(x,y,z,z_{1x})\, z_{,y} - B(x,y,z,z_{1x},z_{2x}) = 0$ (Gomes Teixeina equation) is a resolvent of the

$$\phi_1 = dz - z_x dx - z_y dy \qquad \phi_2 = dz_x - wdx - (Az_x + B)dy$$

($z_{,xx}$ is replaced by w in B). In this case $<\theta,\pi> = -1$, $<\pi,\pi> = -2A$ and

$$\theta' = dz_x - Adz - (w - Az_x)dx - Bdy$$

Then if f is an integrating factor of $dz_x - Adz$ the preceeding equation can be written in canonical form.

B_2 and B_3 transformations

One can show that just as there is a correspondence between the Cartan characteristics on the solutions of the resolvents e, e' related by a B_1 transformation, that the Cartan characteristics correspond on resolvents related by B_2 or B_3 transformations. There are also interesting interrelations between the transformations. Thus if E(M) admits resolvents e, f, e' of types II, I and II respectively then the transformation B_3 which gives a correspondence between e and e' can be decomposed into two B_2 transformations.

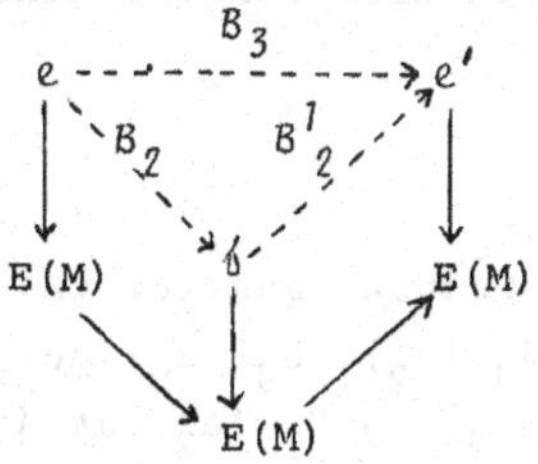

It is also possible to pass from a II to a II by a transformation which is not a B_3,

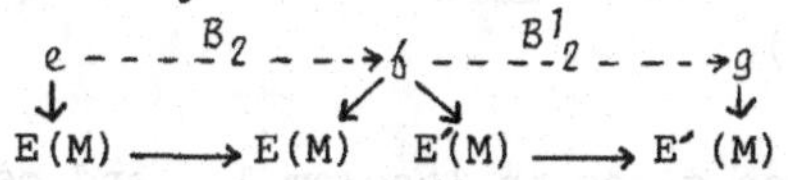

that is f is a resolvent of type II for two different systems E(M), E^1(M). Further relationships of a more complicated nature also apparently exist if we extend the discussion to canonical forms of class less than 5.

Example 9

Consider [16], [12]

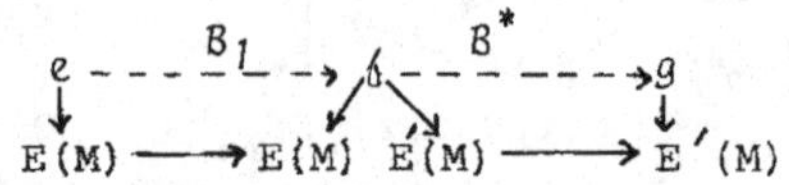

given by

$$
\begin{aligned}
e &\equiv z_{xy} - \exp z \quad \text{(Liouville equation)} \\
B_1 &: z' = z_x,\ z'_y = \exp z,\ z'_x = w \\
s &\equiv z'_{xy} - z'_y z' \\
B^* &: z'' = z'_x - \tfrac{1}{2} z'^2,\ z'_y = 0,\ z'_x = w - z'z'_x \\
g &\equiv z''_{xy}
\end{aligned}
$$

Thus from the general solution of $z'_{,xy} = 0$ by inverting the sequence of transformations (B_1, B^*) one is able to solve the Liouville equation. Compare this with example 5 where the Liouville equation was solved by using a B_3 transformation.

In the case of a B_3 *auto-Bäcklund transformation* that is the transformation gives a correspondence between solutions to the same equation it is possible to exploit the existence of a parameter in the transformation, which usually arise from a symmetry of the equation (see however §2) to produce new solutions by purely *algebraic* means. Thus we have the following classical theorem for the sine-Gordon equation due to Bianchi.

Theorem 1.4

Given a solution s_1 of the sine-Gordon equation and solutions s_2, s_3 obtained from s_1 by transforms $B_3(a_2)$, $B_3(a_3)$ respectively, then a fourth solution s_4 obtained from s_2 by $B_3(a_2)$ or from s_3 by $B_3(a_2)$ is given, for distinct a_2, a_3, by the algebraic formula

$$\tan\left(\frac{s_4 - s_1}{4}\right) = \frac{\sin\left(\frac{a_3 + a_2}{2}\right)}{\sin\left(\frac{a_3 - a_2}{2}\right)} \tan\left(\frac{s_3 - s_2}{4}\right)$$

Proof: Use the auto-$B_3(a)$ transformation given in Theorem 1.3 and perform the manipulations given in the *Lamb diagram*,

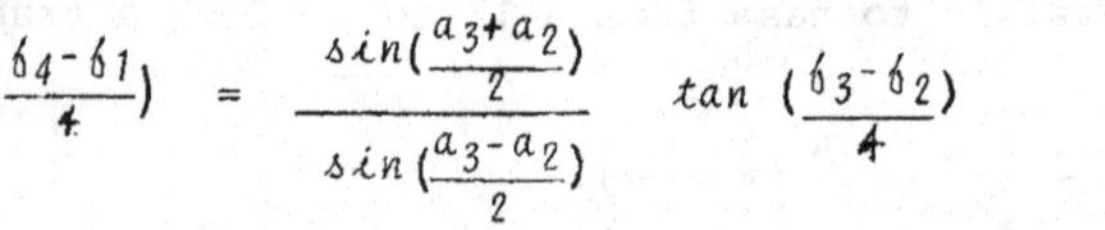

to obtain the result.

By combining Lamb diagrams and starting from the zero solution of the sine-Gordon equation one can produce *a tower of soliton solutions* [19]. The type of B_2 and B_3 transformations considered here can after a contact transformation be brought into the form

B_2 : $w'_x = X(x,y,z,zx,zy,w)$, $w'_y = Y(x,y,z,zx,zy,w)$ not resolvable w.r.t. z_x, z_y; B_3 : $w'_x = X(x,y,z,zx,zy,w)$, $w'_y = Y(x,y,z,z_x,z_y,w)$, resolvable w.r.t. z_x and z_y. It is clear that θ_2 in (1.15) on (1.16) is just the *Wahlquist-Estabrook prolongation form* [20] for the associated resolvent of second order. It is also clear that an inverse problem resulting from restricting θ_2 to a solution manifold is completely integrable and corresponds to a B_3 transformation.

The extension to higher order B_3 transformations is obvious [21]. Let $M = J^r(2,1) \times W$ and on $M^{(1)} = J^{r+1}(2,1) \times W$, $E(M^{(1)}) = \{\pi^*\phi_2, p_1^*\,\Omega^{r+1}(2,1)\}$ $\pi^*: M^{(1)} \longrightarrow M$, $p_1^* : M^{(1)} \longrightarrow J^{r+1}(2,1)$ where

$$\phi_2 = dw - Xdx - Ydy, \quad Z, Y \in \Lambda\,(M)$$

The corresponding B_3 transformation is

$$w'_x = X \qquad w'_y = Y$$

and the resolvent equation is

$$e = D_y^{r+1} X - D_x^{r+1} Y + [X, Y]$$

Finally notice that $\phi_2 = dw - Xdx - Ydy - Z_x dz_x - Z_y dz_y$ corresponds to a generalisation of the Wahlquist - Estabrook prolongation form for the case when either the scattering problem (for which we replace M by a vector bundle and search for a ϕ_2 in which the *vectors* w enter linearly into X and Y) or an associated non linear equation has different independent variable from the original equation.

References for section 1.

[1] Bäcklund A.,V., Math. Ann. 17,285 (1880)

[2] Bäcklund A.,V., Math. Ann. 19,387(1883)

[3] Lamb,G.L.article in "Bäcklund Transformations" (ed. R.M.Miura, Springer Verlag,Berlin-Heidelberg-New York) S.L.N.maths.515,1976

[4] Skyrme,T.H.R. ,Proc. Roy. Soc. 262A,237(1961)

[5] Rogers,C. S.L.N.maths. 515,106(1976)

[6] Loewner,C. NASA Tech. Note 2¼65(1950)

[7] Forsyth,A.R. "Theory of Differential Equations",5&6,Dover Publications(1959).

[8] Lie,S. "Sophus Lies's 1880 Transformation Group paper" eds.N. Wallach&R.Hermann Math.Sci.Press Brookline (1976).

[9] Backlund A.V.,Math.Ann.,13,69.(1877)

[10] Gardener R.B.,Trans.Amer.Math.Soc.,126,514.(1967)

[11] Ibragimov N,Anderson R.L.,J.Math.Anal&Appl.,59,145.(1977)

[12] Gardener R.B.,"Constructing Backlund Transformations" Invited address at the Berlin conference on Differenential Geometry and Global Analysis July 1977.

[13] Clairin,Ann.Sci de l'Ecole Normal Superieure,19(3),supplement(1902)

[14] Liouville J.,J.de Math.Pures et Appl.Paris,18(1),72.(1853)

[15] Eisenhart L.P.,"A treatise on the differential geometry of curves and surfaces" Dover N.Y.

[16] Goursat E.,Memor.Sci.Math.6,Gautier Villars,Paris 1925.

[17] Cartan E.,Les systemes Differentielles Exterieures at Leurs Applications Geometrique,Hermann,Paris,1946.

[18] Hermann R.,"The geometry of non-linear differential equations, Backlund transformations and solitons" 12A&B. Math.Sci.Press Brookline Massachusettes 1976.

[19] Scott A,Chu F.Y,McLaughlin D.W.,IEEE ,1443,Oct 1973.

[20] Wahlquist H,Estabrook F., J.Math.Phys.,16,1(1975) and 17,1993(1976)

[21] Pirani F,Robinson D.C,Shadwick W, Local Jet Bundle formulation of Backlund Transformations,Reidel 1979.

§ 2 Recent Examples

In the first section of this paper we were concerned primarily with classical concepts and some of the original examples which motivated them. However, in more recent times, additional new examples have arisen in theoretical physics and in this section we wish to look at a few of these. It should always be remembered that it is the examples of today that are our inspiration and the examples of yesteryear merely a guide. We will look at four examples only but many others exist and more will surely be discovered. Each is chosen to illustrate some explicit facet, either the relevance of a particular formalism or a new idea extending classical notions. Very few details will be given and our attention will be confined mainly to examples of auto-Bäcklund transformations from an evolution equation to itself in order to divert attention away from the associated linear scattering problems as their discussion would take us too far away from our central theme.

2.1 The Massive Thirring Model:

The two dimensional massive Thirring model [1] is a model of a self interacting electron. The particle is represented by a classical complex values wave function ψ_a (a=1,2) satisfying the field equations

$$i\psi_{1,\xi} = 2(\psi_2 + |\psi_2|^2\psi_1) \qquad (2.1.1a)$$

$$i\psi_{2,\eta} = -2(\psi_1 + |\psi_1|^2\psi_2) \qquad (2.1.1b)$$

where ξ and η are the light cone coordinates $\xi=\frac{1}{2}(x+t)$ and $\eta = \frac{1}{2}(x-t)$. We will use this example to illustrate the standard Wahlquist-Estabrook [2] [3] [4][5] approach to the problem of determining a Bäcklund transformation directly from equations (2.1.1). Define P to be the manifold with local coordinates $(\xi, \eta, \psi_1, \psi_2, \psi_1^*, \psi_2^*)$. The differential 2-forms α_1 and α_2 defined by

$$\alpha_1 = id\psi_1\wedge d\eta - 2(\psi_2 + |\psi_2|^2\psi_1)d\xi\wedge d\eta \qquad (2.1.2a)$$

$$\alpha_2 = id\psi_2\wedge d\xi - 2(\psi_1 + |\psi_1|^2\psi_2)d\xi\wedge d\eta \qquad (2.1.2b)$$

together with their complex conjugates α_1^* and α_2^* generate an exterior system $E(P)\subset\Lambda(P)$ the 2-dimensional solutions of which, parametrised by ξ and η, are completely equivalent, to those of the original equations (2.1.1) The method now proceeds by constructing a fibre bundle $\pi_M:M\to P$ over the base space P. A 1-form ω on M is called a Wahquist-Estabrook prolongation form for E(P) iff $d\omega=\gamma\wedge\omega \bmod(\pi_M^*E(P))$. If the local coordinates in the fibre are denoted by ζ and we consider the 1-form

$$\omega = d\zeta - F(\xi,\eta,\psi_1,\psi_2,\psi_1^*,\psi_2^*,\zeta)d\xi - G(\xi,\eta,\psi_1,\psi_2,\psi_1^*,\psi_2^*,\zeta)d\eta \quad (2.1.3)$$

then the condition $d\omega = \gamma\wedge\omega \bmod(\pi_M^* E(P))$ becomes

$$= f^a \alpha_a + g^a \alpha_a + \gamma\wedge\omega \quad (2.1.4)$$

where $f^a, g^a \in \Lambda^0(M)$ and $\gamma\in\Lambda^1(M)$. If the 1-form γ is taken to be of the special form $(\gamma_1 d\xi + \gamma_2 d\eta)$ we may simplify (3.1.4) and obtain the equations

$$F,_{\psi_1} = F,_{\psi_1^*} = G,_{\psi_2} = G,_{\psi_2^*} = 0 \quad (2.1.5a)$$

together with the principal equation of the Wahlquist-Estabrook method

$$[F,G] = (F,_\eta - G,_\xi) + 2i(\psi_2 + |\psi_2|^2\psi_1)G,_{\psi_1} + 2i(\psi_1 + |\psi_1|^2\psi_2)F,_{\psi_2} - 2i(\psi_2^* + |\psi_2|^2\psi_1^*)G,_{\psi_1^*} - 2i(\psi_1^* + |\psi_1|^2\psi_1^*)F,_{\psi_2^*} \quad (2.1.5b)$$

where the bracket $[A,B]$ of two functions of ζ is defined by

$$[A,B] = A\, B,_\zeta - B\, A,_\zeta \quad (2.1.6)$$

and generalised in the obvious manner if dimension of the fibre is greater than 1. The bracket (2.1.6) satisfies all the normal properties of a Lie bracket such as the Jacobi identity.

If we take F and G to have the form

$$F = X_0 + X_1\psi_2 + X_2\psi_2 + X_3\psi_2\psi_2^* \quad (2.1.7)$$

$$G = X_4 + X_5\psi_1 + X_6\psi_1^* + X_7\psi_1\psi_1^* \quad (2.1.7b)$$

then equation (2.15a) is automatically satisfied. If we further assume that the X_i have no explicit ξ or η dependence and are simply functions of the fibre coordinate ζ then substitution into (2.1.5b) yields the following bracket constraints on the quantities X_i.

$$[X_0,X_4] = 0 \quad [X_0,X_5] = -2iX_1 \quad [X_0,X_6] = 2iX_2 \quad [X_0,X_7] = 0$$

$$[X_1,X_4] = 2iX_5 \quad [X_1,X_5] = 0 \quad [X_1,X_6] = 2i(X_3+X_6) \quad [X_1,X_7] = 2iX_1$$

$$[X_2,X_4] = -2iX_6 \quad [X_2,X_5] = -2i(X_3+X_7) \quad [X_2,X_6] = 0 \quad [X_2,X_7] = 2iX_2$$

$$[X_3,X_4] = 0 \quad [X_3,X_5] = 2iX_5 \quad [X_3,X_6] = -2iX_6 \quad [X_3,X_7] = 0$$

(2.1.8)

The bracket relations together with any others which result from the application of the Jacobi identity constitute a *Wahlquist-Estabrook prolongation structure* for E(P). A representation for the X_i is sought by completing the prolongation structure (2.1.8) into a Lie algebra for which standard methods are available. A one dimensional representation of the above structure is found to be given by

$$
\begin{array}{lll}
X_0 = -2i\lambda^2 y\partial_y & X_1 = -2i\lambda y^2\partial_y & X_2 = 2i\lambda\partial_y \\
X_3 = 2iy\partial_y & X_4 = -2i\lambda^{-2}y\partial_y & X_5 = -2i\lambda^{-1}y\partial_y \\
X_6 = 2i\lambda^{-1}\partial_y & X_7 = 2iy\partial_y &
\end{array} \tag{2.1.9}
$$

where y is the complex valued coordinate in the fibre, $\partial_y = \frac{\partial}{\partial y}$ and λ is an arbitrary complex constant. The origin of the complex parameter λ will be considered in a moment but let us first summarise our results so far. We have shown that we can construct a complex line bundle π_M: $M\to P$ with local coordinates $(\xi,\eta,\psi_1,\psi_2,\psi_1^*,\psi_2^*,y,y^*)$. On M we have the exterior system $E(M) = (\Pi_M^* E(P), \omega,\omega^*)$. The prolongation forms ω,ω^* when restricted to a solution manifold of $\Pi_M^*(P)$ yield additional equations compatible with those defining the massive Thirring model. These extra equations constitute the *inverse scattering problem* for the Thirring equations. The 1-form ω corresponding to the 1-dimensional representation (2.1.9) is given by

$$
\omega = dy+2i(-\lambda\psi_2^*+(\lambda^2-\psi_2\psi_2^*)y+\lambda\psi_2 y^2)d\xi+2i(-\lambda^{-1}\psi_1^*+(\lambda^{-2}-\psi_1\psi_1^*)y+\lambda^{-1}\psi_1 y^2)d\eta \tag{2.1.10}
$$

From the Wahlquist-Estabrook point of view an *auto-Bäcklund transformation* is a map $B:M\to P$ with the property $B^*(\ E(P))\subset E(M)$. For this case B_λ is given in local coordinates by $B_\lambda:(\xi,\eta,\psi_1,\psi_2,\psi_1^*,\psi_2^*,z,z^*)\mapsto(\xi,\eta,\psi_1',\psi_2',\psi_1'^*,\psi_2'^*)$ where ψ_1' and ψ_2' are defined by

$$
\psi_1' = \frac{(\lambda^*+\lambda zz^*)}{(\lambda+\lambda^* zz^*)}\psi_1 + z\frac{(\lambda^{-2}-\lambda^{*-2})}{(\lambda^{*-1}+\lambda^{-1}zz^*)} \tag{2.1.11a}
$$

$$
\psi_2' = \frac{(\lambda+\lambda^* zz^*)}{(\lambda^*+\lambda zz^*)}\psi_2 + z\frac{(\lambda^2-\lambda^{*2})}{(\lambda^*+\lambda zz^*)} \tag{2.1.11b}
$$

Details may be found in the work of Holod [6] and Morris [7]. For this Bäcklund transformation to be nontrivial we clearly require $\mathrm{Im}\lambda\neq 0$. This model provides an example of a situation in which the parameter in the Bäcklund transformation does *not* derive from a symmetry of the original equation which is a widely held misconception. The occurrence of the parameter λ is a result of a symmetry of the prolongation structure (2.1.8). It is possible for more than one equation to have the same prolongation structure. In this case it is easy to show that not only the Thirring equations (2.1.1) but also the equations

$$iu_{,\xi} = 2(v + vuw) \tag{2.1.12a}$$
$$ip_{,\eta} = -2(w + uvp) \tag{2.1.12b}$$
$$iv_{,\eta} = -2(u + upv) \tag{2.1.12c}$$
$$iw_{,\eta} = 2(p + puv) \tag{2.1.12d}$$

have the prolongation (2.1.8). The equations (2.1.12) admit a 1-parameter symmetry group with complex parameter μ defined as follows. If we denote the vector $(u,p,v,w)^t$ by V and the manifold with local coordinates (ξ,η,V) by $\bar{P}$ the group action is defined locally on $\bar{P}$ by

$$G\mu \;:\; (\xi,\eta,V) \longmapsto (\mu^2\xi, \mu^{-2}\eta, U_\mu V) \quad \text{where}$$
$$U_\mu V = \mathrm{diag}(\mu^{-1},\mu^{-1},\mu,\mu)V \tag{2.1.13}$$

We could have constructed a Bäcklund transformation $\bar{B}\mu$ for these equations and in that case we would have $\bar{G}\mu \circ \bar{B}\mu = \bar{B}_1 \circ \bar{G}\mu$ where $\bar{G}\mu$ is the prolongation of $G\mu$ to the product bundle $\bar{P} \times C^1$ given by trivial action on C^1. Only if $\mu \varepsilon R$ does this symmetry continue to hold for the Thirring equation in which case it becomes the normal Lorentz group symmetry of that model. The massive Thirring model was originally constructed as a quantum field theory model and quantised with canonical anti-commutation relations on the electron field. Consequently, the classical limit of that quantum system is not the classical model we have just considered. Rather, it is a system in which the fields are not complex valued but in fact take values in a *Grassman algebra* A. In order to define the equations A must also be equipped with a discrete anti-linear operator $J:A \to A$ sending $a \to \bar{a}$ such that $\overline{ab} = \bar{b}\bar{a}$ and $J^2 = 1$. J replaces the classical operation of complex conjugation and the equations take the form

$$i\phi_{1,\xi} = 2(\phi_2 + \bar{\phi}_2\phi_2\phi_1) \tag{2.1.14a}$$
$$i\phi_{2,\eta} = -2(\phi_1 + \bar{\phi}_1\phi_1\phi_2) \tag{2.1.14b}$$

where the fields ϕ_a are A-valued and great care must be taken to maintain the orderings of expressions when manipulating the equations. Following the method used in the classical case we define a generalised manifold P_G with local coordinates $(\xi,\eta,\phi_1,\phi_2,\bar{\phi}_1,\bar{\phi}_2)$. It is much more natural in this case, where the local coordinates are a mixture of real and Grassman algebra valued quantities, to set up the equations using a generalised form of jet bundle language. However, it would take too long to develop here and so we continue on the normal Wahlquist-Estabrook path. It is possible to define a space of A-valued p-forms on P_G denoted by $\Lambda^p(P_G)\otimes A$. Then the forms α_1 and α_2 defined by

$$\alpha_1 = i d\phi_1 \wedge d\eta - 2(\phi_2 + \bar{\phi}_2\phi_2\phi_1) d\xi\wedge d\eta \quad (2.1.15a)$$

$$\alpha_2 = i d\phi_2 \wedge d\xi - 2(\phi_1 + \bar{\phi}_1\phi_1\phi_2) d\xi\wedge d\eta \quad (2.1.15b)$$

together with their J-conjugates $\bar{\alpha}_1$ and $\bar{\alpha}_2$ span a closed ideal E_G in $\Lambda(P_G)\otimes A$.

For this case we construct a fibre bundle $\pi_{M_G} : M_G \to P_G$ with fibre A. In analogy with the classical case it can be shown that the A-valued 1-form ω defined by

$$\begin{aligned}\omega = \; & dy + [-2\lambda + (4i - 2\lambda\bar{y}y)\phi_2 + 2i\bar{\phi}_2\phi_2 y] d\xi \\ & - [2\lambda^{-1}y + (-4\lambda^{-1} + 2i\bar{y}y)\phi_1 + 2i\bar{\phi}_1\phi_1 y] d\eta\end{aligned} \quad (2.1.16)$$

is a prolongation form for $E(P_G)$ and we may define $E(M_G)$ as in the classical case. An auto-Bäcklund transformation is then defined to be a map $B : M_G \to P_G$ having the property $B^* E(P_G) \subset E(M_G)$. In terms of the local coordinates $(\xi,\eta,\phi_1,\phi_2,\bar{\phi}_1,\bar{\phi}_2,y,\bar{y})$ on M_G the map B_λ defined by B_λ: $(\xi,\eta,\phi_1,\phi_2,\bar{\phi}_1,\bar{\phi}_2,z,z) \mapsto (\xi,\eta,\phi_1',\phi_2',\bar{\phi}_1',\bar{\phi}_2')$ where ϕ_1' and ϕ_2' are defined by

$$\phi_1' = -\phi_1 + z + \frac{i\lambda}{2}\bar{z}z\phi_1 \quad (2.1.17a)$$

$$\phi_2' = \phi_2 + i\lambda z + \frac{i\lambda}{2}\bar{z}z\phi_2 \quad (2.1.17b)$$

has the property $B^* E(P_G) \subset E(M_G)$ and therefore defines a Bäcklund transformation for the anti-commuting massive Thirring model. Further details may be found in reference [3].

2.2 *The O(3) σ-model*

The two-dimensional σ-model [9] [10] describes a dynamical system of three dimensional spin vectors q of unit length $\langle q,q\rangle = 1$ with a Lagrangian density $\mathcal{L} = \langle\partial\mu q, \partial\mu q\rangle$. The resulting field equations are then given by

$$\Box_2 q + \langle\partial\mu q, \partial\mu q\rangle\, q = 0, \quad q(x,t)\in S^2 \quad (2.2.1)$$

where $g_{\mu\nu} = \mathrm{diag}(1,-1)$ and $\Box_2 = \partial_t^2 - \partial_x^2$. One can, of course, eliminate the constraint on q by expressing it in polar coordinates by

$$q = (\cos\phi\sin\theta,\ \sin\phi\sin\theta,\ \cos\theta)$$

in which case equation (2.2.1) takes the form

$$\Box_2\phi = -2(\partial\mu\phi\partial\mu\theta)\cot\theta \quad (2.2.2a)$$

$$\Box_2\theta = (\partial\mu\phi\partial\mu\phi)\sin\theta\cos\theta \quad (2.2.2b)$$

and these equations may be analysed in the Wahlquist-Estabrook manner in which we treated the Thirring model. However in order to illustrate the advantages of the jet bundle theory language, we will retain the constraint $\langle q,q\rangle = 1$ as one of our dynamical equations, use the jet bundle approach, and obtain results which would have been difficult to extract by the standard method. It is more convenient for us to start, as we did in the Thirring model, by expressing the equations in the light cone coordinates $\xi = \frac{1}{2}(x+t)$ and $\eta = \frac{1}{2}(x-t)$.
Equation (2.2.1) becomes

$$q,_{\xi\eta} + \langle q,_\xi, q,_\eta\rangle q = 0 \qquad \langle q,q\rangle = 1$$

Define the ten dimensional submanifold $\Sigma \subset J^2$ (2,3) by the equations

$$\langle q,q\rangle = 1 \quad \langle q,q_1\rangle = 0 \quad \langle q,q_2\rangle = 0 \quad q_{12} + \langle q_1,q_2\rangle = 0$$

$$\langle q,q_{11}\rangle + \langle q_1,q_1\rangle = 0 \quad \langle q,q_{22}\rangle + \langle q_2,q_2\rangle = 0 \quad \langle q_{11},q_2\rangle = 0 \quad \langle q_{22},q_1\rangle = 0$$

$(uq_2) = (\xi,\eta,q,q_1,q_2,q_{11},q_{12},q_{22})$ is a local coordinate system on J^2 (2,3), and let $i:\Sigma\to J^2(2,3)$ be an embedding of Σ in $J^2(2,3)$. The restriction $i^*\Omega^2$ (2,3) of the contact module to Σ, which we denote by $E(\Sigma)$, defines a completely integrable system on Σ. Consider a fibre bundle $\pi_M\colon M \to \Sigma$ with base Σ. A 1-form ω on M is called a Wahlquist-Estabrook prolongation form for $E(\Sigma)$ iff $d\omega = \gamma\wedge\omega \bmod(\pi_M^*E(\Sigma))$. Define the exterior differential system $E(M) = \{\Pi_M^*E(\Sigma),\omega\}$ on M which is generated by the six 2-forms $\delta_1 = dq_{11}\wedge d\xi, \delta_2 = dq_{22}\wedge d\eta$ (2.2.3a) and the nine 1-forms

$$\kappa_1 = dq - q_1 d\xi - q_2 d\eta,\ \kappa_2 = dq_1 - q_{11}d\xi + \langle q_1,q_2\rangle q d\eta,\ \kappa_3 = dq_2 + \langle q_1,q_2\rangle q\, d\xi - q_{22}d\eta \quad (2.2.3b)$$

If the local coordinates in the fibre are denoted by ζ then local coordinates on M are (uq_2,ζ). A canonical form for the Wahlquist-Estabrook prolongation form is then

$$\omega = d\zeta - Fd\xi - Gd\eta \qquad F,G \in \Lambda^0(M) \quad (2.2.4)$$

and the condition $d\omega = \gamma\wedge\omega \bmod(\pi_M^*E(\Sigma))$ becomes

$$d\omega = \alpha^a{}_b\,\kappa_a{}^b + \beta^a\delta_a + \gamma\wedge\omega \quad (2.2.5)$$

If the 1-forms $\alpha^a{}_b$ and γ are taken to be of the special form $(ad\xi + bd\eta)$ we then obtain the analogue of (2.1.5) in this formalism. They are

$$F,_{q_{22}} = 0 = G,_{q_{11}} \tag{2.2.6a}$$

$$[F,G] = F,_\eta - G,_\xi + \langle q_{22}, F_{q_2}\rangle - \langle q_{11}, G,_{q_1}\rangle + \langle q_2, F,_q\rangle - \langle q_1, G,_q\rangle$$
$$+ \langle q_1, q_2\rangle\langle q, G_{q_2}\rangle - \langle q_1, q_2\rangle\langle q, Fq_1\rangle \tag{2.2.6b}$$

In the case of the Thirring model M was the complex line bundle $P \times C^1$ the simplest possible situation. However, this case provides an example of a more general situation in which M is not a vector bundle.

The three 1-forms ω defined by

$$\omega = dY - (-q_1 - q\langle q_1, Y\rangle + \langle q_1, Y\rangle Y)d\xi - (q_2 - q\langle q_2, Y\rangle - Y\langle q_2, Y\rangle)d\eta \tag{2.2.7}$$

regarded as 1-forms on the vector bundle $\Pi_M : M \to \Sigma$ with fibre $\Pi_M^{-1}(x) = R^3\ x\epsilon\Sigma$ do not provide a Wahlquist-Estabrook prolongation form for the exterior system $\Pi_M^* E(\Sigma)$ on M. However, if we define the sub-bundle $\Pi_N : N \to \Sigma$ with $\pi_N^{-1}(x) = \{Y\epsilon R^3 | \langle Y,Y\rangle=1, \langle(\beta\circ i)(x), Y\rangle=0\}$ R^3 where β is the target map on J^2 (2,3), then$\omega$is a Wahquist-Estabrook prolongation form for the exterior system $\pi_N^* E(\Sigma)$ on the circle bundle N. A mapping $\mathcal{B}$: $N \to \Sigma$ is the auto-Bäcklund transformation if $\mathcal{B}^* E \subset E(N)$ where $E(N) = \{\Pi_N^* E(\Sigma), \omega\}$. Consider a fibre bundle $\pi_{N_2} : N_2 \to J^2$ (2,3) in which N is embedded. Choose local coordinates (uq_2, z) on N_2 with the properties $\langle z,z\rangle=1, \langle z,q\rangle=0$ which coordinatise the fibre over the point (uq_2) of J^2 (2,3). Define the map $b : N_2 \to J^2$ (2,3) to be the transformation given in local coordinates by

$$\xi' = \xi,\ \eta' = \eta,\ q' = z,\ q_1' = -q_1 - q\langle q_1, z\rangle + z\langle q_1, z\rangle, q_2 = q_1 - q\langle q_2, z\rangle - z\langle q_2, z\rangle$$

$$q'_{11} = -q_{11} - 2q_1\langle q_1, z\rangle + (2\langle q_1, z\rangle^2 + \langle q_{11}, z\rangle - \langle q_1, q_2\rangle)(z - q)$$

$$q'_{22} = q_{22} - 2q_2\langle q_2, z\rangle + (2\langle q_2, z\rangle^2 - \langle q_{22}, z\rangle - \langle q_2, q_2\rangle)(z + q)$$

$$q'_{12} = -q_{12} - q\langle q_1, q_2\rangle + z(\langle q_1, q_2\rangle - 2\langle q_1, z\rangle\langle q_2, z\rangle)$$

We then define $\mathcal{B} = b|_N$ and note that $\mathcal{B} : N \to \Sigma$. One easily shows that $\mathcal{B}^* E(\Sigma) \subset E(N)$ and that therefore $\mathcal{B}$ provides an auto-Bäcklund map for the O(3) σ-model. It is clear that a similar analysis can also be made for the O(p,q) σ-model by considering n-vectors $\mathcal{Q}$, which take their values in R^{p+q} equipped with the inner product

$$\langle P, Q\rangle = P_1 Q_1 + \ldots + P_p Q_p - P_{p+1} Q_{p+1} - \ldots - P_{p+q} Q_{p+q}\ .$$

2.3 *The Ernst Equation:*

Both of the examples considered so far have had trivial explicit coordinate dependence. For our third example we consider the Ernst equation

$$(ReE)\, \nabla^2 E = \nabla E . \nabla E \qquad (2.3.1)$$

where E is a complex valued function and the Laplacian and gradient operators are defined on R^3. This equation occurs in the formulation of the axially symmetric gravitational field problem [11] and E is a function which depends only upon axial polar coordinates r and z. Therefore, despite the symmetric appearance of (2.3.1) in which there is no explicit coordinate dependence in cartesian coordinates, once we choose to express the equation in axial polars such a dependency immediately appears. If we adopt the complex coordinates

$$\xi = \tfrac{1}{2}(r + iz) \qquad \eta = \tfrac{1}{2}(r - iz)$$

then (2.3.1) reduces to the form

$$E_{,\xi\eta} + \tfrac{1}{2}r^{-1}(E_{,\xi} + E_{,\eta}) = T^{-1}E_{,\xi}E_{,\eta} \qquad (2.3.2)$$

in which we have written $T = ReE$ and $r = (\xi+\eta)$. This equation and its complex conjugate may then be represented by the four 2-forms

$$\alpha = -dt\wedge d\xi + \tfrac{1}{2}(tu-tw-r^{-1}(t+w))\,d\eta\wedge d\xi \qquad (2.3.3a)$$

$$\beta = -du\wedge d\eta + \tfrac{1}{2}(tu-vu-r^{-1}(u+v))d\xi\wedge d\eta \qquad (2.3.3b)$$

$$\gamma = -dv\wedge d\xi + \tfrac{1}{2}(vw-vu-r^{-1}(u+v))d\eta\wedge d\xi = \bar{\beta} \qquad (2.3.3c)$$

$$\delta = -dw\wedge d\eta + \tfrac{1}{2}(vw-tw-r^{-1}(t+w))d\xi\wedge d\eta = \bar{\alpha} \qquad (2.3.3d)$$

where we have introduced the notation

$$t = T^{-1}E_{\xi} - r^{-1} \qquad w = \bar{t} = T^{-1}\bar{E}_{\eta} - r^{-1} \qquad (2.3.4a)$$

$$u = T^{-1}E_{\eta} - r^{-1} \qquad v = \bar{u} = T^{-1}\bar{E}_{\xi} - r^{-1} \qquad (2.3.4b)$$

The 2-forms α,β,γ and δ generate an exterior system E(P) on the manifold P with local coordinates (ξ,η,t,u,v,w).

In order to construct an auto-Bäcklund for the Ernst equation we must construct a prolongation 1-form on an appropriate fibre bundle. By following the Wahlquist-Estabrook approach Harrison [12] was able to show that the 1-form

$$\omega = d\zeta - \tfrac{1}{2}(t(\zeta+\zeta^2) - v(\zeta+\lambda))\,d\xi - \tfrac{1}{2}(w(\zeta+\zeta^2\lambda^{-1}) - u(\zeta+\lambda^{-1})) \qquad (2.3.5)$$

defined on a complex line bundle $\pi_N : N \to P$ with fibre coordinate ξ and depending on the 1-parameter family of functions $\lambda(k)$ defined by

$$\lambda(k) = \left[\frac{ik-\eta}{ik+\xi}\right]^{\frac{1}{2}} \qquad k\varepsilon R \qquad (2.3.6)$$

a prolongation form. To determine an auto-Bäcklund we must find a map $B : N \to P$ having the basic property $B^* E \subset E(N) \bmod(\pi_N^* E)$. By directly implementing this condition Harrison was able to show that the map $B : N \to P$ given in local coordinates by $B : (\xi,\eta,t,u,v,w,z) \mapsto (\xi,\eta,t',u',v',w')$ with

$$t' = -z(z+\lambda)^{-1}(\ (\lambda z + 1)t + r^{-1}(1-\lambda^2)) \qquad (2.3.7a)$$

$$u' = -(z+\lambda)^{-1}(z^{-1}(\lambda z+1)u + r^{-1}(\lambda^2-1)) \qquad (2.3.7b)$$

$$v' = -(\lambda z+1)^{-1}(z^{-1}(z+\lambda)v + r^{-1}(1-\lambda^2)) \qquad (2.3.7c)$$

$$w' = -z(\lambda z+1)^{-1}((z+\lambda)w + r^{-1}(\lambda^2-1)) \qquad (2.3.7d)$$

had that property and therefore defined an auto-Bäcklund map for the Ernst equation (2.3.2). This is the first example of a Bäcklund map determined solely by the Wahlquist-Estabrook geometric method. In view of the complexity of the mapping it is doubtful that it could have been found by other means.

Despite the tremendous difficulty resulting from the occurrence of the coordinate dependent function $\lambda(k)$ Harrison [13] has managed to use the Bäcklund map (2.3.7) to construct solutions to the Einstein field equations which are believed to have been previously unknown. It is also possible to construct a generalised Lax representation for the Ernst system but still containing the function $\lambda(k)$ and k playing the role of a scattering parameter [12][14][15]. However, the continuing appearance of $\lambda(k)$ will always lead to difficulty. Morris and Dodd [16] have proposed an alternative attack on the problem which eliminates the problem of complicated coordinate dependence from the start. By using the generalised prolongation method developed in ref [17] directly on equation (2.3.1) without the assumption that E depends only on axial polar coordinates, and refraining from implementing that constraint until a latter phase, they are able to show that the Ernst equation may be associated with an operator bundle in three dimensions. The generalised scattering problem obtained is

$$T(r\zeta,_r + 2s\zeta,_s - s\zeta,_3) = (irT,_r X_0 - r\phi,_r Y_- + s(iT,_3 X_0 + \phi,_3 Y_+))\zeta \qquad (2.3.8a)$$

$$T(r^{-1}\zeta_{,r} + s^{-1}\zeta_{,3}) = (ir^{-1}T, X_0 - r^{-1}\phi_{,r}Y_+ + s^{-1}(iT_{,3}X_0 - \phi_{,3}Y_-))\zeta \qquad (2.3.8b)$$

where s is a complex valued independent variable, $\phi = \text{Im } E$ and X_0, Y_+ and Y_- are any three matrices which satisfy the commutation relations

$$[X_0, Y_+] = -iY_+ \quad [X_0, Y_-] = iY_- \quad [Y_+, Y_-] = -2iX_0$$

Equations (2.3.8) may be given an interpretation as a Bäcklund transformation between a two dimensional surface and a three dimensional one as considered in the first section. It is not known at this time whether it is possible to determine an auto-Bäcklund map within this formulation. The coordinates still appear in the problem but in a greatly simplified form. The most likely way in which the problem (2.3.8) will be solved is as a system describing a generalised monodromy preserving deformation rather than as an inverse scattering problem. The relationship between the Harrison and Morris & Dodd problems is also unknown except that we might note that the latter problem probably corresponds to an infinite dimensional realisation of the prolongation structure of the Ernst equation and the former to a finite one. Further details may be found in ref [16].

2.4 *A Functional Bäcklund Transformation:*

Despite its generality the formalism of jet bundles is not adequate to describe all of the known examples we would like our theory to encompass. The nonlocal equation

$$u_t + 2uu_x + K[u_x] = 0 \qquad (2.4.1)$$

where K is an integral operator on R with kernel $K(x-z)$ is known, from the result of numerical studies [18], to have solution with soliton type collisional properties for a wide variety of different kernels. This suggests that it may be possible, for special kernels, to determine inverse scattering transformations for such equations.

Equation (2.4.1) has the form

$$u_t + 2uu_x + W_x = 0 \qquad (2.4.2)$$

and if we regard u and W to be independent we can introduce a geometric framework for (2.4.2) as follows. Consider the two equations

$$V_{1t} = (i(au+w_1)-(\lambda+a)\partial_x + i\partial_{xx})V_1 \qquad (2.4.3a)$$

$$V_{2t} = (i(au+w_2)-(\lambda+a)\partial_x + i\partial_{xx})V_2 \qquad (2.4.3b)$$

in which a,b and λ are complex constants and u and w_1 and w_2 functions of x and t. If we ask under what circumstance this flow preserves the relationship

$$V_{1x} = i(u - \lambda/2)V_1 + bV_2 \qquad (2.4.3c)$$

we readily discover that the necessary constraints on u, w_1 and w_2 are

$$i(w_1 - w_2) = 2u_x \qquad (2.4.4)$$

and

$$u_t + 2uu_x + W_x = 0 \text{ where } W = -\tfrac{1}{2}(w_1 + w_2) \qquad (2.4.5)$$

We see therefore that equations (2.4.3) almost provide an inverse scattering equation for (2.4.2) but we are lacking an equation relating V_1 and V_2 analogous to the normal equation for V_{2x}.

As a step towards the determination of a compatible relationship between V_1 and V_2 we introduce the parameterisations

$$V_1 = exp[\tfrac{1}{2}(r - iv)] \qquad V_2 = exp[\tfrac{1}{2}(r + iv)]$$

where r and v are complex valued. Substitutions of these forms into (.4.4) yields the equations

$$v_t = -(\lambda + a)v_x + iv_{xx} + 2iu_x + ir_x v_x \qquad (2.4.6a)$$

$$r_t = -(\lambda + a)r_x + ir_{xx} - 2iW + i\tfrac{1}{2}(r_x^2 - v_x^2) + 2aiu \qquad (2.4.6b)$$

$$r_x = i(v_x + 2u - \lambda) + 2be^{iv} \qquad (2.4.6c)$$

The form of (2.4.6a&b) suggests that we might find a solution of the form $r = L[v]$ where L is a linear operator akin to K. We obtain from that supposition the equation

$$2(L[u_x] + W - au) = (L[r_x v_x] - \tfrac{1}{2}(r_x^2 - v_x^2)) \qquad (2.4.7)$$

from which we see that the choice

$$W = [-L + a\int_{-\infty}^{x} dx]u_x \qquad (2.4.8)$$

can be made provided the operator L satisfies the algebraic requirement

$$L[\sigma L[\sigma]] = \tfrac{1}{2}(L[\sigma]^2 - \sigma^2) \qquad (2.4.9)$$

The equation $r = L[v]$ becomes in terms of V_1 and V_2 the relationship

$$(1 + iL)\log V_1 = (-1 + iL)\log V_2 \qquad (2.4.10)$$

and (2.4.10) together with (2.4.3) defines a linear inverse scattering problem for equation (2.4.1) with K given by

$$K = (-L + a\int_{-\infty}^{x} dx\,)\partial_x$$

However, like a linear equation with nonlinear boundary conditions, the linearity may be more illusory than real.

We are now in possession of an inverse scattering problem for the functional equation

$$u_t + 2uu_x + T[u_{xx}] = 0 \quad \text{where } T = (-L + a\int_{-\infty}^{x} dx) \qquad (2.4.11)$$

and L has the algebraic property expressed in equation (2.4.9).

If we introduce the potential w defined by $u = w_x$ equation (2.4.11) may be integrated to the form

$$w_t + w_x^2 + T[w_{xx}] = 0 \qquad (2.4.12)$$

If w is any solution of equation (2.4.12)
let us define w' by $w = w + v$ where v is a solution to equations (2.4.6) with $r=L[v]$ and $u = w_x$. Direct substitution shows that

$$w'_t + w'^2_x + T[w'_{xx}] = r_x - av_x + T[v_{xx}] = \partial_x(\,L - a\int_{-\infty}^{x} dx + T)v_x = 0 \qquad (2.4.13)$$

and consequently the mapping $w \rightarrow w'$ defines an auto-Bäcklund transformation for equation (2.4.11). Equations (2.4.6) with $r=L[v]$ may be expressed in terms of w and w' as

$$L[w'_x - w_x] = i(\,(w'_x - w_x) + 2u - \lambda) + 2be^{i(w'-w)} \qquad (2.4.14a)$$

$$w'_t - w_t = -(\lambda + a)(w'_x - w_x) + i(w'_{xx} - w_{xx}) + 2iu_x + i(w'_x - w)L[w'_x - w_x] \qquad (2.4.14b)$$

We are left with the problem of identifying the operator L or equivalently T. One solution is given by

$$L[\delta] = \frac{-P}{\pi}\int_{-\infty}^{\infty} \frac{\delta(z)}{\xi - z}\,dz = -H[\delta]$$

otherwise known as the Hilbert transform. With the choice a=0 equation (2.4.11) then becomes

$$u_t + 2uu_x + H[u_{xx}] = 0 \qquad (2.4.15)$$

which is the noted Benjamin-Ono equation [19][20] which describes the propagation of internal waves in stratified fluids of infinite depth. Once again, as in the past, we have the remarkable coincidence of a physical interpretation of a special equation with soliton properties. A more general solution is given by the choice of kernel

$$T_a[\delta] = P\int_{-\infty}^{\infty}(-\tfrac{1}{2}a\ \coth\tfrac{\pi a}{2}(\xi-z) + \tfrac{1}{2}a\ \mathrm{sgn}(\xi-z))\,\delta(z)dz$$

in which case equation (2.4.11) becomes the equation of Joseph [21] for the propagation of internal waves in stratified fluids of finite depth a^{-1}. As $a\to 0$, $T_a \to H$ and the Joseph equation reduces to the Benjamin-Ono equation. Further details of this example can be found in refs. [22] [23].

References for section 2.

[1] Thirring W,Ann.Phys.,3,91.(1958)
[2] Wahlquist H,Estabrook F.,J.Math.Phys.,16,1.(1975)
[3] Wahlquist H,Estabrook F.,J.Math.Phys.,17,1993.(1976)
[4] Wahlquist H,Estabrook F,Hermann R.,"Differential Geometric Prolongations and Backlund Transformations" in the Proc. Ames Sem. on Differential and Algebraic Geometry for Control Engineers (Brookline:Math.Sci.Press 1977)
[5] Dodd R,Gibbon J.,Proc.R.Soc.Lond.A.359,411.(1978)
[6] Holod P.,"Pseudopotentials and Backlund Transformations for the Thirring equation" Institute for theoretical physics(Kiev)preprint 78-100.(1978)
[7] Morris H.,J.Phys.A.,12,131.(1979)
[8] Morris H.,J.Math.Phys.,19,85-(1978)
[9] Pohlmeyer K.,Comm.Math.Phys.,46,207.(1976)
[10] Luscher M,Pohlmeyer K.,Nuclear Physics B137,46.(1978)
[11] Ernst F.,Phys.Rev.,167,1175.(1968)
[12] Harrison K.,Phys.Rev.Lett 41,1197.(1978)
[13] Harrison K.,"A large family of vacuum solutions of the equations of general relativity" submitted to Phys.Rev.Lett. June 1979.
[14] Maison D.,Phys.Rev.Lett.,41,521.(1978)
[15] Maison D.,J.Math.Phys.,20,871.(1979)
[16] Morris H,Dodd R.,"A 2-connection and operator bundles for the Ernst equation for axially symmetric gravitational fields" Physics Letters (to appear)
[17] Morris H.,"Inverse scattering problems in higher dimensions: Yang-Mills fields and the supersymmetric sine-Gordon equation" J.Math.Phys. (to appear)
[18] Fornberg B,Whitam G.,Phil. Trans.R.Soc.Lond.,289,373.(1978)
[19] Benjamin T.,J.Fluid Mech.,29,559.(1967)
[20] Ono H.,J.Phys.Soc.Japan,39,1082.(1975)
[21] Joseph R.,J.Phys.A10,L225.(1977)
[22] Satsuma J,Ablowitz M,Kodama Y.,"On an internal wave equation describing a stratified fluid with finite depth" Physics Letters (to appear)
[23] Bock T,Kruskal M.,Physics Letters,74A,173.(1979)

Generalised Bäcklund Transformations for Integrable Evolution Equations associated with Nth order Scattering Problems.

R.K.Dodd & H.C.Morris.
School of Mathematics,Trinity College,Dublin 2,Republic of Ireland.

Calogero and Degasperis have shown how it is possible by using their *generalised Wronskian technique* to obtain transformations which relate equations solvable by the same scattering problem. In particular for the Nth order Schrödinger equation [1] and the 2nd order Zakharov-Shabat scattering problem [2] they have obtained *generalised auto-Bäcklund transformations* which enable new solutions of an equation to be obtained from a given solution of the same equation. The generalisation consists in the fact that whereas a Bäcklund transformed solution can contain, asymptotically as $t\to\infty$, only one extra soliton, the generalised Bäcklund transformed solution can contain an arbitrary number.

We present here the generalised Bäcklund transformation for a subclass of the equations solvable by the Nth order Zakharov-Shabat scattering problem [3],[4]. The generalised Bäcklund transformations (or rather one half of them) are obtained by using a method which depends on a generalisation [5] of the squared eigenfunctions [6] associated with a linear scattering problem.

§ 1 *Bäcklund transformations for the Nth order Z-S problem.*

Consider the scattering problem,

$$V_{,x} = (\xi R + P)V \qquad -\infty < x < \infty \tag{1.1}$$

where V is an N-column vector, $P=(p_{ij})$, with $p_{ij}\to 0$ as $|x|\to\infty$ and

$\int_{-\infty}^{\infty}|P_{ij}|dx < \infty$. R is a constant diagonal matrix, diag $R = i\beta_j\delta_{jk}(\beta_1<\beta_2<\ldots<\beta_N)$ [3],[4]. We do not consider the case when $\beta_i=\beta_j$ in this paper. Fundamental matrix solutions to (1.1) are denoted by $\Phi(x,t),\Psi(x,t)$ which have the asymptotic behaviour $E(x)=\mathrm{diag}(e^{i\beta_1\xi x},e^{i\beta_2\xi x},\ldots,e^{i\beta_N\xi x})$ as $x\to-\infty, x\to+\infty$ respectively. Since Φ and Ψ are linearly independent,

$$\Phi(x,t,\xi) = \Psi(x,t,\xi)A(\xi,t) \tag{1.2}$$

which introduces $A(\xi,t)$, the scattering matrix.

Using the approach suggested in [5] consider the systems,

$$V'_{,x} = (\xi R + P')V' \tag{1.3a}$$

$$V_{,x} = (\xi R + P)V \tag{1.3b}$$

with associated fundamental matrix solutions Φ' and Φ. It is straightforward to show from (1.3) that,

$$A^{-1}(A'-A) = \int_{-\infty}^{\infty}\Phi^{-1}(P'-P)\Phi'\,dx \tag{1.4}$$

where A' and A are the scattering matrices corresponding to (1.3a) and (1.3b) respectively. It is also possible to obtain from (1.3) the following formula,

$$A^{-1}(CA'-AC) = \int_{-\infty}^{\infty}\Phi^{-1}(CP'-PC)\Phi'\,dx \tag{1.5}$$

where C is an arbitrary diagonal matrix. Then the ij th entry in (1.4) can be written as

$$(A^{-1}(A'-A))_{ij} = \int_{-\infty}^{\infty}(P'-P).U_{ij}\,dx \tag{1.6a}$$

where U_{ij} is given by

$$U_{ij} = \sum_{k,l=1}^{n}(\Phi^{-1t}_{ki}\,\Phi'^{t}_{jl}) = (\Phi^{-1t}\Delta^{ij}\Phi'^{t}) = (\Phi'\Delta^{ji}\Phi^{-1})^{t} \tag{1.6b}$$

and Δ^{ij} is the matrix with 1 in the ijth entry and 0 elsewhere and t is the transpose operation. A similar result holds for (1.5).

We now show by construction that there exists an operator Λ, such that for the special class of potentials diag$P$=0=diagP',

$$\langle \Lambda H_0, U_{ij0}\rangle = \xi\langle H_0, U_{ij0}\rangle \tag{1.7a}$$

where $\langle H,U\rangle = \int_{-\infty}^{\infty}H.U\,dx$ and $A_0=A-A_D$ with $A_D=\mathrm{diag}A$ (1.7b)

From (1.3) we obtain

$$U_{ij,x} = \xi\,(U_{ij}R-RU_{ij}) + U_{ij}P'^{t} - P^{t}U_{ij} \tag{1.8}$$

It follows that provided $H(x=\pm\infty) = 0$,

$$\langle H_{,x} + HP' - PH + [H,R]\,,\ U_{ij}\rangle = 0 \tag{1.9}$$

Integrating (1.8) we also have the relation,

$$F_D,U_{ij} \;=\; \int_{-\infty}^{\infty} F_D.\Delta^{ij}\,dx + \langle \int_x^{\infty} F_D dyP' - P\int_x^{\infty} F_D dy, U_{ij}\rangle \qquad (1.10)$$

where $F_D(x=\pm\infty)=0$. Combining (1.9) and (1.10) produces the result,

$$\langle \Lambda H_O, U_{ijO}\rangle = \langle S_{O,x} + (S_O P' - PS_O)_O + \int_x^{\infty}(S_O P' - PS_O)_D dyP' - P\int_x^{\infty}(S_O P' - PS_O)_D dy, U_{ijO}\rangle$$

$$= \xi\langle H_O, U_{ijO}\rangle \qquad (1.11)$$

where

$$S_{Oij} = \frac{H_{Oij}}{i(\beta_i - \beta_j)} \qquad (1.12)$$

Finally from (1.4),(1.5) we have the relationships,

$$(A^{-1}(A'-A) - \Omega(\xi)(CA'-AC))_O = \sum_{i,j=1}^{N}\left(\int_{-\infty}^{\infty}(P'-P) - \Omega(\xi)(CP'-PC).U_{ijO}dx\right).\Delta_O^{ij}$$

$$(1.13)$$

where $\Omega(\xi) = f(\xi)/g(\xi)$ and f and g are entire functions.

Then since $(CP'-PC)_O = (CP'-PC)$ if diagP=0=diagP',the generalised Bäcklund transformations for this system are defined by,

$$g(\Lambda)(P'-P) - f(\Lambda)(CP'-PC) = 0 \qquad (1.14a)$$

The corresponding change in the scattering data involves the off diagonal "reflection coefficients" and is given by

$$(A^{-1}(g(\xi)(A'-A) - f(\xi)(CA'-AC)))_O \;=\; 0 \qquad (1.14b)$$

We omit the corresponding change in the diagonal elements from this short paper. Finally as an example of formula (1.14a) we produce the generalised Bäcklund transformations for the A.K.N.S. system [6].

Generalised Bäcklund transformations for the A.K.N.S.system.

This is defined by (1.1) with

$$P = \begin{bmatrix} 0 & q \\ r & 0 \end{bmatrix} \qquad R = \begin{bmatrix} -i & 0 \\ 0 & i \end{bmatrix} \qquad C = \begin{bmatrix} 1 & 0 \\ 0 & -1 \end{bmatrix}$$

Define $H_O = 2i\begin{bmatrix} 0 & -g \\ f & 0 \end{bmatrix}$ so that $S_O = \begin{bmatrix} 0 & g \\ f & 0 \end{bmatrix}$ and (1.11) becomes

$$\Lambda H_O = \begin{bmatrix} 0 & g_{,x} \\ f_{,x} & 0 \end{bmatrix} + \begin{bmatrix} 0 & q\int_x^{\infty}(rg-q'f)dy - q'\int_x^{\infty}(qf-r'g)dy \\ r\int_x^{\infty}(qf-r'g)dy - r'\int_x^{\infty}(rg-q'f)dy, & 0 \end{bmatrix}$$

$$= \xi H_O \qquad (1.15)$$

Define

$$D\begin{bmatrix} f & 0 \\ 0 & g \end{bmatrix} = \begin{bmatrix} 0 & 1 \\ -1 & 0 \end{bmatrix}\Lambda\begin{bmatrix} 0 & 1 \\ -1 & 0 \end{bmatrix}\begin{bmatrix} f & 0 \\ 0 & g \end{bmatrix} = \xi\begin{bmatrix} f & 0 \\ 0 & g \end{bmatrix} \qquad (1.16)$$

Clearly

$$D^k = \begin{bmatrix} 0 & 1 \\ -1 & 0 \end{bmatrix}\Lambda^k\begin{bmatrix} 0 & -1 \\ 1 & 0 \end{bmatrix}$$

so that the Bäcklund transformation equation (1.14a) for this system can be written,

$$\begin{bmatrix} r'-r & 0 \\ 0 & -(q'-q) \end{bmatrix} = -\Omega(D)\begin{bmatrix} r+r' & 0 \\ 0 & q+q' \end{bmatrix} \tag{1.17}$$

From (1.15) we have,

$$L\begin{bmatrix} f \\ g \end{bmatrix} = \left[-\tfrac{1}{2}i\begin{bmatrix} 1 & 0 \\ 0 & -1 \end{bmatrix}\frac{\partial}{\partial x} + \begin{bmatrix} r\int_x^\infty dyq + r'\int_x^\infty dyq' & -r\int_x^\infty dyr' - r'\int_x^\infty dyr \\ q\int_x^\infty dyq' + q'\int_x^\infty dyq & -q\int_x^\infty dyr - q'\int_x^\infty dyr' \end{bmatrix}\right]\begin{bmatrix} f \\ g \end{bmatrix}$$

$$= \xi\begin{bmatrix} f \\ g \end{bmatrix} \tag{1.18}$$

The Bäcklund transformation for the A.K.N.S. system can therefore be written,

$$\begin{bmatrix} r'-r \\ -(q'-q) \end{bmatrix} + \Omega(L)\begin{bmatrix} r + r' \\ q + q' \end{bmatrix} = 0 \tag{1.19}$$

References:

[1] Calogero F,Degasperis A.,Nuovo Cimento 39B,1.(1977)

[2] Calogero F,Degasperis A.,Nuovo Cimento 32B,201.(1976)

[3] Newell A.C.,"The general Structure of integrable evolution equations" to appear Proc.Roy.Soc.1978.

[4] Newell A.C.,"Near Integrable Systems,nonlinear tunnelling, and solitons in slowly changing media" in Nonlinear evolution equations solvable by the spectral transform" Pitman Research Notes in Mathematics 26. Edited by F.Calogero.(1978)

[5] Dodd R.K. J.Phys A.11,81.(1978)

[6] Kaup D.J.J.Math.Anal & Appl.,54,849.(1976)

MEROMORPHIC FORMS SOLUTIONS OF COMPLETELY INTEGRABLE PFAFFIAN SYSTEMS WITH REGULAR SINGULARITIES

by Raymond GERARD

This lecture is a part of a joint paper in preparation with J.P. RAMIS on a general Residue theory associated to singular linear connections.

Here I will restrict myself to some concrete results concerning the theory of linear Pfaffian systems.

1. NOTATIONS.

Let us denote by :

$\mathcal{O}^m$: the set of holomorphic maps from $\mathbb{C}^n$ to $\mathbb{C}^m$,

Ω^q : the set of holomorphic forms of degree q ,

$$(\Omega^q)^m = \underbrace{\Omega^q \times \Omega^q \times \dots \times \Omega^q}_{m}$$

$$X_i = \{x = (x_1 , x_2 ,\dots, x_n) \in \mathbb{C}^n \mid x_i = 0\} ,$$

$$X = \bigcup_{i=1}^{p} X_i \qquad 0 < p \leq n ,$$

$\Omega^q \langle X_i \rangle$: the set of q-differential forms having a pole of logarithmic type on X_i , this means that each $\alpha \in \Omega^q \langle X_i \rangle$ can be written in the form

$$\alpha = \frac{dx_i}{x_i} \wedge \alpha_1 + \alpha_2 ,$$

where α_1 is a holomorphic $(q-1)$-form, which does not contain dx_i , α_2 is a holomorphic q-form which does not contain dx_i .

$\Omega^q \langle X \rangle$ defined in the same way.

Let Z be an analytic subset of C^n, we shall assume that $Z = \bigcup_{i=1}^{t} Z_i$, where for each $i = 1, 2, \dots, t$, Z_i is irreducible without singularities.

Let us introduce :

$\Omega^q(* Z)$: the set of q-differential forms having poles on Z,

$\Omega^q\langle X\rangle(* Z)$: the set of q-forms having at most a logarithmic pole on X and poles on Z.

Then the following sets are well-defined : $\Omega^q(* X)$, $\Omega^q(* X \cup Z)$, $\Omega^q\langle X \cup Z\rangle$ (if $X \cup Z$ is normal crossing).

Consider in C^n the following linear Pfaffian system :

$$(1) \qquad \frac{\partial z}{\partial x_i} + zB_i(x) = 0 \qquad i = 1, 2, \dots, n$$

$$x = (x_1, x_2, \dots, x_n), \quad z = (z_1, z_2, \dots, z_m).$$

$B_i(x)$ is a $m \times m$ -square matrix.

We assume that :

$$B_i(x) = \frac{A_i(x)}{x_i} \qquad \text{for } i = 1, 2, \dots, p \le n,$$

where $A_i(x)$ is a $m \times m$ -square matrix holomorphic in C^n (or in a neighbourhood of the origin of C^n).

In the following we shall assume that (1) is completely integrable : this means that for all i, j :

$$(I) \qquad \frac{\partial B_i}{\partial x_j} - \frac{\partial B_j}{\partial x_i} = [B_i, B_j] = B_i B_j - B_j B_i .$$

The system (1) can be written in the form

$$dz + z\omega = 0,$$

where $\omega = \sum_{i=1}^{p} A_i(x) \frac{dx_i}{x_i} + \sum_{i=p+1}^{n} B_i(x)\, dx_i$, and the integrability condition is $d\omega = \omega \wedge \omega$.

To this Pfaffian system we associate the map

$$\nabla : \mathcal{O}^m \longrightarrow (\Omega^1 \langle X \rangle)^m$$

$$v \longrightarrow \nabla v = dv + v\omega .$$

This map has the following properties :

1) C-linear ,

2) satisfies Leibnitz formula :

$\nabla(fv) = dfv + f\nabla v$ for all $f \in \mathcal{O}$ and $v \in \mathcal{O}^m$.

In fact our Pfaffian defines a linear connection on $\mathcal{O}^m$.

Now let us extend the operator ∇ to differential forms by the formula

$$\nabla \alpha = d\alpha + (-1)\alpha \wedge \omega$$

for each $\alpha \in (\Omega^q)^m$ (resp. $(\Omega^q \langle X \rangle)^m$ ...) .

Now it is easy to see that if

- $\alpha \in (\Omega^q)^m$, then $\nabla\alpha \in (\Omega^{q+1} \langle X \rangle)^m$
- $\alpha \in (\Omega^q \langle X \rangle)^m$, then $\nabla\alpha \in (\Omega^{q+1} \langle X \rangle)^m$
- $\alpha \in (\Omega^q \langle Z \rangle)^m$, then $\nabla\alpha \in (\Omega^{q+1} \langle X \cup Z \rangle)^m$
- $\alpha \in (\Omega^q (* Z))^m$, then $\nabla\alpha \in (\Omega^{q+1} \langle X \rangle (* Z))^m$
- $\alpha \in (\Omega^q (* X \cup Z))^m$, then $\nabla\alpha \in (\Omega^{q+1} (* X \cup Z))^m$.

But now we can consider $\nabla \circ \nabla = \nabla^2$.

And the complete integrability condition is equivalent to $\nabla^2 = 0$.

In fact, for each q-form α of any type,

$$\begin{aligned}\nabla(\nabla\alpha) &= \nabla(d\alpha + (-1)^q \alpha\wedge\omega) \\ &= d(d\alpha + (-1)^q \alpha\wedge\omega) + (-1)^{q+1}(d\alpha + (-1)^q \alpha\wedge\omega) \wedge \omega \\ &= (-1)^q\, d\alpha\wedge\omega + (-1)^{2q}\, \alpha\wedge d\omega + (-1)^{q+1}\, d\alpha\wedge\omega + (-1)^{2q+1}\, \alpha\wedge\omega\wedge\omega \\ &= \alpha \wedge (d\omega - \omega\wedge\omega) .\end{aligned}$$

Assuming that the Pfaffian system (1) is completely integrable, we have several complexes : the so called De Rham complexes associated with ∇ :

$$(\Omega^\bullet < X > , \nabla) \; ; \; (\Omega^\bullet < X > (*Z) , \nabla)$$
$$(\Omega^\bullet(* X \cup Z) , \nabla) \; ; \; (\Omega^\bullet(* X) , \nabla) .$$

In the one variable case, this complexes are very short $X = \{x = 0\} = 0$; ∇ is given by $\nabla y = dy + y\, A(x) \frac{dx}{x}$; we have always $\nabla^2 = 0$, and for example

$$(\Omega^\bullet < X > , \nabla) \text{ is the complex } 0 \longrightarrow \mathcal{O}^m \xrightarrow{\nabla} (\Omega^1 < X >)^m \longrightarrow 0$$
$$(\Omega^\bullet (* X) , \nabla) \text{ is the complex } 0 \longrightarrow \mathcal{O}^m \longrightarrow (\Omega^1 (* X))^m \longrightarrow 0 .$$

Solving $\nabla y = 0$ or $\nabla y = f$ is equivalent to finding the Kernel of ∇ and the cokernel of ∇ .

This means for example for the first complex $(\Omega^\bullet <X> , \nabla)$ to know the cohomology of this complex which is

1) $H^0(\Omega^\bullet <X>) = \operatorname{Ker} \nabla$

2) $H^1(\Omega^\bullet <X>) = \dfrac{(\Omega^1 <X>)^m}{\nabla(\mathcal{O}^m)}$,

$H^0(\Omega^\bullet <X>)$ is the vector space over C of holomorphic solutions of $\nabla = 0$.

To find the cohomology space $H^1(\Omega^\bullet <X>)$, we have to solve the following problem, find for each one form

$$\varphi(x) \frac{dx}{x} \; , \qquad \varphi \in \mathcal{O}^m$$

an element $z \in \mathcal{O}^m$ such that $\nabla z = \varphi(x) \frac{dx}{x}$, which means that

$$x \frac{dz}{dx} + z\, A(x) = \varphi(x) .$$

Very simple computations will show you that :

1) If $A(0)$ has no eigenvalue which is a negative or zero integer, then the solution exist and is unique. In this case $H^1(\Omega^\bullet <X>) = \{0\}$.

2) In the other case we can prove easily that $H^1(\Omega^\bullet <X>)$ is of finite

dimension and to compute this dimension, we are back to a classical problem of differential systems with regular singularities.

2. THE PROBLEMS.

In the several variable case, the complexes defined above are much longer, so we have more cohomology vector spaces and the problems are very clear :

Study the cohomology of the De-Rham complexes associated with the linear connexion ∇ .

3. RECALL OF SOME CLASSICAL MATERIAL [1].

Let $\nabla = d\bullet + (-1)^{\bullet}\bullet\wedge\omega$;

$$\omega = \sum_{i=1}^{p} A_i(x) \frac{dx_i}{x_i} + \sum_{i=p+1}^{n} B_i(x)\, dx_i \ ,$$

let us write $x = (x_1, x_2, \dots, x_p)$, $y = (x_{p+1}, \dots, x_n)$, then

$$\omega = \sum_{i=1}^{p} A_i(x, y) \frac{dx_i}{x_i} + B(x, y) \ .$$

Set by definition

$$\mathrm{Res}_{X_i}(\nabla) = A_i(x, y)|_{X_i} \ .$$

Then we have the following properties :

1) $[\mathrm{Res}_{X_i}(\nabla) , \mathrm{Res}_{X_j}(\nabla)]_{X_i \cap X_j} = 0$ all i , j

2) $\nabla_{X_i}(\mathrm{Res}_{X_i} \nabla) = 0$ all i ,

where for all $k = 1, 2, \dots, p$

$$\nabla_{X_k} = d + (-1)^{\bullet}\bullet \wedge \omega_{X_k} \ ,$$

with

$$\omega_{X_k} = \sum_{\substack{i=1 \\ i\neq k}}^{p} A_i|_{X_k} \frac{dx_i}{x_i} + B(x, y)|_{X_k} .$$

The connexion ∇_{X_k} is integrable.

3) For all i, the eigenvalues of $\mathrm{Res}_{X_i} \nabla$ are constant.

Later on, we have also to use the integrable linear connexion

$$\nabla_{[kX]} = \nabla_{[k_1X_1 + k_2X_2 + \ldots + k_pX_p]}$$

defined by the one-form

$$\sum_{i=1}^{p} (A_i - k_i I) \frac{dx_i}{x_i} + B(x, y) .$$

4. THE SIMPLEST CASE.

Let us assume here that $p = 1$ so that

$$\omega = A(x, y) \frac{dx}{x} + B(x, y) ,$$

$$x = (x_1) , \quad y = (x_2, \ldots, x_n)$$

and

$$X = \{x = 0\} .$$

We are considering here the complex $\Omega^{\bullet}(*X)$.

PROPOSITION 1. If $\varphi \in (\Omega^q(*X))^m$ is ∇-closed and has a pole of order r on $x = 0$, then φ is ∇-cohomologous to a form of the type

$$\frac{dx}{x^r} \wedge \psi_1 + \frac{\psi_2}{x^{r-1}} ,$$

where ψ_1 is a holomorphic $(q-1)$-form, ψ_2 is a holomorphic q-form; both of them do not contain dx.

Proof. φ is of the form : $\varphi = \frac{dx}{x^r} \wedge \varphi_1 + \frac{\varphi_2}{x^r}$,

φ_1 , φ_2 do not contain dx . Then

$$\nabla \varphi = - \frac{dx}{x^r} \wedge d\varphi_1 - r \frac{dx \wedge \varphi_2}{x^{r+1}} + \frac{d\varphi_2}{x^r}$$

$$+ (-1)^q \left(\frac{dx}{x^r} \wedge \varphi_1 + \frac{\varphi_2}{x^r}\right) \wedge \left(A(x , y) \frac{dx}{x} + B(x , y)\right)$$

Let us denote for each differential form h , $d_x h$ the differential with respect to dx , $d_y h$ the differential with respect to the variables $y = (x_2 , \dots , x_n)$, and if $d_x h = dx \wedge k$,

$$k = \frac{d_x h}{dx} .$$

Then

$$\nabla\varphi = dx \wedge \left[- \frac{d_y\varphi_1}{x^r} - \frac{r\varphi_2}{x^{r+1}} + \frac{1}{x^r} \frac{d_x\varphi_2}{dx} + \right.$$

$$\left. + (-1)^q \left(\frac{\varphi_1 \wedge B}{x^r} + (-1)^q \frac{\varphi_2 \ A}{x^{r+1}}\right)\right] + \frac{d_y\varphi_2}{x^r} + (-1)^q \frac{\varphi_2 \wedge B}{x^r}$$

and $\nabla\varphi = 0$ gives you

$$\begin{cases} \varphi_2(A - rI) + x[- d_y \varphi_1 + \frac{d_x\varphi_2}{dx} + (-1)^q \varphi_1 \wedge B] = 0 \\ d_y\varphi_2 + (-1)^q \varphi_2 \wedge B = 0 . \end{cases}$$

If for each form h , we denote by h° the restriction of h to $x = 0$, we have

$$\begin{cases} \varphi_2^\circ (A^\circ - rI) = 0 \\ d_y \varphi_2^\circ + (-1)^q \varphi_2^\circ \wedge B^\circ = 0 \end{cases}$$

If $A^\circ = \text{Rés}_X \nabla$ has not the eigenvalue r , then $\varphi_2^\circ = 0$ and the lemma is proved.

$$\varphi = \frac{dx}{x^r} \wedge \varphi_1 + \frac{\tilde{\varphi}_2}{x^{r-1}} .$$

Assume that A° has the eigenvalue r , then $\varphi_2^\circ \in \mathrm{Ker}(A^\circ - rI)$, and consider the system

$$(S_o) \begin{cases} d_y u + (-1)^{q-1} u \wedge B^\circ = \varphi_2^\circ \\ u(A^\circ - rI) = 0 . \end{cases}$$

The connection $\nabla_y = d_y + (-1)^{\bullet} \bullet \wedge B^\circ$ is completely integrable, it is the restriction of ∇ to $x = 0$, so we have $d_y B^\circ = B^\circ \wedge B^\circ$.

<u>Let us study the system</u> S_o :

First

$$(S_o) \begin{cases} \nabla_y u = \varphi_2^\circ \\ u(A^\circ - rI) = 0 \end{cases}$$

$\nabla_y \varphi_2^\circ = 0$ implies that it does exist $u(y)$ holomorphic such that $\nabla_y u = \varphi_2^\circ$.

Let us compute

$$\nabla_y(u(A_o - rI)) = d_y u (A^\circ - rI) + (-1)^{q-1} u \wedge d_y A^\circ + (-1)^{q-1} u(A^\circ - rI) \wedge B^\circ$$

or $d_y u + (-1)^{q-1} u \wedge B^\circ = \varphi_2^\circ$, so

$$\begin{aligned} \nabla_y(u(A_o - rI)) &= (\varphi_2^\circ + (-1)^q u \wedge B^\circ)(A^\circ - rI) + (-1)^{q-1} u \wedge (d_y A^\circ + (A^\circ - rI) \wedge B^\circ) \\ &= (-1)^{q-1} u \wedge \left[d_y A^\circ + (A^\circ - rI)B^\circ - B^\circ(A^\circ - rI)\right] \\ &= (-1)^{q-1} u \wedge \left[d_y A^\circ + [A^\circ, B^\circ]\right] \\ &= 0 \end{aligned}$$

because of the integrability condition which implies

$$d_y A^\circ + [A^\circ, B^\circ] = 0 .$$

Finally

$$\nabla_y(u(A_o - qI)) = 0 ,$$

but by the Poincaré lemma (Frobenius theorem) for ∇_y we have

$$u(A_o - qI) = \nabla_y v \quad , \quad v = v(y) .$$

Compute now

$$\nabla\left(\frac{u}{x^r} + \frac{dx\wedge v}{x^{r+1}}\right) = \frac{du}{x^r} - r\,\frac{dx\wedge u}{x^{r+1}} - \frac{dx\wedge d_y v}{x^{r+1}} + (-1)^{q-1}\left(\frac{u}{x^r} + \frac{dx\wedge v}{x^{r+1}}\right)\wedge\left(A^\circ \frac{dx}{x} + B^\circ\right)$$

$$+ (-1)^{q-1}\left(\frac{u}{x^r} + \frac{dx\wedge v}{x^{r+1}}\right)\wedge\left(A^1\, dx + xB^1\right)$$

$$= \frac{du + (-1)^{q-1}\, u\wedge B^\circ}{x^r} + \frac{dx}{x^{r+1}}\wedge u(A^\circ - rI) - \frac{dx}{x^{r+1}}\left[d_y v + (-1)^{q-2}\, v\wedge B^\circ\right]$$

$$+ (-1)^{q-1}\,\frac{dx\wedge v}{x^r}\wedge B^1 + (-1)^{q-1}\,\frac{u\wedge A^1 dx}{x^r} + (-1)^{q-1}\,\frac{u\wedge B^1}{x^{r-1}}$$

$$= \frac{\nabla_y u}{x^r} + \frac{dx}{x^{r+1}}\wedge\left[u(A^\circ - rI) - \nabla_y v\right] + (-1)^{q-1}\,\frac{dx\wedge v\wedge B^1}{x^r}$$

$$+ \frac{dx}{x^r}\wedge uA^1 + (-1)^{q-1}\,\frac{u\wedge B^1}{x^{r-1}}$$

$$= \frac{\varphi_2^\circ}{x^r} + (-1)^{q-1}\,\frac{dx\wedge v\wedge B^1}{x^r} + \frac{dx}{x^r}\wedge uA^1 + (-1)^{q-1}\,\frac{u\wedge B^1}{x^{r-1}} .$$

Then

$$\varphi = \frac{dx\wedge\varphi_1}{x^r} + \frac{\varphi_2^\circ}{x^r} + \frac{\varphi_2^1}{x^{r-1}}$$

$$= \frac{dx}{x^r}\wedge\left(\varphi_1 + (-1)^q\, v\wedge B^1 - uA^1\right) + \frac{\varphi_2^1 + (-1)^q\, uB^1}{x^{r-1}} + \nabla\left(\frac{u}{x^r} + \frac{dx\wedge v}{x^{r+1}}\right)$$

which proves the proposition 1.

PROPOSITION 2. <u>Each ∇-closed form of degree q of the type</u>

$$\varphi = \frac{dx}{x^r}\wedge\varphi_1 + \frac{\varphi_2}{x^{r-1}}$$

<u>is ∇-cohomologous to a form of the type</u>

$$\psi = \frac{dx}{x^{r-1}} \wedge \psi_1 + \frac{\psi_2}{x^{r-2}} .$$

Write $\nabla\varphi = 0$.

$$\nabla\varphi = -\frac{dx}{x^r} \wedge d\varphi_1 - (r-1)\frac{dx \wedge \varphi_2}{x^r} + \frac{d\varphi_2}{x^{r-1}} + (-1)^q\left(\frac{dx}{x^r}\wedge\varphi_1 + \frac{\varphi_2}{x^{r-1}}\right) \wedge \left(A\frac{dx}{x} + B\right) ,$$

which give us

$$-\frac{d_y\varphi_1}{x^r} - (r-1)\frac{\varphi_2}{x^r} + \frac{1}{x^r}\frac{d_x\varphi_2}{x^r} + (-1)^q\frac{\varphi_1 \wedge B}{x^r} + \frac{\varphi_2 \wedge A}{x^r} = 0$$

and

$$\frac{d_y\varphi_2 + (-1)^q\,\varphi_2 \wedge B}{x^{r-1}} = 0 .$$

Finally

$$-\nabla_y\,\varphi_1^\circ + \varphi_2^\circ(A^\circ - (r-1)I) = 0$$

$$\nabla_y\,\varphi_2^\circ = 0 .$$

Consider now the system

$$(S)\quad \begin{aligned} -\nabla_y\,u + v(A^\circ - (r-1)I) &= \varphi_1^\circ \\ \nabla_y\,v &= \varphi_2^\circ \end{aligned}$$

Remark. Here it is necessary to assume that $q > 1$. In fact if $q = 1$, then φ_2° is a 1-form and v is a 0-form, and the first equation has no meaning. When $q = 1$ replace the first equation by $v(A^\circ - (r-1)I) = \varphi_1^\circ$ and do exactly the same as in the following. Now let us solve the system (S) . As $\nabla_y\,\varphi_2^\circ = 0$, there exist a $v(y)$ such that $\nabla_y\,v = \varphi_2^\circ$.

Now we have to find a $u(y)$ satisfying the first equation, this means find a $u(y)$ such that :

$$\nabla_y\,u = v(A^\circ - (r-1)I) - \varphi_1^\circ .$$

As above, let us compute :

$$\nabla_y(v(A^\circ - (r-1)I) - \varphi_1^\circ)$$

$$= d_y\, v(A^\circ - (r-1)I) + (-1)^{q-1}\, v \wedge d_y A^\circ - d_y\, \varphi_1^\circ$$

$$+ (-1)^{q-1}\, (v(A^\circ - (r-1)I - \varphi_1^\circ) \wedge B_o$$

$$= [\varphi_2^\circ + (-1)^q\, v \wedge B^\circ](A^\circ - (r-1)I) + (-1)^{q-1}\, v \wedge d_y A^\circ$$

$$- [d_y\, \varphi_1^\circ + (-1)^{q-1}\, \varphi_1^\circ \wedge B_o] + (-1)^{q-1}\, v(A^\circ - (r-1)I) \wedge B^\circ$$

$$= \varphi_2^\circ(A^\circ - (r-1)I) - \nabla_y\, \varphi_1^\circ + (-1)^{q-1}\, v \wedge [d_y\, A^\circ - B^\circ A^\circ + A^\bullet B^\bullet]$$

$$= 0 .$$

So there does exist a $u(y)$ such that

$$- \nabla_y\, u + v(A^\circ - (q-1)I) = \varphi_1^\circ .$$

Now let us compute

$$\nabla\left(\frac{dx \wedge u}{x^r} + \frac{v}{x^{r-1}}\right) = - \frac{dx}{x^r} \wedge d_y u - (r-1)\, \frac{dx \wedge v}{x^{r\bullet}} + \frac{d_y v}{x^{r-1}}$$

$$+ (-1)^{q-1}\, \left(\frac{dx \wedge u}{x^r} + \frac{v}{x^{r-1}}\right) \wedge \left(A\, \frac{dx}{x} + B\right)$$

$$= - \frac{dx}{x^r} \wedge (d_y u + (r-1)v) + \frac{d_y \varphi}{x^{r-1}} + (-1)^{q-1}\left(\frac{dx \wedge u}{x^r} + \frac{v}{x^{r-1}}\right) \wedge \left(A^\circ\, \frac{dx}{x} + B^\circ\right)$$

$$+ (-1)^{q-1}\left(\frac{dx \wedge u}{x^r} + \frac{v}{x^{r-1}}\right) \wedge (A^1\, dx + xB^1)$$

$$= \frac{dx}{x^r} \wedge (- d_y u + (-1)^{q-1}\, u \wedge B^\circ + vA^\circ - (r-1)v) + \frac{\nabla_y\, v}{x^{r-1}}$$

$$+ (-1)^{q-1}\, \frac{dx \wedge u \wedge B^1}{x^{r-1}} + \frac{dx \wedge vA^1}{x^{r-1}} + (-1)^{q-1}\, \frac{v \wedge B_1}{x^{r-2}} ,$$

$$\nabla\left(\frac{dx\wedge u}{x^r}+\frac{v}{x^{r-1}}\right)=\frac{dx}{x^r}\wedge\varphi_1^{\circ}+\frac{\varphi_2^{\circ}}{x^{r-1}}+\frac{dx}{x^{r-1}}\wedge(vA^1+(-1)^{q-1}\,u\wedge B^1)+\frac{B^1\wedge v}{x^{r-2}}$$

$$=\frac{dx}{x^r}\wedge\varphi_1^{\circ}+\frac{\varphi_2^{\circ}}{x^{r-1}}+\frac{dx}{x^{r-1}}\wedge W_1+\frac{W_2}{x^{r-2}}$$

$$\varphi=\frac{dx}{x^r}\wedge\varphi_1+\frac{\varphi_2}{x^{r-2}}=\frac{dx}{x^r}\wedge\varphi_1^{\circ}+\frac{\varphi_2^{\circ}}{x^{r-1}}+\frac{dx}{x^{r-1}}\wedge\varphi_1^1+\frac{\varphi_2^1}{x^{r-2}}.$$

Replacing $\frac{dx}{x^r}\wedge\varphi_1^{\circ}+\frac{\varphi_2^{\circ}}{x^{r-1}}$ by $\nabla\left(\frac{dx\wedge u}{x^r}+\frac{v}{x^{r-1}}\right)-\frac{dx}{x^{r-1}}\wedge\omega_1-\frac{\omega_2}{x^{r-2}}$, we have proved the lemma 2.

Remark. In all the computations, it was not necessary to assume that r was a positive integer.

Finally the two lemmas imply the following theorem :

THEOREM. If $q>1$, then each $\varphi\in(r^q(*X))^m$ which is ∇-closed is ∇-cohomologous to a holomorphic ∇-closed form.

This means the folowwing : if $\nabla\varphi=0$, φ having pole on X, then $\varphi=\hat{\varphi}+\nabla(\psi)$, where $\hat{\varphi}$ is holomorphic, $\nabla\hat{\varphi}=0$.

So to find the q-forms with poles on X solutions of $\nabla=0$, it is enough to find holomorphic forms which are solutions.

Then each meromorphic form solution φ is of the form $\hat{\varphi}+\nabla\psi$, where $\hat{\varphi}$ is a holomorphic form satisfying $\nabla\hat{\varphi}=0$ and ψ is an arbitrary $(q-1)$-form having pole on X.

5. A MORE GENERAL CASE.

Now let be given a linear connexion ∇ having regular singularities along the X_i' s . This means that :

$$\nabla = d. + (-1)^{\bullet} \bullet \wedge \omega$$

where

$$\omega = \sum_{i=1}^{p} A_i(x, y) \frac{dx_i}{x_i} + B(x, y) .$$

And let $Z = \bigcup_{i=1}^{t} Z_i$, Z_i irreductible without singularities.

And assume moreover that $\bigcup_{i=1}^{p} X_i \bigcup_{1}^{t} Z_i$ are normal norming. This is not a restriction for a general theory, because of Hironaka's theorem.

Notations.

$$(k) = (k_1, k_2, \ldots, k_p) \in \mathbb{Z}^p$$

$$(1) = (1, 1, \ldots, 1)$$

$$(1_i) = (0, \ldots, 0, 1, 0, \ldots, 0)$$

$\uparrow$ i^{th} place

and denote by

$$-\Omega^q <(k)X> = \Omega^q <X> \otimes_{\mathcal{O}} \mathcal{O}(((k)-(1))X) ,$$

where $\mathcal{O}(((k)-(1))X)$ are the meromorphic functions having on X_i a pole of order at most $k_i - 1$.

$$-\Omega^q <(k)X> (*Z) = \Omega^q <(k)X> \otimes_{\mathcal{O}} \mathcal{O}(*Z) ,$$

where $\mathcal{O}(*Z)$ are the meromorphic functions having at most a pole on Z .

- $\hat{\Omega}^q <(k)X> (*Z)$ means the same object, but in the formal sens which means that all the coefficients of the forms occuring are not necessarly convergent, but formal power series.

PROPOSITION 3. Assume that for each i , $\mathrm{Res}_{X_i} \nabla$ has not the eigenvalue k_i . Then let $\varphi \in (\Omega^q(* X \cup Z))^m$ having a pole of order at most k_i on X_i $(i = 1, 2, \dots, p)$ and assume that $\nabla\varphi$ has a pole of order at most $k_i - 1$ on X_i , then :

$$\varphi \in \Omega^q <(k)X>(*Z) .$$

Proof.

Remark 1. Using instead of ∇ the connexion $\nabla_{[((k)-(1))X]}$ defined by

$$\omega = \sum_{i=1}^{p} (A_i - (k_i - 1)I) \frac{dx_i}{x_i} + B(x, y) ,$$

it is enough to prove the proposition for $k_i = 1$, all $i = 1, 2, \dots, p$.

Remark 2.

$$\bigcap_{i=1,\dots,p} \Omega^q <X_i>(* X_i' \cup Z) = \Omega^q <X>(*Z) ,$$

and it is enough to prove the proposition when $p = 1$.

$$\text{Let} \quad \nabla = d + (-1)^{\bullet} \bullet \wedge (A(x, y) \frac{dx}{x} + B(x, y) ,$$

and write

$$A = A^\circ(y) + xA^1$$

$$B = B^\circ(y) + xB^1$$

and

$$\varphi = \frac{dx}{x} \wedge \varphi_1 + \frac{\varphi_2}{x}$$

with

$$\nabla\varphi \in \Omega^{q+1}(*Z) ,$$

$$\varphi_1 \in \Omega^{q-1}(*Z) , \quad \varphi_2 \in (\Omega^q(*Z))^m .$$

Compute

$$\nabla\varphi = \frac{dx}{x^2} \wedge \left(\varphi_2(A^\circ - I) + x(-dy\varphi_1 + \frac{d_x\varphi_2}{dx} + \varphi_2 A^1 + (-1)^q \varphi_1 \wedge B)\right)$$

$$+ \frac{1}{x}(d_y\varphi_2 + (-1)^q \varphi_2 \wedge B) .$$

Set

$$\alpha = \varphi_2(A^\circ - I) + x(-d_y\varphi_1 + \frac{d_y\varphi_2}{dx} + \varphi_2 A^1 + (-1)^q \varphi_1 \wedge B) = \varphi_2(A^\circ - I) + x\eta .$$

$\nabla\varphi \in \Omega^{q+1}(*Z)$ implies that $\gamma = \frac{\alpha}{x^2} \in (\Omega^{q-1}(*Z))^m$, then $(A^\circ - I)$ being invertible :

$$\varphi_2 = \alpha(A^\circ - I)^{-1} - x\eta(A^\circ - I)^{-1}$$

$$= x^2\gamma(A^\circ - I)^{-1} - x\eta(A^\circ - I)^{-1}$$

$$= x\,\hat{\varphi}_2 \ , \quad \hat{\varphi}_2 \in \Omega^q(*Z) ,$$

and so

$$\varphi = \frac{dx}{x} \wedge \varphi_1 + \hat{\varphi}_2 ,$$

and the proposition is proved.

PROPOSITION 4. Assume that for all $i = 1, 2, \dots, p$, $\mathrm{Res}_{X_i} \nabla$ has not the eigenvalue $k_i - 1$. Let $\varphi \in (\Omega^q<(k)X>(*Z))^m$, then it does exist

$$\psi \in (\Omega^q((k)-(1))X)(*Z))^m$$

and

$$u \in (\Omega^{q-1}(((k)-(1))X)(*Z))^m$$

such that $\varphi = \psi + \nabla u$. If moreover $\nabla\varphi \in (\Omega^{q+1}(((k)-(1))X)(*Z))^m$, then

$$\psi \in (\Omega^q<((k)-(1))X> (*Z))^m .$$

With the same remark as for proposition 3, it is enough to prove this proposition for $k_i = 1$ for all i , and the statement is :

Assume $\operatorname{Res}_{X_i} \nabla$ has not the eigenvalue 0 and let $\varphi \in (\Omega^q\langle X\rangle(*Z))^m$, then it does exist $\psi \in (\Omega^q(*Z))^m$, $u \in (\Omega^{q-1}(*Z))^m$ such that $\varphi = \psi + \nabla u$. Moreover if $\nabla\varphi \in (\Omega^{q+1}(*Z))^m$, then $\psi \in (\Omega^q\langle(0)X\rangle(*Z))^m$.

<u>Proof</u>. By induction on p.

Assume that the result is true for all forms of degree $0, 1, 2, \dots, q$ and let us prove the result for $0, 1, 2, \dots, q$ and p. The result is trivial for degree $0, 1, 2, \dots, q$ and $p = 0$, there is nothing to prove.

Let $\varphi \in (\Omega^q\langle X\rangle(*Z))^m$, then φ can be written in an unique way in the form

$$\varphi = \frac{dx_p}{x_p} \wedge \varphi_{p_1} + \varphi_{p_2} .$$

φ_{p_1} and φ_{p_2} do not contain dx_p and have logarithmic poles along X'_p.

$$\varphi_{p_1} = \varphi^{\circ}_{p_1}(y) + x\varphi^1_{p_1}(x, y) .$$

Set

$$u = \varphi^{\circ}_{p_1} A^{\circ-1}_p \qquad A^{-1}_{p_o} = \operatorname{Res}_{X_p} \nabla$$

$$A_p = A^{\circ}_p + xA^1_p \qquad \omega = A_p \frac{dx_p}{x_p} + B_p ,$$

then

$$\nabla u = \nabla(\varphi^{\circ}_{p_1} A^{-1}_{p_o}) = dy\, \varphi^{\circ}_{p_1} A^{-1}_{p_o} + (-1)^q \varphi^{\circ}_{p_1} A^{-1}_{p_o} \wedge (A_p \frac{dx_p}{x_p} + B_p)$$

$$= \frac{dx_p}{x_p} \wedge \varphi^{\circ}_{p_1} + \left[d_y\varphi^{\circ}_{p_1} A^{-1}_{p_o} + (-1)^q \varphi^{\circ}_{p_1} A^{-1}_{p_o} \wedge (A^1_p \frac{dx_p}{x_p} + B_p)\right]$$

Then

$$\varphi - \nabla u = \psi \in (\Omega^q\langle X'_p\rangle(*Z))^m , \quad u \in (\Omega^{q-1}\langle X'_p\rangle(*Z))^m ,$$

and by our induction assumption, we have

$$u = v + \nabla w \qquad v \in (\Omega^{q-1}(*Z))^m$$
$$\qquad\qquad\quad w \in (\Omega^{q-2}(*Z))^m$$

and

$$\psi = \xi + \nabla\zeta \qquad \xi \in (\Omega^{q}(*Z))^m$$
$$\qquad\qquad\quad \zeta \in (\Omega^{q-1}(*Z))^m$$

It follows that

$$\varphi - \nabla(v + \nabla w) = \xi + \nabla\zeta$$

and

$$\varphi = \xi + \nabla(v + \zeta)$$

and the proposition is proved.

Now let us state

THEOREM. Let $k^+ = (k_1^+, \dots, k_p^+) \in \mathbb{Z}^p$ and $k^- = (k_1^-, \dots, k_p^-) \in \mathbb{Z}^p$ such that for all $i = 1, 2, \dots, p$, $\mathrm{Res}_{X_i} \nabla$ has no eigenvalue which is an integer greater than k_i^+ or smaller than k_i^-. Then

i) $(k) \in \mathbb{Z}^p$, $(k) \geq k^+$.

The natural map $\Omega^\bullet \langle (k^+)X \rangle (*Z) \longrightarrow \Omega^\bullet \langle (k)X \rangle (*Z)$ is a quasi-isomorphism.

ii) The natural map $\Omega^\bullet \langle (k^+)X \rangle (*Z) \longrightarrow \Omega^\bullet(* X \cup Z)$ is a quasi-isomorphism.

iii) Let $(k) \in \mathbb{Z}^p$, $(k) \leq (k^-)$, then the natural map $\Omega^\bullet \langle (k)X \rangle (*Z) \longrightarrow \Omega^\bullet \langle (k)X \rangle (*Z)$ is a quasi-isomorphism.

iv) Let $(k) \leq (k^-)$, then the complex $\hat{\Omega} \langle (k)X \rangle (*Z)$ is acyclic (has 0-cohomology).

v) Let $(k) \leq (k^-)$, then the complex $\Omega^\bullet \langle (k)X \rangle (*Z)$ is acyclic.

Remarks. When $(k^+) = (0)$, the assertion ii) is equivalent to a result due to Deligne ([1], p. 80).

Idea of the proof.

The proposition 4 gives you easily by induction the assertion i) and iii) . The statement ii) follows by going to the limit. The statement iv) is a consequence of iii) and of the fact that if $\varphi \in (\Omega_X^q<(k)Y>)^m$ for all $k \in \mathbb{Z}^p$, then $\varphi = 0$.

The statement v) is the most difficult to prove : the idea is to show that the computation used in proving iv) (which gives you formal objects that these objects are in fact convergent).

The basic tool is that each formal solution of an integrable Pfaffian system with regular singularities is convergent.

BIBLIOGRAPHY

-:-:-:-

[1] P. DELIGNE. Equations différentielles à points singuliers. Lecture Notes in Mathematics, n° 163, Springer-Verlag.

[2] R. GERARD et J.P. RAMIS. Théorie des résidus associée à une connexion linéaire avec singularités régulières. Applications. (to appear)

FAR FIELDS, NONLINEAR EVOLUTION EQUATIONS, THE BÄCKLUND TRANSFORMATION AND INVERSE SCATTERING

Alan Jeffrey

Department of Engineering Mathematics,
University of Newcastle upon Tyne,
NE1 7RU, England.

I. INTRODUCTION

The main objectives of this paper are (i) to discuss the important notion of a far field, (ii) to examine the origin of some nonlinear evolution equations exhibiting soliton behaviour, and (iii) to comment on the relationships that exist between the Bäcklund transformation, the Riccati equation, inverse scattering theory and conservation laws. These topics have been examined previously by other authors and we refer to the collected papers on Bäcklund transformations edited by R. M. Miura [1], to the collected papers on the reductive perturbation method for nonlinear wave propagation organised by T. Taniuti [2], to the review paper on nonlinear wave propagation by Jeffrey [3] and to the review on solitons by A. C. Scott, F. Y. F. Chu and D. W. McLaughlin [4] for further information and references.

II. FAR FIELDS

There are many different types of higher order equations and systems of equations that characterise nonlinear wave propagation in $\mathbb{R} \times t$, either with or without dispersion. A simplification frequently takes place in the representation of the solutions to initial value problems to such equations after a suitable lapse of time or, equivalently, suitably far from the origin, particularly when the initial data is localised and so has compact support. These simplified forms of solution are often asymptotic solutions, and are appropriately called far fields.

Perhaps the simplest examples of these are the types of far field behaviour exhibited by the ordinary linear wave equation and by a homogeneous quasilinear hyperbolic system with n dependent variables. Thus the wave equation

$$\frac{1}{c^2}\frac{\partial^2 u}{\partial t^2} = \frac{\partial^2 u}{\partial x^2} \qquad (c = \text{const.}) \tag{2.1}$$

may be written either in the form

$$\left(\frac{\partial}{\partial t} + c\frac{\partial}{\partial x}\right)\left(\frac{\partial u}{\partial t} - c\frac{\partial u}{\partial x}\right) = 0 , \tag{2.2a}$$

or as

$$\left(\frac{\partial}{\partial t} - c\frac{\partial}{\partial x}\right)\left(\frac{\partial u}{\partial t} + c\frac{\partial u}{\partial x}\right) = 0 . \tag{2.2b}$$

Then, if $u^{(\pm)}$ is the solution of

$$\frac{\partial u^{(\pm)}}{\partial t} \mp c\,\frac{\partial u^{(\pm)}}{\partial x} = 0\ , \tag{2.3}$$

it follows that $u^{(+)}$ is a degenerate solution of (2.2a) and $u^{(-)}$ is a degenerate solution of (2.2b). The general solution of (2.3) is then

$$u^{(\pm)} = f^{(\pm)}\,(x \mp ct)\ , \tag{2.4}$$

with $f^{(\pm)}$ arbitrary C^1 functions.

These travelling wave solutions are such that $u^{(+)}$ propagates to the right and $u^{(-)}$ to the left with speed c. We thus have the situation that $u^{(\pm)}$ are special simple types of solution to the wave equation (2.1), in the sense that they only satisfy a first order partial differential equation, whereas the wave equation itself is of second order. Such special solutions become of considerable interest when the initial data $f_o^{(\pm)}$ is differentiable with compact support, so that $f_o^{(\pm)}(x)\varepsilon C_o^1$. Then, if the support of the initial data lies in $|x| < d$, after an elapsed time d/c the interaction between waves moving to the left and right ceases and only the solutions $u^{(-)}$ and $u^{(+)}$ are observed to the left and right of the origin, respectively. These are the far fields of the wave equation (2.1). Since $u^{(+)}$ is transported along the $C^{(+)}$ characteristics $x-ct = \xi$ and $u^{(-)}$ along the $C^{(-)}$ characteristics $x+ct = \eta$, and neither family of characteristics intersects itself, the far fields of the wave equation will propagate indefinitely after the interaction has finished.

The situation is different in the case of the homogeneous quasilinear hyperbolic system

$$\frac{\partial U}{\partial t} + A(U)\,\frac{\partial U}{\partial x} = 0\ , \tag{2.5}$$

in which U is an n × 1 vector with elements $u_1, u_2, \ldots, u_n$ and $A(U) = [a_{ij}(u_1, u_2, \ldots, u_n)]$ is an n × n matrix with elements, depending on the elements of U. System (2.5) will be hyperbolic [3,5] when the n eigenvalues of A are all real and the corresponding set of eigenvectors, either left or right, are linearly independent and so span the space E_n associated with A.

If, now, we seek a special solution of (2.5) in which n-1 elements of U are functions of only the one remaining element, say u_1, we may set $U = \tilde{U}(u_1)$. Direct substitution into (2.5) shows that

$$\left(\frac{\partial u_1}{\partial t}\, I + \frac{\partial u_1}{\partial x}\, A(\tilde{U})\right) \frac{d\tilde{U}}{du_1} = 0\ , \tag{2.6}$$

in which I is the unit matrix.

A non-trivial solution of this form only exists when

$$\left|\frac{\partial u_1}{\partial t}\, I + \frac{\partial u_1}{\partial x}\, A\right| = 0\ , \tag{2.7}$$

showing that if λ is an eigenvalue of A,

$$\frac{\partial u_1}{\partial t} \Big/ \frac{\partial u_1}{\partial x} = -\lambda(\tilde{U}) . \tag{2.8}$$

Since system (2.5) is hyperbolic there are n real eigenvalues $\lambda^{(1)}, \lambda^{(2)}, \ldots, \lambda^{(n)}$ of A, from which it follows that when (2.5) is totally hyperbolic (the $\lambda^{(i)}$ are all distinct) there are n different solutions $u_1^{(i)}$ satisfying

$$\frac{\partial u_1^{(i)}}{\partial t} + \lambda^{(i)}(\tilde{U}) \frac{\partial u_1^{(i)}}{\partial x} = 0 \tag{2.9}$$

for $i = 1,2,\ldots,n$. Like the solutions to (2.3), the solutions of (2.9) are, of course, simple wave solutions, and for initial data having compact support they represent the far field solutions after the interaction has finished. The characteristic curves $C^{(i)}$ in this case are given by solving

$$C^{(i)} : \frac{dx}{dt} = \lambda^{(i)}(\tilde{U}) \tag{2.10}$$

for $i = 1,2,\ldots,n$.

The characteristics comprising each family $C^{(i)}$ are again straight lines, but now they are no longer parallel within the family as the gradient of a characteristic depends on the value of the solution that is transported along it. This leads to a breakdown of differentiability when members of a family of characteristics $C^{(i)}$ intersect, and to the formation of a discontinuous solution at some finite elapsed time $t_c^{(i)}$. Thus the simple waves $U = \tilde{U}(u_1^{(i)})$ corresponding to the solutions $u_1^{(i)}$ of (2.9) can only form far fields in the time interval between the end of the interaction period for initial data with compact support and the breakdown time $t_c = \min\{t_c^{(1)}, t_c^{(2)}, \ldots, t_c^{(n)}\}$. The determination of such breakdown times has been discussed in detail by Jeffrey [3,5].

III. REDUCTIVE PERTURBATION METHOD

The far field equations discussed so far are very special, since the equations that gave rise to them involved neither dissipation nor dispersion, and one was, in fact, linear. In more general situations both dissipation and dispersion may be present, and typical of the far field equations that then result are the following nonlinear evolution equations:

Burgers' Equation (dissipative)

$$\frac{\partial v}{\partial t} + v \frac{\partial v}{\partial x} = \nu \frac{\partial^2 v}{\partial x^2} , \quad (\nu > 0) \tag{3.1}$$

KdV Equation (weakly dispersive)

$$\frac{\partial v}{\partial t} + v \frac{\partial v}{\partial x} + \mu \frac{\partial^3 v}{\partial x^3} = 0 , \; (\mu > 0) \tag{3.2}$$

Nonlinear Schrödinger Equation (strongly dispersive)

$$i \frac{\partial v}{\partial t} + \frac{1}{2} \frac{\partial^2 v}{\partial x^2} + a|v|^2 v = 0 . \tag{3.3}$$

An important scalar equation that has either (3.2) or (3.3) as a far field equation, depending on the circumstances, is the Boussinesq equation

$$\frac{\partial^2 u}{\partial t^2} - c^2 \frac{\partial^2 u}{\partial x^2} - \mu \frac{\partial^4 u}{\partial t^2 \partial x^2} = \frac{1}{2} \frac{\partial^2}{\partial x^2} (u^2) , \tag{3.4}$$

in which $u(x,t)$ is a one-dimensional field, c is the phase velocity in the long wave limit and μ is the dispersion parameter. This occurs in the study of water waves, and we refer to the book by Whitham [7] for the details of how it arises in that context.

A very general quasilinear system that contains as special cases many of the systems that are of physical interest has the form

$$\frac{\partial U}{\partial t} + A(U) \frac{\partial U}{\partial x} + B + \left(\sum_{\beta=1}^{s} \prod_{\alpha=1}^{p} \left\{ H_\alpha^\beta \frac{\partial}{\partial t} + K_\alpha^\beta \frac{\partial}{\partial x} \right\} \right) U = 0 . \qquad (p \geq 2) \tag{3.5}$$

Here U is an $n \times 1$ vector with elements $u_1, u_2, \ldots, u_n$, the matrices A, H_α^β, K_α^β are all $n \times n$ matrices depending on U and B is an $n \times 1$ vector depending on U. When wave propagation is involved it is weakly dispersive when $B = 0$ and strongly dispersive when $B \neq 0$.

We now outline the so-called reductive perturbation method due to T. Taniuti and C. C. Wei [8], referring either to that paper or to the review by A. Jeffrey and T. Kakutani [9] for the full details.

Considering the weakly dispersive case ($B = 0$) we apply the Gardiner-Morikawa transformation

$$\xi = \varepsilon^a (x - \lambda t) , \quad \tau = \varepsilon^{a+1} t , \quad a = 1/(p-1) \text{ for } p \geq 2 \tag{3.6}$$

to system (3.5) where λ is taken to be a real eigenvalue of A. It is not necessary that all of the eigenvalues of A are real, but when they are, and the corresponding eigenvectors span the space E_n associated with A, the first order system comprising the first order derivatives in (3.5) will be hyperbolic.

Set

$$U = U_o + \varepsilon U_1(\xi, \tau) + \ldots , \tag{3.7}$$

where U_o is a constant solution of the homogeneous form of (3.5) (i.e., $B = 0$). Then, rewriting the system in terms of derivatives with respect to ξ and τ, and equating like powers of ε, we obtain the results

$$O(\varepsilon^{a+1}) : (-\lambda I + A_o) \frac{\partial U_1}{\partial \xi} = 0 , \tag{3.8a}$$

$$O(\varepsilon^{a+2}) : (-\lambda I + A_o) \frac{\partial U_2}{\partial \xi} + \frac{\partial U_1}{\partial \tau} + \{U_1 \cdot (\nabla_u A)_o\} \frac{\partial U_1}{\partial \xi}$$

$$+ \sum_{\beta=1}^{s} \prod_{\alpha=1}^{p} \left(-\lambda H_{\alpha o}^{\beta} + K_{\alpha o}^{\beta}\right) \frac{\partial^p U_1}{\partial \xi^p} = 0 . \tag{3.8b}$$

Here A_o, $H_{\alpha o}^{\beta}$, $K_{\alpha o}^{\beta}$ and $(\nabla_u A)_o$ indicate quantities appropriate to the solution $U = U_o$, while ∇_u denotes the gradient operator with respect to the elements of U.

Then if ℓ and r denote the left and right eigenvectors of A_o corresponding to the eigenvalue λ, so that

$$\ell(A_o - \lambda I) = 0 \text{ and } (A_o - \lambda I)\, r = 0 , \tag{3.9}$$

equation (3.7) may be solved in the form

$$U_1 = r\phi_1(\xi, \tau) + V_1(\tau) \tag{3.10}$$

with ϕ one of the elements of U_1 and V_1 an arbitrary vector function of τ.

The compatibility condition for (3.8) when solving for $\partial U_2/\partial \xi$ is:

$$\ell \frac{\partial U_1}{\partial \tau} + \ell[V_1 \cdot (\nabla_u A)_o] \frac{\partial U_1}{\partial \xi} + \ell \sum_{\beta=1}^{s} \prod_{\alpha=1}^{p} \left(-\lambda H_{\alpha o}^{\beta} + K_{\alpha o}^{\beta}\right) \frac{\partial U^p}{\partial \xi^p} = 0 . \tag{3.11}$$

Then taking the boundary condition $U \to U_o$ as $x \to \infty$, so that we may set $V_1 \equiv 0$, we find that ϕ satisfies the nonlinear evolution equation

$$\frac{\partial \phi_1}{\partial \tau} + c_1 \phi_1 \frac{\partial \phi_1}{\partial \xi} + c_2 \frac{\partial^p \phi_1}{\partial \xi^p} = 0 , \tag{3.12}$$

where

$$c_1 = \ell \cdot \{r \cdot \nabla_u A)_o r\} / (\ell \cdot r)$$

and

$$c_2 = \ell \cdot \sum_{\beta=1}^{s} \prod_{\alpha=1}^{p} \left(-\lambda H_{\alpha o}^{\beta} + K_{\alpha o}^{\beta}\right) r / (\ell \cdot r) .$$

When $p = 2$ we see that equation (3.12) becomes Burgers' equation and when $p = 3$ the KdV equation. Equation (3.12) thus governs the far field behaviour of the homogeneous form of system (3.5) that is associated with the eigenvalue λ. There will be such a far field for each real eigenvalue λ of A.

Special cases arise when $c_1 = 0$, for then equation (3.12) becomes linear showing that the coordinate transformation (3.6) that has been used is no longer valid since when $c_1 = 0$ equation (3.12) cannot represent a far field. This problem is resolved when both dependent and independent variables are scaled. We refer to reference [9] for the details since here it will suffice merely to mention that this becomes necessary when

$$c_1 \propto (\nabla_u)\cdot r \quad \text{and} \quad (\nabla_u \lambda)\cdot r = 0 . \tag{3.13}$$

This latter condition is the exceptional condition identified by Lax (see [3,5]) which is in effect a weak nonlinearity condition for the associated hyperbolic mode of the reduced order system

$$\frac{\partial U}{\partial t} + A(U)\frac{\partial U}{\partial x} = 0 . \tag{3.14}$$

Typically, when a suitable scaling is employed, in place of the KdV equation in the dispersive case we find as the far field equation the modified KdV equation

$$\frac{\partial \phi_1}{\partial \tau} + \phi_1^q \frac{\partial \phi_1}{\partial \xi} + \mu \frac{\partial^3 \phi_1}{\partial \xi^3} = 0 . \tag{3.15}$$

We refer again to [9] for the details of such a derivation, and to [3] for details of some of the properties and consequences of the exceptional condition in relation to hyperbolic systems.

In concluding this section we remark that although in what follows we shall be referring to properties of exact solutions of some far field equations, it should be remembered that these far field equations are in the main only asymptotic approximations to the solution that is of interest.

IV. KRYLOV-BOGOLIUBOV-MITROPOLSKY METHOD

To provide an example of the derivation of the nonlinear Schrödinger equation, let us illustrate how the Krylov-Bogoliubov-Mitropolsky (KBM) method may be used in conjunction with the Boussinesq equation (3.4). In what follows we base our approach on the one described by A. Jeffrey and T. Kawahara [10]. For an application of this method to plasma physics we refer the reader to the papers by D. Montgomery and D. A. Tidman [11], D. A. Tidman and H. M. Stainer [12] and T. Kakutani and N. Sugimoto [13].

Our starting point is then the Boussinesq equation

$$\frac{\partial^2 u}{\partial t^2} - c^2\frac{\partial^2 u}{\partial x^2} - \mu\frac{\partial^4 u}{\partial x^2 \partial t^2} = \frac{1}{2}\frac{\partial^2 (u^2)}{\partial x^2} , \tag{4.1}$$

and wave modulation in the form of a perturbation solution

$$u = \sum_{n=1}^{\infty} \varepsilon^n u_n . \tag{4.2}$$

For the lowest order starting solution we take a monochromatic plane wave solution

$$u = Ae^{i\theta} + A^{*}e^{-i\theta} \tag{4.3}$$

where a star denotes the complex conjugate, A is the complex amplitude and θ is the phase function $\theta = kx - \omega t$. For solution (4.3) to be non-trivial we find the condition

$$D(k,\omega) = c^2k^2 - \omega^2 - \mu k^2\omega^2 = 0 , \tag{4.4}$$

which is just the linear dispersion relation.

Now let us seek a perturbation solution of the form

$$u = \varepsilon u_1 + \sum_{n=2}^{\infty} \varepsilon^n u_n(A,A^*,\theta) , \tag{4.5}$$

where $u_2, u_3, \ldots$ have only an implicit dependence on x,t through A, A^* and θ. If the complex amplitude varies only slowly with respect to x and t we may write

$$\frac{\partial A}{\partial t} = \sum_{n=1}^{\infty} \varepsilon^n a_n(A,A^*) , \tag{4.6a}$$

$$\frac{\partial A}{\partial x} = \sum_{n=1}^{\infty} \varepsilon^n b_n(A,A^*) , \tag{4.6b}$$

and also the complex conjugate of these expressions. The change of phase is taken into account through the complex amplitude which is phase dependent, while a_i, b_i are independent of θ.

Expressions for derivatives with respect to t and x then follow directly where, for example,

$$\frac{\partial}{\partial t} = \frac{\partial A}{\partial t}\frac{\partial}{\partial A} + \frac{\partial A^*}{\partial t}\frac{\partial}{\partial A^*} + \frac{\partial \theta}{\partial t}\frac{\partial}{\partial \theta}$$

$$\equiv (\varepsilon a_1 + \varepsilon^2 a_2 + \ldots)\frac{\partial}{\partial A} + (\varepsilon a_1^* + \varepsilon^2 a_2^* + \ldots)\frac{\partial}{\partial A^*} - \omega\frac{\partial}{\partial \theta} ,$$

with corresponding expressions for $\partial/\partial x$ and higher derivatives. Now define the operators

$$L\left(\frac{\partial}{\partial x}, \frac{\partial}{\partial t}\right) \equiv \frac{\partial^2}{\partial t^2} - c^2\frac{\partial^2}{\partial x^2} - \mu\frac{\partial^4}{\partial x^2 \partial t^2} , \tag{4.7a}$$

$$N\left(\frac{\partial}{\partial x}, \frac{\partial}{\partial t}\right) \equiv \frac{1}{2}\frac{\partial^2}{\partial x^2} , \tag{4.7b}$$

so that the Boussinesq equation (4.1) becomes

$$L[u] = N[u^2] . \tag{4.8}$$

Then the derivative operators themselves may be written

$$L \equiv L_o + \varepsilon L_1 + \varepsilon^2 L_2 + \dots , \tag{4.9a}$$

$$N \equiv N_o + \varepsilon N_1 + \varepsilon^2 N_2 + \dots , \tag{4.9b}$$

where the coefficients are given by the derivatives with respect to A, A^*, θ together with the unknowns a_i, b_i, a_i^*, b_i^* and their derivatives with respect to A and A^*. Substitution of (4.5) and (4.9a,b) into (4.8) followed by equating like powers of ε gives:

$$O(\varepsilon) \ : \quad L_o u_1 = 0 \tag{4.10a}$$

$$O(\varepsilon^2) \ : \quad L_o u_2 + L_1 u_1 = N_o u_1^2 \tag{4.10b}$$

$$O(\varepsilon^3) \ : \quad L_o u_3 + L_1 u_2 + L_2 u_1 = N_o [2u_1 u_2] + N_1 u_1^2 \tag{4.10c}$$

$$\dots$$

where

$$L_o \equiv \omega^2 \frac{\partial}{\partial\theta^2} - c^2 k^2 \frac{\partial}{\partial\theta^2} - \mu k^2 \omega^2 \frac{\partial^4}{\partial\theta^4} , \tag{4.11a}$$

$$L_1 \equiv -2\omega \left(\frac{\partial}{\partial\theta} - \mu k^2 \frac{\partial^3}{\partial\theta^3} \right) a_1 \frac{\partial}{\partial A}$$

$$-2k \left(c^2 \frac{\partial}{\partial\theta} - \mu\omega^2 \frac{\partial^3}{\partial\theta^3} \right) b_1 \frac{\partial}{\partial A} + \text{complex conjugate} , \tag{4.11b}$$

$$\dots$$

$$N_o \equiv \frac{1}{2} k^2 \frac{\partial^2}{\partial\theta^2} , \tag{4.12a}$$

$$N_1 \equiv k b_1 \frac{\partial^2}{\partial A \partial\theta} + \text{complex conjugate} , \tag{4.12b}$$

$$\dots$$

and the unknowns a_i, b_i are to be determined from the non-secularity conditions.

The lowest order equation (4.10a) yields the linear dispersion relation (4.4). Using the lowest order solution (4.3) in the higher order equations gives to order $O(\varepsilon^2)$,

$$\mu k^2 \omega^2 \left(\frac{\partial^4}{\partial\theta^4} + \frac{\partial^2}{\partial\theta^2} \right) u_2$$

$$= i \left(\frac{\partial D}{\partial\omega} a_1 - \frac{\partial D}{\partial k} b_1 \right) e^{i\theta} + 2k^2 A^2 e^{2i\theta} + \text{complex conjugate} , \tag{4.13}$$

where

$$\frac{\partial D}{\partial k} = 2k(c^2 - \mu\omega^2) \text{ and } \frac{\partial D}{\partial \omega} = -2\,(1 + \mu k^2)\omega \ .$$

To order $O(\varepsilon^3)$ we have:

$$\mu k^2\omega^2\left(\frac{\partial^4}{\partial\theta^4} + \frac{\partial^2}{\partial\theta^2}\right) u_3$$

$$= \left[\left\{-2\omega\left(\frac{\partial}{\partial\theta} - \mu k^2 \frac{\partial^2}{\partial\theta^2}\right) a_1 \frac{\partial}{\partial A} - 2k\left(c^2 \frac{\partial}{\partial\theta} - \mu\omega^2 \frac{\partial^3}{\partial\theta^3}\right) b_1 \frac{\partial}{\partial A} + \text{complex conjugate}\right\} u_2 + \left\{i\left(\frac{\partial D}{\partial\omega} a_2 - \frac{\partial D}{\partial k} b_2\right)\right.\right.$$

$$- \frac{1}{2}\left(\frac{\partial^2 D}{\partial\omega^2} a_1 \frac{\partial a_1}{\partial A} - 2\frac{\partial^2 D}{\partial k\partial\omega} b_1 \frac{\partial a_1}{\partial A} + \frac{\partial^2 D}{\partial k^2} b_1 \frac{\partial b_1}{\partial A}\right)$$

$$\left.+ \text{complex conjugate}\right\} e^{i\theta} - 4ikb_1Ae^{2i\theta}$$

$$- k^2\left(\frac{\partial^2 u_2}{\partial\theta^2} + 2i\frac{\partial u_2}{\partial\theta} - u_2\right) Ae^{i\theta}$$

$$+ \text{complex conjugate} \ . \tag{4.14}$$

The occurrence of a term proportional to $\exp[i\theta]$ on the right hand side of (4.13) gives rise to secular terms in the solution u_2. However, if the condition

$$a_1 + v_g b_1 = 0 \ , \text{ with } v_g = -\frac{\partial D}{\partial k}\Big/\frac{\partial D}{\partial\omega} = \frac{d\omega}{dk} = \frac{\omega^3}{c^2k^3} \tag{4.15}$$

and its complex conjugate are satisfied we may obtain the secular free solution for u_2:

$$u_2 = \frac{1}{6\mu\omega^2} A^2e^{2i\theta} + E(A,A^*)e^{i\theta} + \text{complex conjugate} + F(A,A^*) \ , \tag{4.16}$$

where $E(A,A^*)$ is complex and $F(A,A^*)$ is real.

If we now substitute (4.16) into (4.14), collect the terms proportional to $\exp[i\theta]$ and equate them to zero, and use the lowest order condition (4.15) we obtain the secular free condition

$$i(a_2 + v_g b_2) + \frac{1}{2}\frac{dv_g}{dk}\left(b_1\frac{\partial b_1}{\partial A} + b_1^*\frac{\partial b_1}{\partial A^*}\right)$$

$$+ \frac{k^2}{\partial D/\partial\omega}\left(\frac{1}{6\mu\omega^2} A^2A^* + F(A,A^*)A\right) = 0 \ , \tag{4.17}$$

where

$$\frac{dv_g}{dk} = -\left(\frac{\partial^2 D}{\partial\omega^2} v_g^2 + 2\frac{\partial^2 D}{\partial k\partial\omega} v_g + \frac{\partial^2 D}{\partial k^2}\right) \Big/ \frac{\partial D}{\partial\omega} = -\frac{3\mu\omega^2}{c^4 k^4} . \tag{4.18}$$

The function $E(A,A^*)$ has been eliminated, but we need to determine $F(A,A^*)$. This follows from the secular free condition for the higher order approximation in the Boussinesq equation. We find that the constant terms in L_2u_2 and $N_2u_1^2$ give, after use of (4.15), the non-secularity condition

$$(v_g^2 - c^2)\, b_1 \frac{\partial}{\partial A}\left(b_1 \frac{\partial F}{\partial A} + b_1^* \frac{\partial F}{\partial A^*}\right) + \text{complex conjugate} = b_1\left(b_1^* + \frac{\partial b_1}{\partial A} A^* + \frac{\partial b_1^*}{\partial A} A\right) + \text{complex conjugate} . \tag{4.19}$$

This is satisfied if we choose for F the function

$$F(A,A^*) = \left(\frac{1}{v_g^2 - c^2}\right) = AA^* + \beta \tag{4.20}$$

with β an absolute constant.

Using this form of F in (4.17) gives

$$i(a_2 + v_g b_2) + \frac{1}{2}\frac{dv_g}{dh}\left(b_1 \frac{\partial b_1}{\partial A} + b_1^* \frac{\partial b_1}{\partial A^*}\right) + \frac{k^2}{\partial D/\partial\omega}\left(\left(\frac{1}{v_g^2 - c^2} + \frac{1}{6\mu\omega^2}\right) A^2 A^* + \beta A\right) = 0 . \tag{4.21}$$

As we may write

$$a_1 = \frac{\partial A}{\partial t_1} + O(\varepsilon) , \quad b_1 = \frac{\partial A}{\partial x_1} + O(\varepsilon) \tag{4.22}$$

where $t_1 = \varepsilon t$, $x_1 = \varepsilon x$ are slow variables, the non-secularity condition (4.15) becomes

$$\frac{\partial A}{\partial t_1} + v_g \frac{\partial A}{\partial x_1} \simeq 0 . \tag{4.23}$$

So, in a reference frame moving with the group velocity v_g, the amplitude of A is almost constant. In terms of the variables $t_2 = \varepsilon^2 t$ and $x_2 = \varepsilon^2 x$ we find from (4.6a,b) that

$$a_2 + v_g b_2 = \frac{\partial A}{\partial t_2} + v_g \frac{\partial A}{\partial x_2} + O(\varepsilon)$$

showing that (4.21) is equivalent to

$$i\left(\frac{\partial A}{\partial t_2} + v_g \frac{\partial A}{\partial x_2}\right) + \frac{1}{2}\frac{dv_g}{dk}\frac{\partial^2 A}{\partial x_1^2} + \frac{k^2}{\partial D/\partial\omega}\left[\left(\frac{1}{v_g^2 - c^2} + \frac{1}{6\mu\omega^2}\right) A^2 A^* + \beta A\right] = 0 \ . \qquad (4.24)$$

This is simply the nonlinear Schrödinger equation in a reference frame moving with the group velocity v_g. In references [11,12] the condition $a_1 = b_1 = 0$ was used instead of the result in equation (4.15). This condition was thus a sufficient condition for secularity, but not a necessary one. It leads to the removal of the second derivative term in (4.24), and so to a special case which cannot take full account of amplitude modulation.

V. BÄCKLUND TRANSFORMATIONS, INVERSE SCATTERING AND CONSERVATION LAWS

We now consider the notion of a Bäcklund transformation, and in doing so we base our approach on that of A. Jeffrey and T. Taniuti [14]. Consider a second order partial differential equation

$$F(u_{xx}, u_{xt}, u_{tt}, u_x, u_t, u, x, t) = 0 \ , \qquad (5.1)$$

then the Bäcklund transformation for this has the form [1], see also the papers by R. M. Miura, R. Herrmann, D. W. McLaughlin and A. C. Scott in [15],

$$u_x^{(n)} = P(u^{(n)}, u^{(n-1)}, u_x^{(n-1)}, u_t^{(n-1)}, x, t) \ , \qquad (5.2a)$$

$$u_t^{(n)} = Q(u^{(n)}, u^{(n-1)}, u_x^{(n-1)}, u_t^{(n-1)}, x, t) \ , \qquad (5.2b)$$

in which $u^{(n)}$ and $u^{(n-1)}$ are two solutions of (5.1). We may, in fact, consider a Bäcklund transformation as a transformation from the solution of one equation to the solution of another, as we now illustrate by means of the Liouville equation

$$u_{xt} = e^u \ . \qquad (5.3)$$

The Bäcklund transformation (5.2a,b) for this takes the form

$$u_x = \bar{u}_x - ke^{(u+\bar{u})/2} \ , \qquad (5.4a)$$

$$u_t = -\bar{u}_t - \frac{2}{k} e^{(u-\bar{u})/2} \ , \qquad (5.4b)$$

where k is an arbitrary constant. The equality of mixed derivatives for (5.4a,b), usually called the integrability condition, then shows that the function $\bar{u}$ must be a solution of the equation

$$\bar{u}_{xt} = 0 \ . \qquad (5.5)$$

This in turn shows that the Bäcklund transformation (5.4a,b) relates the solution u of the nonlinear Liouville equation and the solution $\bar{u}$ of the linear wave equation (5.5). When viewed differently, this provides a means of solving a nonlinear equation in terms of the solution of a linear equation and a transformation.

Similarly, the solutions of the modified KdV equation

$$\frac{\partial v}{\partial t} - 6v^2 \frac{\partial v}{\partial x} + \frac{\partial^3 v}{\partial x^3} = 0 , \tag{5.6}$$

and the KdV equation

$$\frac{\partial u}{\partial t} - 6u \frac{\partial u}{\partial x} + \frac{\partial^3 u}{\partial x^3} = 0 , \tag{5.7}$$

are related by the transformation found by Miura

$$u = v^2 + \frac{\partial v}{\partial x} , \tag{5.8}$$

which may also be regarded as a Bäcklund transformation connecting the solutions of (5.6) and (5.7).

The Riccati equation enters here, because if u(x,t) is known, equation (5.8) is simply a Riccati equation for v(x,t) and the Schrödinger equation

$$- \frac{\partial^2 \psi}{\partial x^2} + u\psi = \lambda\psi \tag{5.9}$$

then follows by means of the transformation

$$v = \frac{\partial \psi}{\partial x} / \psi . \tag{5.10}$$

So, in general, the Riccati equation may be derived from the Bäcklund transformation. When the Riccati equation is transformed into a linear equation it follows from a study of the KdV equation that the Riccati equation is the eigenvalue equation for the inverse scattering method [4].

To illustrate ideas further we now derive the Riccati equation and the corresponding linear equation from the Bäcklund transformation for the Sine-Gordon equation

$$\frac{\partial^2 u}{\partial t^2} - \frac{\partial^2 u}{\partial x^2} + \sin u = 0 . \tag{5.11}$$

The Bäcklund transformation is

$$\frac{1}{2} (u_t + \bar{u}_t) = \frac{1}{k} \sin \left(\frac{u-\bar{u}}{2}\right) , \tag{5.12a}$$

$$\frac{1}{2} (u_x - \bar{u}_x) = k \sin \left(\frac{u+\bar{u}}{2}\right) , \tag{5.12b}$$

where k is a constant.

If we now write $f = \tan \left(\frac{u+\bar{u}}{4}\right)$, results (5.12a,b) reduce to the Riccati equations for f :

$$f_t - \frac{\sin u}{2k}(1 - f^2) + \left(\frac{\cos u}{k}\right) f = 0 , \tag{5.13a}$$

$$f_x - \frac{u_x}{2}(1 + f^2) + kf = 0 . \tag{5.13b}$$

The transformation $f = \psi_2/\psi_1$ now converts the Riccati equation to the linear equations for ψ_1 and ψ_2

$$\psi_{1t} = \frac{1}{2k}(\psi_1 \cos u + \psi_2 \sin u) , \tag{5.14a}$$

$$\psi_{2t} = \frac{1}{2k}(\psi_1 \sin u - \psi_2 \cos u) , \tag{5.14b}$$

$$\psi_{1x} + \frac{u_x}{2}\psi_2 = \frac{k}{2}\psi_1 , \tag{5.15a}$$

$$\psi_{2x} - \frac{u_x}{2}\psi_1 = -\frac{k}{2}\psi_2 . \tag{5.15b}$$

The following properties can be established.

(i) Equations (5.15a,b) give the eigenvalue equation for the inverse scattering method for the Sine-Gordon equation (5.11).

(ii) The transformation parameter k in the Bäcklund transformation (5.12a,b) is the eigenvalue.

(iii) Equations (5.14a,b) describe the time evolution of the eigenfunctions ψ_1 and ψ_2.

The Riccati equation and the Bäcklund transformation may, in fact, be derived from the inverse scattering equations. That is, from the eigenvalue equation and from the equations determining the time evolution of the eigenfunctions. To see how this happens let us use the procedure due to Ablowitz, Kaup, Newell and Segur (AKNS) [16].

Consider the eigenvalue problem for the linear operator L

$$L\psi = \zeta\psi , \tag{5.16}$$

and the equation governing the time evolution of ψ

$$i\frac{\partial\psi}{\partial t} = \tilde{A}\psi \tag{5.17}$$

where

$$L \equiv \begin{pmatrix} i\frac{\partial}{\partial x} & -iq(x,t) \\ ir(x,t) & -i\frac{\partial}{\partial x} \end{pmatrix} , \quad \psi = \begin{pmatrix} \psi_1 \\ \psi_2 \end{pmatrix} \tag{5.18}$$

and

$$\tilde{A} = \begin{pmatrix} A(x,t,\zeta) & B(x,t,\zeta) \\ C(x,t,\zeta) & -A(x,t,\zeta) \end{pmatrix} . \tag{5.19}$$

The fact that L, $\tilde{A}$ are not self-adjoint means that the eigenvalue ζ is usually complex, but we take it to be independent of the time t. The functions q and r are solutions of nonlinear equations that do not have an explicit dependence on ζ.

Now differentiation of (5.16) and (5.17) with respect to t and x, respectively, followed by subtraction gives zero when the equality of mixed derivatives is required. As ζ is taken to be independent of the time, $\zeta_t = 0$, which then implies

$$A_x = 2C - rB , \tag{5.20a}$$

$$B_x + 2i\zeta B = iq_t - 2Aq , \tag{5.20b}$$

$$C_x - 2i\zeta C = ir_t + 2Ar . \tag{5.20c}$$

When A, B, C are polynomials involving either ζ or ζ^{-1}, using this fact in (5.20) and equating corresponding terms gives the nonlinear evolution equations for q(x,t) and r(x,t).

<u>*Example 1*</u> If A is quadratic in ζ, so that B, C are linear,

$$A = 2\zeta^2 + qr ,$$

and we have

$$B = 2iq - q_x , \quad C = 2ir + r_x ;$$

when

$$iq_t + q_{xx} - 2q^2r = 0 , \tag{5.21a}$$

$$ir_t - r_{xx} + 2qr^2 = 0 . \tag{5.21b}$$

Then, making the identifications $r = -q^*Q/2$, $q = u$, converts (5.21a,b) into the nonlinear Schrödinger equation.

<u>*Example 2*</u> If A is assumed to be an inverse power of ζ, like

$$A = -\frac{\cos u}{4\zeta} ,$$

then

$$B = q_t/2\zeta , \quad C = -r_t/2\zeta$$

and

$$(\cos u)_x = 2(qr)_t , \quad q_{xt} = q \cos u , \quad r_{xt} = r \cos u . \tag{5.22}$$

Then, setting $r = -q = u_x/2$ converts (5.22) into the Sine-Gordon equation. Substituting for A, B and C into the equations

$$L\psi = \zeta\psi \,, \quad i\psi_t = \tilde{A}\psi \tag{5.23}$$

we find the eigenvalue equation for ζ and the equation governing the time evolution of the eigenfunction for the form of inverse scattering appropriate to the Sine-Gordon equation. These are, in fact, the linear results (5.14a,b) and (5.15a,b) found from the Bäcklund transformation.

The parameter k in the Backlund transformation is related to ζ by the result

$$k = -2i\zeta \,.$$

So, for the Sine-Gordon equation, we have established the process:

Bäcklund transformation → Riccati equation → AKNS equation.

We may also show the converse result:

AKNS equation → Riccati equation → Backlund transformation.

This follows because the transformation $f_1 = \psi_2/\psi_1$ or $f_2 = \psi_1/\psi_2$ reduces the AKNS equations

$$L\psi = \zeta\psi \,, \quad i\psi_t = \tilde{A}\psi$$

to the Riccati equations

$$f_{1x} = 2i\zeta f_1 + r - qf_1^2 \,, \tag{5.24a}$$

$$f_{2x} = -2i\zeta f_2 + q - rf_2^2 \,, \tag{5.24b}$$

and

$$f_{1t} = i(2Af_1 - C + Bf_1^2) \,, \tag{5.25a}$$

$$f_{2t} = -i\,(2Af_2 + B - Cf_2^2) \,. \tag{5.25b}$$

So, for the Sine-Gordon equation, the Bäcklund transformation follows from (5.24a,b) and our previous result

$$f = \tan\left(\frac{u+u}{2}\right) \,. \tag{5.26}$$

VI. CONSERVATION LAWS

We conclude by making an observation about conservation laws. Equations (5.20a,b), which we repeat again here for convenience,

$$A_x = 2C - rB \,, \tag{6.1a}$$

$$B_x + 2i\zeta B = iq_t - 2Aq \tag{6.1b}$$

may be written in the form of the conservation equations

$$\frac{\partial}{\partial t}(qf_1) + i\frac{\partial}{\partial x}(A + Bf_1) = 0 , \tag{6.2a}$$

$$\frac{\partial}{\partial t}(rf_2) + i\frac{\partial}{\partial x}(-A + Bf_2) = 0 . \tag{6.2b}$$

This shows that qf_1 and rf_2 are the densities that are conserved by these laws. Let us now rewrite equations (5.24a,b) in the form

$$qf_1 = \frac{1}{2i\zeta}[(qf_1)^2 - qr + qf_{1x}] , \tag{6.3}$$

and expand qf_1 in terms of powers of ζ^{-1} by setting

$$qf_1 = \sum_{n=1}^{\infty} h_n \zeta^{-n} . \tag{6.4}$$

Then the following recurrence relation follows for the coefficients h_n:

$$h_{n+1} = \frac{1}{2i}\left\{\sum_{k=1}^{n-1} h_k h_{n-k} - (rq)\delta_{no} + q(h_n/q)_x\right\} . \tag{6.5}$$

If we use equation (6.3) that governs the conserved density qf_1 in the conservation equations (6.2a,b) we then arrive at an enumerably infinite set of conserved quantities with respect to each order of ζ^{-1}. This shows that a connection exists between the inverse scattering method, the Bäcklund transformation and the conservation laws, via the Riccati equation. We refer to the basic paper by Miura et al. [17] for further information about conservation laws and the so called associated constants of motion.

REFERENCES

[1] R. M. Miura (Editor), Bäcklund Transformations, Lecture Notes in Mathematics 515, Springer, Berlin, 1974.

[2] T. Taniuti (Editor), Reductive Perturbation Method for Nonlinear Wave Propagation, Progress of Theoretical Physics Supplement, 55 (1974), pp. 306.

[3] A. Jeffrey, Nonlinear wave propagation, ZAMM, 58 (1978), T38-T56.

[4] A. C. Scott, F. Y. F. Chu and D. W. McLaughlin, The soliton: a new concept in applied science, Proc. IEEE, 61 (1973), 1443-1483.

[5] A. Jeffrey, Quasilinear Hyperbolic Systems and Waves, Research Note in Mathematics 5, Pitman Publishing, London, 1976.

[6] T. Taniuti, Reductive perturbation method and far fields of wave equations, Progress of Theoretical Physics Supplement, 55 (1974), 1-35.

[7] G. Whitham, Linear and Nonlinear Waves, Wiley-Interscience, New York, 1974.

[8] T. Taniuti and C. C. Wei, Reductive perturbation method in nonlinear wave propagation - Part I, J. Phys. Soc. Japan, 24 (1968), 941-946.

[9] A. Jeffrey and T. Kakutani, Weak nonlinear dispersive waves: a discussion centred around the KdV equation, SIAM Review, 14 (1972), 582-643.

[10] A. Jeffrey and T. Kawahara, Asymptotic Methods in Nonlinear Wave Theory, Pitman Publishing, London (in preparation).

[11] D. Montgomery and D. A. Tidman, Secular and nonsecular behaviour for the cold plasma equations, Phys. Fluids, 7 (1964), 242-249.

[12] D. A. Tidman and H. M. Stainer, Frequency and wavenumber shifts for nonlinear equations in a "hot" plasma, Phys. Fluids, 8 (1965), 345-353.

[13] T. Kakutani and N. Sugimoto, Krylov-Bogoliubov-Mitropolsky method for nonlinear wave modulation, Phys. Fluids, 17 (1974), 1617-1625.

[14] A. Jeffrey and T. Taniuti, Nonlinear Dispersive and Nondispersive Wave Propagation, Pitman Publishing, London (in preparation).

[15] K. Lonngren and A. C. Scott, Solitons in Action, Academic Press, New York, 1978.

[16] M. J. Ablowitz, D. J. Kaup, A. C. Newell and H. Segur, The inverse scattering transformation: Fourier analysis for nonlinear problems, Studies in Applied Mathematics, 53 (1974), 249-315.

[17] R. M. Miura, C. S. Gardner and M. D. Kruskal, Korteweg-de Vries equation and generalisations II. Existence of conservation laws and constants of motion, J. Math. Phys., 9 (1968), 1204-1209.

CONVERGENCE OF FORMAL POWER SERIES SOLUTIONS OF A SYSTEM OF NONLINEAR DIFFERENTIAL EQUATIONS AT AN IRREGULAR SINGULAR POINT

Yasutaka Sibuya*)
School of Mathematics
University of Minnesota
Minneapolis, Minnesota 55455
U.S.A.

§1. Introduction. We consider a system of differential equations

(1.1) $$x^{p+1}\frac{du}{dx} = E(x,y,u) \quad ,$$

where

(i) p is a positive integer;

(ii) x is an independent variable;

(iii) y is a parameter;

(iv) u and E are n- vectors;

(v) entries of E are holomorphic in a neighborhood of $(x,y,u) = (0,0,0)$.

Assume that there exists a formal solution of system (1.1):

(1.2) $$u = \psi(x,y) = \sum_{h=0}^{\infty} \psi_h(x)y^h \quad ,$$

where the coefficients $\psi_h(x)$ are n- vectors whose entries are holomorphic in a neighborhood of $x = 0$. We shall prove the following theorem.

Theorem 1: If $\psi_o(0) = 0$ and if $E_u(0,0,0) \in GL(n;\mathbb{C})$, then ψ is convergent in a neighborhood of $(x,y) = (0,0)$.

§2. A special case. Let $A(x)$ be an n- by -n matrix whose entries are holomorphic in a neighborhood of $x = 0$. Define a differential operator L by

(2.1) $$L(u) = x^{p+1}\frac{du}{dx} - A(x)u \quad .$$

We consider a system of differential equations

(2.2) $$L(u) = y\ E(x,y,u) \quad ,$$

where we assume (i) ~ (v) of Section 1. We also assume that system (2.2) admits a formal solution

(2.3) $$u = \psi(x,y) = \sum_{h=1}^{\infty} \psi_h(x)y^h \quad ,$$

where the coefficients $\psi_h(x)$ are n- vectors whose entries are holomorphic in a neighborhood of $x = 0$. Before we prove Theorem 1, we shall prove the following theorem.

Theorem 2: If $A(0) \in GL(n;\mathbb{C})$, then ψ is convergent in a neighborhood of $(x,y) = (0,0)$.

*) Partially supported by NSF MCS 79-01998.

§3. <u>A transformation.</u> For a positive number δ_o, set $\mathcal{D}(\delta_o) = \{y; |y| < \delta_o\}$, and denote by $\Omega(\delta_o)$ the set of all mappings from $\mathcal{D}(\delta_o)$ to $\mathbb{C}^n$ which are holomorphic and bounded in $\mathcal{D}(\delta_o)$. For $C \in \Omega(\delta_o)$, set

$$|C|_{\delta_o} = \sup_{\mathcal{D}(\delta_o)} |C(y)| .$$

For a power series $\phi = \sum_{m=0}^{\infty} C_m x^m \quad (C_m \in \Omega(\delta_o))$, set

$$\|\phi\|_{\delta_o,\delta} = \sum_{m=0}^{\infty} |C_m|_{\delta_o} \delta^m ,$$

where δ is a positive constant.

Let $A(x) = \sum_{m=0}^{\infty} A_m x^m$ be the matrix given in Section 2. We set

$$\|A\|_{\delta} = \sum_{m=0}^{\infty} |A_m| \delta^m .$$

Since the matrices A_m are independent of y, the quantity $\|A\|_\delta$ is independent of δ_o.

Let $E(x,y,u) = \sum_{|\Theta|\geq 0} E_\Theta(x,y)u^\Theta$ be the n- vector given in Section 2, where $\Theta = (p_1,\ldots,p_n) \ (p_j \in \mathbb{Z}_+)$, $|\Theta| = \sum_{j=1}^{n} p_j$ and $u^\Theta = u_1^{p_1} \ldots u_n^{p_n}$, the u_j being entries of the vector u. Set

$$E_\Theta(x,y) = \sum_{m=0}^{\infty} E_{\Theta,m}(y)x^m = \sum_{h=0}^{\infty} \tilde{E}_{\Theta,h}(x)y^h .$$

Since entries of $A(x)$ and $E(x,y,u)$ are holomorphic in a neighborhood of $x = 0$ and $(x,y,u) = (0,0,0)$ respectively, there exist three positive numbers δ_o, δ and ρ_o such that

(a) $\|A\|_\delta < +\infty$;

(b) $E_{\Theta,m} \in \Omega(\delta_o)$;

(c) $\|E_\Theta\|_{\delta_o,\delta} < +\infty$ for all Θ ;

(d) $\sum_{|\Theta|\geq 0} \|E_\Theta\|_{\delta_o,\delta}\, \rho_o^{|\Theta|} < +\infty$.

Since ψ given by (2.3) is a formal solution of system (2.2), the coefficients $\psi_h(x)$ satisfy linear differential equations:

$$L(\psi_h) = H_h(x) \qquad h = 1,2,\ldots ,$$

where the quantities H_h are n- vectors whose entries are respectively polynomials in the entries of ψ_ℓ and $E_{\Theta,\ell} (\ell \leq h-1)$. Therefore, if we set $\psi_h(x) = \sum_{m=0}^{\infty} \psi_{h,m} x^m$ we have

(e) $\|\psi_h\|_\delta = \sum_{m=0}^{\infty} |\psi_{h,m}|\delta^m < +\infty$ for all h

for some positive number δ.

We fix δ and ρ_o so that (a) ~ (e) hold with some δ_o. We will replace δ_o by a smaller number later. Note that if $\delta_o' < \delta_o$ then $\Omega(\delta_o) \subset \Omega(\delta_o')$ and $|C|_{\delta_o'} \leq |C|_{\delta_o}$ for $C \in \Omega(\delta_o)$. Hence, if $\delta_o' < \delta_o$ then $\|\phi\|_{\delta_o',\delta} < \|\phi\|_{\delta_o,\delta}$.

Choose a positive integer M so that

(3.1) $$\frac{\|A\|_\delta}{\delta^{p_M}} < 1 \quad ,$$

where p is the integer given in Section 2.

If $\delta_o > 0$ is sufficiently small, we can find $C_o,\ldots,C_{M-1}$ in $\Omega(\delta_o)$ such that

$$\phi(x,y) = \sum_{m=0}^{M-1} C_m(y)x^m$$

satisfies the condition that

(3.2) $$L(\phi) - E(x,y,y\phi) = O(x^M) \quad .$$

Note that $A(0)\in GL(n;\mathbf{C})$.

Set

(3.3) $$u = v + y\phi(x,y)$$

and

(3.4) $$L(v) = yF(x,y,v) \quad ,$$

where

(3.5) $$F(x,y,v) = E(x,y,v + y\phi) - L(\phi) \quad .$$

Let

(3.6) $$F(x,y,v) = \sum_{|\Theta|\geqq 0} F_\Theta(x,y)v^\Theta$$

and

(3.7) $$F_\Theta(x,y) = \sum_{m=0}^{\infty} F_{\Theta,m}(y)x^m \quad .$$

Then, if δ_o and ρ_1 are sufficiently small positive numbers, we have

(b') $F_{\Theta,m} \in \Omega(\delta_o)$;

(c') $\|F_\Theta\|_{\delta_o,\delta} < +\infty$ for all Θ ;

(d') $\displaystyle\sum_{|\Theta|\geqq 0} \|F_\Theta\|_{\delta_o,\delta}\, \rho_1^{|\Theta|} < +\infty$.

Note that two positive numbers δ and ρ_o were chosen so that conditions (a)~(e) are satisfied with some δ_o . We choose δ_o and ρ_1 so that $\rho_1 + \delta_o \|\phi\|_{\delta_o,\delta} < \rho_o$.

Furthermore, (3.2) implies that if $|\Theta| = 0$ we have

(3.8) $$F_\Theta(x,y) = F(x,y,0) = O(x^M) \quad .$$

Set

(3.9) $$\tilde{\psi}(x,y) = \psi(x,y) - y\phi(x,y) = \sum_{h=1}^{\infty} \tilde{\psi}_h(x)y^h \quad .$$

Since $v = \tilde{\psi}$ is a formal solution of system (3.4), it follows from (3.8) that $\tilde{\psi}_h(x) = O(x^M)$ for all h . Therefore, we get

(3.10) $\tilde{\psi}_h(x) \displaystyle\sum_{m=M}^{\infty} \psi_{h,m}\, x^m$ for all h , where the $\psi_{h,m}$ are coefficients of ψ_h .

Since the $\psi_{h,m}$ satisfy (e), we have

(e') $$\|\tilde{\psi}_k\|_\delta = \sum_{m=M}^{\infty} |\psi_{h,m}|\delta^m < +\infty \qquad \text{for all} \quad h \quad .$$

Observe that the convergence of $\tilde{\psi}$ implies the convergence of ψ . Therefore, we shall prove that $\tilde{\psi}$ is convergent in a neighborhood of $(x,y) = (0,0)$.

§4. Lemmas. In this section, we review some of our previous results (cf. Y. Sibuya [5]).

Set $\mathcal{B}(\delta_o,\delta)=\{f=\sum_{m=0}^{\infty}\alpha_m x^m\,;\ \alpha_m\in\Omega(\delta_o)\ \text{and}\ \|f\|_{\delta_o,\delta}<+\infty\}$

and

$$\mathcal{B}(\delta_o,\delta,M)=\{x^M f\,;\ f\in\mathcal{B}(\delta_o,\delta)\}\quad,$$

where we assume (3.1). Define the following mappings:

(4.1) $A:\mathcal{B}(\delta_o,\delta,M)\to\mathcal{B}(\delta_o,\delta,M)$,

(4.2) $P:\mathcal{B}(\delta_o,\delta,M)\to\mathcal{B}(\delta_o,\delta,M+p)$,

(4.3) $T:\mathcal{B}(\delta_o,\delta,M+p)\to\mathcal{B}(\delta_o,\delta,M)$,

where

(4.1') $A(f)(x,y)=A(x)f(x,y)$ for $f\in\mathcal{B}(\delta_o,\delta,M)$,

(4.2') $P(f)=\sum_{m=M+p}^{\infty}\alpha_m x^m$ for $f=\sum_{m=M}^{\infty}\alpha_m x^m$,

(4.3') $T(f)=\sum_{m=M}^{\infty}\frac{\alpha_m}{m}x^m$ for $f=\sum_{m=M}^{\infty}\alpha_m x^{m+p}$

These mappings are $\Omega(\delta_o)$ - linear and

(4.4)
$$\begin{cases}\|A(f)\|_{\delta_o,\delta}\leqq\|A\|_{\delta}\,\|f\|_{\delta_o,\delta}\ ,\\ \|P(f)\|_{\delta_o,\delta}\leqq\|f\|_{\delta_o,\delta}\ ,\\ \|T(f)\|_{\delta_o,\delta}\leqq\dfrac{1}{\delta p_M}\|f\|_{\delta_o,\delta}\ .\end{cases}$$

Hence $\|TPA(f)\|_{\delta_o,\delta}\leqq\dfrac{\|A\|_{\delta}}{\delta p_M}\|f\|_{\delta_o,\delta}$ and

(4.5) $$I-TPA:\mathcal{B}(\delta_o,\delta,M)\to\mathcal{B}(\delta_o,\delta,M)$$

is an isomorphism, where I is the identity map. Observe that

(4.6) $$\|(I-TPA)^{-1}(f)\|_{\delta_o,\delta}\leqq\gamma\|f\|_{\delta_o,\delta}\quad,$$

where

$$\gamma=\frac{1}{1-\dfrac{\|A\|_{\delta}}{\delta p_M}}$$

Set

(4.7) $$\Phi=(I-TPA)^{-1}TP\quad.$$

Then the mapping

(4.8) $$\Phi:\mathcal{B}(\delta_o,\delta,M)\to\mathcal{B}(\delta_o,\delta,M)$$

is $\Omega(\delta_o)$ - linear and

(4.9) $$\|\Phi(f)\|_{\delta_o,\delta}\leqq\gamma_o\|f\|_{\delta_o,\delta}\quad,$$

where

(4.10) $$\gamma_o=\frac{1}{\delta p_M-\|A\|_{\delta}}\quad.$$

We previously proved (cf. Y. Sibuya [5]) the following lemmas

Lemma 1: For $f\in\mathcal{B}(\delta_o,\delta,M)$ we have

(4.11) $$L\,\Phi(f)=f-Q(f)\quad,$$

where L is given by (2.1) and

(4.12) $$Q = (I - P)(I + A\Phi) \quad .$$

Lemma 2: If $f \in \mathcal{B}(\delta_o, \delta, M)$ and $x^{p+1} \frac{df}{dx} \in \mathcal{B}(\delta_o, \delta, M+p)$, then

(4.13) $$\Phi L(f) = f \quad .$$

Lemma 3: If $u \in \mathcal{B}(\delta_o, \delta, M)$ and if $L(u) = f = \sum_{m=M}^{M+p-1} \alpha_m x^m \quad (\alpha_m \in \Omega(\delta_o))$, then $u = 0$ and $f = 0$.

Tp prove Lemma 3, we used the assumption that $A(0) \in GL(n;C)$.

§5. Proof of Theorem 2. In this section, we shall prove the convergence of the formal solution $\tilde{\psi}$ of system (3.4). It has been already proved that F of system (3.4) satisfies the following conditions:

(d') $$\sum_{|\theta| \geq 0} \|F_\theta\|_{\delta_o, \delta} \rho_1^{|\theta|} < +\infty \quad ,$$

where $F = \sum_{|\theta| \geq 0} F_\theta(x,y) v^\theta$; and

(3.8) $$F(x,y,0) = O(x^M) \quad .$$

Set

(5.1) $$\mathcal{B} = \{f \in \mathcal{B}(\delta_o, \delta, M) \ ; \ \|f\|_{\delta_o, \delta} \leq \rho_1\} \quad .$$

Then $F(x,y,f) \in \mathcal{B}(\delta_o, \delta, M)$ for all $f \in \mathcal{B}$. Furthermore, if $\delta_o > 0$ is sufficiently small, we have

(5.2) $$y\Phi F(x,y,f) \in \mathcal{B} \quad \text{for all} \quad f \in \mathcal{B} \quad ,$$

and $y\Phi F(x,y,f)$ admits a Lipschitz constant smaller than one with respect to $f \in \mathcal{B}$, where Φ is the mapping given by (4.7). Therefore, by virtue of the fixed point theorem of Banach, we can find an $f \in \mathcal{B}$ such that

(5.3) $$f = y\Phi F(x,y,f) \quad .$$ (Cf. W.A. Harris [3].)

Then it follows from Lemma 1 that

(5.4) $$L(f) = yF(x,y,f) - yQF(x,y,f) \quad .$$

Since Q is given by (4.12), the quantity $yQF(x,y,f)$ is a polynomial in x of the form:

(5.5) $$yQF(x,y,f) = y \sum_{m=M}^{M+p-1} \alpha_m x^m \quad (\alpha_m \in \Omega(\delta_o)) \quad .$$

Set

(5.6) $$f = \sum_{h=1}^{\infty} f_h(x) y^h$$

and

(5.7) $$f_h(x) = \sum_{m=M}^{\infty} f_{h,m} x^m \quad .$$

Since

$$f_h(x) = \frac{1}{2\pi i} \oint \frac{f(x,y)}{y^{h+1}} dy \quad ,$$

we have

(5.8) $$\|f_h\|_\delta = \sum_{m=M}^{\infty} |f_{h,m}| \delta^m < +\infty \quad \text{for all} \quad h \quad .$$

We shall prove that

(5.9) $$\tilde{\psi} = f$$

as formal power series in y. This will complete the proof of convergence of $\tilde{\psi}$. To prove (5.9), we derive

(5.10) $$L(\tilde{\psi} - f) = y \sum_{m=M}^{M+p-1} \alpha_m x^m + y\{F(x,y,\tilde{\psi}) - F(x,y,f)\} \quad .$$

Note that $\tilde{\psi}$ is a formal solution of system (3.4) and that $L(f)$ is given by (5.4). We must regard (5.10) as a relation of formal power series in y. If $\tilde{\psi} \neq f$, then there would exist a positive integer h_o such that

$$(5.11)\qquad \begin{cases} \tilde{\psi}_h - f_h = 0 & \text{for} \quad h < h_o \ , \\ \psi_{h_o} - f_{h_o} \neq 0 & . \end{cases}$$

Then (5.10) would imply that

$$(5.12)\qquad L(\tilde{\psi}_{h_o} - f_{h_o}) = \sum_{m=M}^{M+p-1} \beta_m x^m$$

for some $\beta_m \in \Omega(\delta_o)$. Since $\tilde{\psi}_{h_o} - f_{h_o} \in B(\delta_o,\delta,M)$, it would follow from Lemma 3 that

$$\tilde{\psi}_{h_o} - f_{h_o} = 0 .$$

This is a contradiction. Thus we proved (5.9).

§6. Proof of Theorem 1. In this section, we shall prove Theorem 1 by using Theorem 2. Since ψ (given by (1.2)) is a formal solution of system (1.1), we have

$$(6.1)\qquad x^{p+1} \frac{d}{dx} (\psi_o + \psi_1 y) - E(x,y,\psi_o + \psi_1 y) = O(y^2) .$$

Let us change u by the transformation

$$(6.2)\qquad u = v + \psi_o + \psi_1 y$$

and derive

$$(6.3)\qquad x^{p+1} \frac{dv}{dx} = G(x,y,v) ,$$

where

$$(6.4)\qquad G(x,y,v) = E(x,y,v + \psi_o + \psi_1 y) - x^{p+1} \frac{d}{dx} (\psi_o + \psi_1 y) .$$

Hence, (6.1) implies that

$$(6.5)\qquad G(x,y,0) = O(y^2) .$$

Observe also that

$$(6.6)\qquad G_v(x,y,0) = E_u(x,y,\psi_o + \psi_1 y) .$$

Set

$$(6.7)\qquad G(x,y,v) = G(x,y,0) + G_v(x,y,0)v + H(x,y,v) .$$

Then if we set

$$(6.8)\qquad v = yw ,$$

we get

$$(6.9)\qquad H(x,y,yw) = O(y^2) .$$

Now change system (6.3) by transformation (6.8) to derive

$$(6.10)\qquad x^{p+1} \frac{dw}{dx} = \frac{1}{y} G(x,y,0) + G_v(x,y,0)w + \frac{1}{y} H(x,y,yw) .$$

If we set

$$(6.11)\qquad L(w) = x^{p+1} \frac{dw}{dx} - G_v(x,0,0)w$$

and

$$(6.12)\qquad \tilde{G}(x,y,w) = \frac{1}{y^2} G(x,y,0) + \frac{1}{y} \{G_v(x,y,0) - G_v(x,0,0)\}w + \frac{1}{y^2} H(x,y,yw)$$

system (6.10) becomes

$$(6.13)\qquad L(w) = y\,\tilde{G}(x,y,w) .$$

Observe that

$$(6.14)\qquad w = \sum_{h=2}^{\infty} \psi_h(x) y^{h-1} = \sum_{h=1}^{\infty} \psi_{h+1}(x) y^h$$

is a formal solution of system (6.13) and that

$$(6.15)\qquad G_v(0,0,0) = E_u(0,0,0) \in GL(n;\mathbb{C}) .$$

Here we used the assumption that $\psi_o(0) = 0$. By applying Theorem 2 to system(6.13), we conclude that the formal solution (6.14) is convergent. This completes the proof of Theorem 1.

Remark: We can also prove Theorem 1 by applying to system (1.1) the theory of existence and uniqueness of asymptotic solutions (cf. R. Gérard and Y. Sibuya [1,2]) .

§7. An application. Let us consider a Pfaffian system

$$(7.1)\qquad \begin{cases} x^{p+1}\dfrac{\partial u}{\partial x} = E(x,y,u) , \\ y^{q+1}\dfrac{\partial u}{\partial y} = F(x,y,u) , \end{cases}$$

where

(i) p and q are positive integers;
(ii) x and y are independent variables;
(iii) u, E and F are n- vectors;
(iv) entries of E and F are holomorphic in a neighborhood of $(x,y,u)=(0,0,0)$;
(v) system (7.1) is completely integrable, i.e. E and F satisfy the condition:

$$(7.2)\qquad \begin{aligned} & y^{q+1}\frac{\partial E}{y}(x,y,u) + E_u(x,y,u)F(x,y,u) \\ & = x^{p+1}\frac{\partial F}{\partial x}(x,y,u) + F_u(x,y,u)E(x,y,u) . \end{aligned}$$

Under this situation, by utilizing the theory of asymptotic solutions, R. Gérard and Y. Sibuya [1,2] proved the following theorem.

Theorem 3: If $E(0,0,0)=0$, $F(0,0,0)=0$, $E_u(0,0,0)\in GL(n;\mathbb{C})$ and $F_u(0,0,0)\in GL(N;\mathbb{C})$, then system (7.1) admits one and only one solution $u=\phi(x,y)$ such that

(α) $\phi(0,0) = 0$; and

(β) ϕ is holomorphic in a neighborhood of $(x,y) = (0,0)$.

Theorem 3 follows from Theorem 1 in the following way:

Since $F(0,0,0) = 0$ and $F_u(0,0,0) \in GL(n;\mathbb{C})$, we can construct a formal power series in y :

$$(7.3)\qquad \psi(x,y) = \sum_{h=0}^{\infty} \psi_h(x)y^h$$

in such a way that

(1) the ψ_h are holomorphic in a neighborhood of $x = 0$;

(2) $\psi_o(0) = 0$;

(3) $u = \psi$ satisfies the differential equation

$$y^{q+1}\frac{\partial u}{\partial y} = F(x,y,u)$$

formally.

In R. Gérard and Y. Sibuya [2], it was proved that $u = \psi$ is also a formal solution of the differential equation

$$x^{p+1}\frac{\partial u}{\partial x} = E(x,y,u) .$$

In fact, it follows from (7.2) that, if we set

$$v(x,y) = x^{p+1}\frac{\partial \psi}{\partial x} - E(x,y,\psi) ,$$

we get

$$y^{q+1}\frac{\partial v}{\partial y} = F_u(x,y,\psi)v$$

as formal power series in y. Since $F_u(0,0,0)\in GL(n;\mathbb{C})$, we have $v = 0$ as a formal power series in y.

Hence, the convergence of ψ follows from Theorem 1, and the proof of Theorem 3 is completed.

Remark: The proof of Theorem 1 in the present paper is based on the idea due to W.A. Harris, Jr., Y. Sibuya and L. Weinberg [4]. Previously, the linear case was investigated in a similar manner (cf. Y. Sibuya [5,6]) .

References:

1. R. Gérard and Y. Sibuya, Étude de certains systèmes de Pfaff au voisinage d'une singularité, C.R. Acad. Sc. Paris, 284 (1977) 57-60.
2. R. Gérard and Y. Sibuya, Étude de certains systèmes de Pfaff avec singularités, Séries de Math. pures et appl., IRMA, Strasbourg, 1978.
3. W.A. Harris, Jr., Holomorphic solutions of nonlinear differential equations at singular points, SIAM Studies in Applied Math., 5 (1969) 184-187.
4. W.A. Harris, Jr., Y. Sibuya and L. Weinberg, Holomorphic solutions of linear differential systems at singular points, Arch. for Rat. Mech. and Anal., 35 (1969) 245-248.
5. Y. Sibuya, Convergence of power series solutions of a linear Pfaffian system at an irregular singularity, Keio Engineering Reports, 31 (1978), No. 7, 79-86 .
6. Y. Sibuya, A linear Pfaffian system at an irregular singularity, to appear in Tôhoku Math. J..

NON-LINEAR WAVE EQUATIONS AS HAMILTONIAN SYSTEMS

L.J.F. Broer

Department of Physics, Technical University, Eindhoven,
P.O. Box 513, the Netherlands.

1. Introduction

In physics one often is confronted with equations which are supposed to describe systems in which dissipative effects can be neglected. Usually then an energy relation can be found. In this connection it would be desirable to bring the equations in Hamiltonian or Lagrangian form. The known properties of this kind of equations, e.g. the use of Poisson brackets, Noether's theorem and the use of multipliers to introduce auxillary conditions in a variational formulation of the problem, then may entail some other advantages.

As the original equations often are derived in a quite different way the required transformation is not always quite obvious. Moreover, not every formal solution of this problem need to be satisfactory from a physical point of view. The reason for this is essentially that the basic idea is to look for an analogy between the system under consideration and classical analytical mechanics, the oldest and in some sense still the best understood branch of theoretical physics. This analogy implies that we prefer equations for which the energy law follows from Noether's theorem applied to the variation of time.

In the following sections a few examples of these problems will be shown, omitting many details and proofs. The nature of the above-meant advantages will, it is hoped, also become more clear in this way.

2. The electromagnetic field

This well-known subject, which has been studied in great detail, seems to offer the best introduction. We first consider the equations in vacuum. The system of units can be chosen in such a way that the equations read:

$$\underline{B}_t + \text{curl } \underline{E} = 0 \qquad \text{div } \underline{B} = 0 \tag{2.1}$$

$$\underline{E}_t - \text{curl } \underline{B} = 0 \qquad \text{div } \underline{E} = 0 \tag{2.2}$$

Considering vector functions with divergence zero only it is observed that the left pair of equations is a Hamiltonian set as it stands. The Hamiltonian is

$$H_1 = \int d\underline{x}\tfrac{1}{2}(\underline{E} \text{ curl } \underline{E} + \underline{B} \text{ curl } \underline{B}) \tag{2.3}$$

and $\underline{E}$, $\underline{B}$ are conjugate variables.

Simple as this is, it is not what we look for. In the first place, H in (2.3) is not the energy but one of the additional conserved quantities termed Zilch by their discoverer Lipkin [1]). Of course the energy

$$H = \int d\underline{x}\tfrac{1}{2}(E^2 + B^2) \tag{2.4}$$

is another conserved quantity of (2.1) and (2.2). It is connected now with the canonical variation:

$$\delta\underline{E} = -\varepsilon\underline{B}, \qquad \delta\underline{B} = \varepsilon\underline{E}$$

which generates a special rotation in E (6).

A second problem is that we want to be able to generalise our equations, e.g allowing for the interaction between field and matter. When this is done by adding some terms to (2.3) both the equations (2.1) and (2.2) are influenced. But it is known that the equations (2.1) are identities, valid in every situation. Therefore we must look for a transformation to other variables.

This is most conveniently done in the Lagrangian way. The equations (2.1) can be satisfied by expressing the fields in terms of the potentials $\underline{A}$ and ϕ. Taking an arbitrary (differentiable) functional as the electromagnetic Lagrangian L_{em} and adding the definitions of the fields by means of multiplyers we write:

$$L = L_{em}\{\underline{E},\underline{B}\} - \int d\underline{x}\,[\underline{D}(\underline{E} + \underline{A}_t + \text{grad }\phi) - \underline{H}(\underline{B} - \text{curl }\underline{A})] \tag{2.5}$$

Performing the variations we obtain (2.1) and:

$$\underline{D} = \frac{\delta L_{em}}{\delta \underline{E}}, \qquad \underline{H} = -\frac{\delta L_m}{\delta \underline{B}} \tag{2.6}$$

$$\underline{D}_t - \text{curl } \underline{H} = 0, \qquad \text{div } \underline{D} = 0 \tag{2.7}$$

Suppose now we deal with local interaction with a homogeneous medium at rest. We put:

$$L_{em} = \int d\underline{x}\ L = \int d\underline{x}\ [\tfrac{1}{2}E^2 - \tfrac{1}{2}B^2 + W(\underline{E},\underline{B})] \tag{2.8}$$

and

$$\underline{D} = \underline{E} + \frac{\partial W}{\partial \underline{E}} = \underline{E} + \underline{P}, \qquad H = \underline{B} - \frac{\partial W}{\partial \underline{B}} = \underline{B} - \underline{M} \tag{2.9}$$

In this way we recover the usual Maxwell equations for a non-linear, non-dispersive medium.

As L_{em} is invariant for variations of t and $\underline{x}$ there are, according to Noether's theorem, four corresponding conserved densities. These are found to be

$$U = \underline{D}.\underline{E} - L \tag{2.10}$$

and

$$\underline{U} = \underline{D}\times\underline{B} \tag{2.11}$$

The first is undisputed as an admissable expression for the energy density. The second is sometimes considered as the electromagnetic momentum. Often, however, one prefers

$$\underline{U}' = \underline{E}\times\underline{H} \tag{2.12}$$

as a momentum vector. This is the Poynting vector, which also is the flux vector associated with (2.10):

$$U_t + \text{div } \underline{U}' = 0$$

There has been a considerable discussion whether (2.11) or (2.12) is the more suitable expression for the electromagnetic momentum. Although $\underline{U}'$ is not a conserved vector it is usually considered as the favorite. One of the arguments is that the material, having finite mass and rigidity, could store up a variable amount of momentum. Therefore it seems of interest to consider the coupling of (2.8) and (2.9) to, similarly formulated, variational equations for the motion of a mechanical continuum. An example will be considered in the next section.

3. Ideal fluid dynamics

The isentropic motion of an ideal fluid is described by the Euler equation:

$$\underline{v}_t + (\underline{v}\ \text{grad})\underline{v} + \text{grad}\ I = 0 \tag{3.1}$$

and the continuity equation:

$$\rho_t + \text{div}\ (\rho\underline{v}) = 0 \tag{3.2}$$

$I(\rho)$ is the specific entalpy. It is related to the specific energy $U(\rho)$ and the pressure by the relations

$$I = \frac{\partial}{\partial\rho}(\rho U),\quad p = \rho^2\frac{\partial U}{\partial\rho},\quad \rho\, dI = dp$$

The equations (3.1) and (3.2) obviously are not of Hamiltonian or Lagrangian form. As the fluid is a purely mechanical system one would expect that a transformation to such a form must be possible.

Some variational principles or Hamiltonian formulations have been found by trial and error methods, either for (3.1) and (3.2) or for special cases, e.g. incompressible fluids or irrotational flow. One of these ad-hoc results will be given in the next section. These variational principles often are of the nature of equations (2.1) to (2.3). They do not lend themselves readily to generalisations, e.g. coupling with an electromagnetic field and they do not enlighten the fact that fluid dynamics must be based on a structure quite analogeous to classical mechanics.

A more satisfactory treatment has been indicated by Eckart [2]). This is based on the use of coordinates m^α, moving with the fluid. The basic dependent variables then are the space coordinates $x^i(m^\alpha,t)$. This amounts to a mapping:

$$x^i = x^i(m^\alpha,t) \tag{3.3}$$

of the physical space on an auxillary space in which an exact replica of the fluid is at rest with unit density. The m^α can be considered as Cartesian coordinates in this space. The specific volume of the fluid in physical space then is:

$$V = \det x^i_{,\alpha} \tag{3.4}$$

In a mechanical system the Lagrangian is the difference of kinetic and deformation energy. Therefore we try:

$$L_{fl}\{\underline{x}\} = \int d\underline{m}\ L_{fl} = \int d\underline{m}\ [\tfrac{1}{2}x^i_{,t}\, x^i_{,t} - U(V)] \tag{3.5}$$

In this way we are sure that time variations will yield the energy equation. The equation of motion is obtained by variation of x^i:

$$x^i_{,tt} - [\frac{dU}{dV} \cdot X^\alpha_i]_{,\alpha} = 0 \tag{3.6}$$

where X^α_i stands for the minor $\frac{\partial V}{\partial x^i_{,\alpha}}$. We note that (3.5) contains only differentials of m^α. Therefore Noether's theorem will give us another set of conservation laws:

$$(x^i_{,t} x^i_{,\alpha})_{,t} - [V \frac{dU}{dV} + L_{fl}]_{,\alpha} = 0 \tag{3.7}$$

In these equations $\frac{\partial}{\partial t}$ is at constant m^α. Therefore

$$x^i_{,t} = v^i \tag{3.8}$$

is the material velocity.

Using (3.3), (3.4) and (3.8) it is possible to transform (3.6) and (3.7) into equations in x-space. The result is that (3.7) yields the Euler equation (3.1) whereas (3.6) transforms into the momentum equation.

$$(\rho \underline{v})_{,t} + \text{div} (\rho \underline{v}\, \underline{v}) + \text{grad } p = 0 \tag{3.9}$$

which follows directly from (3.1) and (3.2). (The continuity equation is an identity now.) This idea is due to Eckart [2]), the transformations have been considered in more detail by Kobussen [3]).

The transformation can be performed in two ways. Either one attacks (3.6) and (3.7) directly or one first transforms (3.5) into an expression for $L_{fl}\{\underline{m}\}$. In the latter case one can also consider the associated Hamiltonian forms of the equation. It can be shown then [4]) that the transformation, which essentially is the inversion $x^i(m^\alpha) \to m^\alpha(x^i)$, is canonical. The results obtained in this way are rather complicated. They implicate that, in $\underline{x}$ space, the functions $\underline{m}(\underline{x},t)$ essentially play the role of a vector potential from which ρ and $\underline{v}$ can be derived. Therefore one could look for a multiplyer form of the theory in the same way as was done in (2.5) for the electromagnetic field. This can be done, the most important difference is that now the definitions are not linear. This essentially is the cause of the complications met by Eckart and Kobussen.

A convenient starting point for the formulation of the theory in local coordinates is:

$$L_{fl}\{m\} = \int d\underline{x}\ [\rho(\tfrac{1}{2}v^2 - U(\rho)) + \lambda_\alpha(m^\alpha_{,t} + v^i m^\alpha_{,i}) + \mu(\rho - \det m^\alpha_{,i})] \tag{3.10}$$

The first auxillary condition or definition simply means that the m^α are material coordinates, moving with the fluid . Here $\frac{\partial}{\partial t}$ is at constant $\underline{x}$. The condition is the counterpart of (3.8). It can be shown that the results of Eckart and Kobussen can be deduced from (3.10) and that a number of the ad-hoc variational principles mentioned above follow rather straightforwardly from (3.10) [5].

We now turn to the coupling problem mentioned at the end of section 2. Combining (2.5) and (3.10) we write:

$$L = L_{em} + L_{fl} + L_{int} \tag{3.11}$$

where L_{int} is a functional of $\underline{E}$, $\underline{B}$, ρ and $\underline{v}$. We will not bother here about the exact expression. Physically it results from corrections to $W(\underline{E},\underline{B})$ due to the variation in density and the motion of the matter through the fields.

These effects entail that we now must write

$L_{em} + L_{int} = \frac{1}{2}E^2 - \frac{1}{2}B^2 + W(\rho, \underline{E}-\underline{v}\times\underline{B}, \underline{B} + \underline{v}\times\underline{E})$

when $|\underline{v}| << c = 1$.

From (3.11) we can derive the equations of motion and the conservation laws for variation of t and $\underline{x}$ in the same manner as before. The latter exercise yields the conserved densities:

$$U = \underline{D}.\underline{E} - L + \rho(\tfrac{1}{2}v^2 + U(\rho)) + U_{int} \tag{3.12}$$

and

$$\underline{U} = \underline{D}\times\underline{B} + \rho\underline{v} + \underline{U}_{int} \tag{3.13}$$

The second terms in (3.12) ena (3.13) are purely mechanical. Therefore these densities can be considered as the total energy and momentum of the system. These quantities are separated in the same way as the Lagrangian (3.11) and it seems natural to consider $\underline{D}\times\underline{B}$, according to (2.11), as the contribution of the field.

However, if we take (3.13) on its own without looking at (3.11), we note that $\underline{U}_{int}$ contains some terms which are independent of the mechanical variables. This is due to the fact that there is, upon expanding, a contribution to L_{int} which is linear in the velocity. According to Noether's theorem U_{int} has a term

$\rho \frac{\partial L_{int}}{\partial \underline{v}} = \rho_o \frac{\partial L_{int}}{\partial \underline{v}} + (\rho-\rho_o) \frac{\partial L_{int}}{\partial \underline{v}}$

These terms could be joined to the first terms as well. In this way we could also write:

$$\underline{U} = \underline{E}\times\underline{H} + \rho\underline{v} + \underline{U}'_{int} \tag{3.14}$$

A special case has been worked out a few years ago [6].

This separation has the disadvantage that the separate terms can not be derived from the terms of L, nor from any other separation of the Lagrangian. It does not correspond to the, universily accepted, separation (3.12) of the energy and $\underline{E}\times\underline{H}$ is in general not a conserved density for the pure electromagnetic equations. Nevertheless many arguments in the electromagnetic momentum discussion implicate the use of (3.14). This is due to the fact that it has a positive side too. Suppose we deal with wave solutions of fairly small apmlitude around some equilibrium states $\rho = \rho_0$, $\underline{v} = \underline{E} = \underline{B} = 0$. Then it turns out that, in (3.12), the main terms are of second order, whereas the interaction term is of third order. This is very satisfactory. In (3.13) however $\underline{U}_{int}$ is of second order but (3.14) $\underline{U}'_{int}$ is a third order term. Therefore, in the discussions mentioned above, it is often neglected implicitely. It seems that this discussion still can not be considered as closed.

The root of this trouble is the fact that in equilibrium $\rho = \rho_0 \neq 0$. It can be shown that this is also the cause of the difficulties met in exactly defining the pressure of sound. After all this quantity is nothing but the time average of the momentum flux in (3.9).

4. Surface waves

In this section we consider the classical problem of irrotational gravity waves, propagating in the x-direction only, on a layer of an incompressible ideal fluid. We consider only layers of constant unperturbed depth. Units are chosen to that this depth, the fluid density and the gravitational constant are equal to one. The velocity potential $\Phi(x,y,t)$ of the fluid then has to satisfy the conditions:

$$\Phi_{xx} + \Phi_{yy} = 0 \tag{4.1}$$

$y = 0$:

$$\Phi_y = 0 \tag{4.2}$$

$y = 1+\eta(x,t)$:

$$\eta_t = \Phi_y - \eta_x \Phi_x \tag{4.3}$$

$$\Phi_t = -\tfrac{1}{2}(\Phi_x^2 + \Phi_y^2) - \eta \tag{4.4}$$

For reasons explained later on it is desirable to obtain a Hamiltonian formulation for this problem. As it clearly is a purely mechanical system we want the Hamiltonian to be equal to the total energy:

$$H = \int_{-\infty}^{\infty} dx \left[\int_{0}^{1+\eta(x,t)} dy \; \tfrac{1}{2}(\Phi_x^2 + \Phi_y^2) + \tfrac{1}{2}\eta^2 \right] \tag{4.5}$$

It remains then to find one pair of conjugate variables. Other pairs then can be found by canonical transformations.

The first step is to note that, on account of (4.1) and (4.2), H is a functional of η and the surface value of Φ:

$$\phi(x,t) = \Phi(x,1+\eta(x,t),t) \tag{4.6}$$

Although the functional $H\{\phi,\eta\}$ cannot be written down explicitely it is possible to prove the canonical theorem [7]):

The equations

$$\eta_t = \frac{\delta H}{\delta \phi}, \quad \phi_t = -\frac{\delta H}{\delta \eta} \tag{4.7}$$

are equivalent to the set (4.3), (4.4) under the conditions (4.1) and (4.2).

There is no difficulty to extend the proof to three-dimensional motions and sloping bottom. As the considerations indicated in section 3. were not used the theorem is an example of the ad-hoc results mentioned there.

A special case is that of small amplitude waves. Then one takes the linearized conditions (4.3) and (4.4) at y = 1. The resulting potential problem then is easily soluble and one finds:

$$H_{lim} = \int_{-\infty}^{\infty} dx(\tfrac{1}{2}\phi_x R\phi_x + \tfrac{1}{2}\eta^2) \qquad (4.8)$$

where R is an integral operator with the spectrum:

$$\hat{R}(\kappa) = \frac{\text{tgh}\kappa}{\kappa} = 1 - \tfrac{1}{3}k^2 + \ldots \qquad (4.9)$$

The reason for this work was, in part, rather practical. The original equations are so complicated that even for a modern computer, it seems hardly possible to obtain e.g. solutions for an initial value problem over large intervals of time. Therefore in practice approximate equations are used. Now it is not always easy to be sure that an approximate equation really is of the same character as the original. A safer procedure seems to look for an approximate Hamiltonian H_a first. In this way we are sure that the simplified equations at any rate are again a "good" dynamical system. Moreover, when H_a can be chosen positive definite, the system is stable.

As an example we look briefly at equations for fairly long, fairly low waves. That is, we require only first order effects of the small parameters wave amplitude/depth and (depth/wave length)^2. The latter corresponds to κ^2 in (4.9) which makes clear that k itself would not do. Formally this means that we suppose that η, ϕ_x and $\frac{\partial^2}{\partial x^2} = -\kappa^2$ all are of order ε. It then is required that $H - H_a = O(\varepsilon^4)$. It can be shown that:

$$H_a = \int dx\, [\tfrac{1}{2}\phi_x R\phi_x + \tfrac{1}{2}\eta^2 + \tfrac{1}{2}\eta\phi_x^2] \qquad (4.10)$$

would be a satisfactory approximation. The operator R is rather cumbersome to deal with. We could truncate the series in (4.9) and write:

$$H_b = \int dx\, [\tfrac{1}{2}\phi_x^2 - \tfrac{1}{6}\phi_{xx}^2 + \tfrac{1}{2}\eta^2 + \tfrac{1}{2}\eta\phi_x^2] \qquad (4.11)$$

Equations (4.7) then reduce to:

$$\begin{aligned} \eta_t &= -(v + \eta v + \tfrac{1}{3}v_{xx})_x \\ v_t &= -(\tfrac{1}{2}v^2 + \eta)_x \end{aligned} \qquad (4.12)$$

where $\phi_x = v$ is, in this approximation, the fluid velocity at the surface. Equations (4.12) are the renowned Boussinesq equations, usually derived by expanding Φ in a power series in y.

It is seen now that H_b is not positive definite. Small wave perturbations (e.g. rounding-off noise) could cause instability. In fact, computer solutions of (4.12) for initial value problems sometimes seem to have run into considerable trouble

A way out is to approximate R by the much simpler operator R' with

$$\hat{R}' = \frac{1}{1+\frac{K^2}{3}} \tag{4.13}$$

The first equation in (4.12 then is modified into: $\eta_t = -(R'v+\eta v)_x$

It seems that this integro-differential equation is not more difficult to handle on a computer than the classical Boussinesq equation. Preliminary experiments indicate that the long-time behaviour of inital value problems is greatly improved by this modification.

Formally the stability is not yet ensured. For large amplitudes the last term in H might cause trouble. This can be avoided by a further modification. Adding a fourth order term we take:

$$H_m = \int dx\, [\tfrac{1}{2}\phi_x\, R'\phi_x + \tfrac{1}{2}(\eta+\tfrac{1}{2}\phi_x^2)^2]$$

and obtain:

$$\eta_t = -(R'v + \eta v + \tfrac{1}{2}v^3)_x$$
$$v_t = -(\eta + \tfrac{1}{2}v^2)_x$$

It is doubtfull however whether this precaution is really necessary as the existence of the solutions of the exact equations for large-amplitudes is not warranted anyhow.

Once this method is available it is not difficult to find other Hamiltonian approximations for long low waves which are stable for short-wave perturbations. Some examples are given in the papers cited under [7].

Relatively simple results are obtained by methods indicated in section 3. The mass coordinate now is : $m(x,t) = \int^x h\, dx$

Applying the inverse of the Eckart-Kobussen transformation and taking $\xi(m,t) = x(m,t)-m$ as the new variable we find:

$$\xi_{tt} - \tfrac{1}{3}\xi_{ttxx} - (\xi_m - \tfrac{3}{2}\xi_m^2)_m = 0$$

This equation has the same qualities as the modified Boussinesq equation as regards stability and degree of approximation. It does not involve any integral operator and moreover has a simple Lagrangian:

$$L = \int dm\, [\tfrac{1}{2}\xi_t^2 + \tfrac{1}{6}\xi_{tm}^2 - \tfrac{1}{2}\xi_m^2 + \tfrac{1}{2}\xi_m^3]$$

In practice some dozens of Boussinesq-like approximate equations are known, usually derived by various, slightly differing, expansion and truncation procedures. Only some of these approximations really are Hamiltonian systems, not all of them stable. It seems that the criterion of a positive approximate Hamiltonian is a useful one to bring some order in this chaos.

A problem similar to the long water waves is that of ion-acoustic waves in a cold plasma. A canonical theorem for these waves has been proved and a stable approximation bases on an approximate Hamiltonian has been given [8].

An important further approximation is concerned with waves running in one direction This leads to an equation of the first order in $\frac{\partial}{\partial t}$. The classical example is the Korteweg-de Vries (KdV) equation as an approximation for uni-directional solutions of the Boussinesq equations.

Suppose, in (4.12), that $\eta \approx v$. Adding the equations we obtain the KdV equation:

$$\eta_t = -(\eta + \tfrac{3}{4}\eta^2 + \tfrac{1}{6}\eta_{xx})_x \tag{4.14}$$

This is certainly not a standard example of exact reasoning. In order to gain more insight into the connections between these second-and first order approximations it would be useful to know under wich conditions first order equations are equivalent to Hamiltonian systems. This is the subject of the next section.

5. First order equations and hidden Hamiltonians

When a first order equation:

$$u_t = F(u, u_x, u_{xx} \ldots)$$

is equivalent to a Hamiltonian system it is obviously required that the function space u(x) can be split up into two parts from which p(x) and q(x) can be constructed. As an example we take the KdV equation (4.13) which can be reduced to:

$$u_t + uu_x + u_{xxx} = 0 \tag{5.1}$$

Let p(x), q(x) be arbitrary (sufficiently differentiable) even functions. Then we can write

$$u = q + p_x \tag{5.2}$$

Putting this into (5.1) and sorting out even and odd parts we obtain:

$$q_t = \frac{\delta H}{\delta p} = -\left(\frac{\delta H}{\delta p_x}\right)_x \qquad p_t = -\frac{\delta H}{\delta q} \tag{5.3}$$

where

$$H = \int dx\left(\tfrac{1}{6}q^3 + \tfrac{1}{2}qp_x^2 - \tfrac{1}{2}q_x^2 - \tfrac{1}{2}p_{xx}^2\right) \tag{5.4}$$

which, on account of (5.2) also can be written as

$$H = \int dx\left(\tfrac{1}{6}u^3 - \tfrac{1}{2}u_x^2\right) \tag{5.5}$$

Then (5.1) takes the form:

$$u_t + \frac{\partial}{\partial x}\frac{\partial H}{\partial u} = 0 \tag{5.6}$$

This result is a special case of a more general property [9]):

When A is an antihermitian and non-singular linear operator an equation of the form:

$$Au_t = \frac{\delta H}{\delta u} \tag{5.7}$$

can be considered as a transformed Hamiltonian equation. H is then called the hidden Hamiltonian of (5.7) . For the KdV equation (5.1) we found:

$$A = \left(-\frac{\partial}{\partial x}\right)^{-1}, \quad H = \tfrac{1}{6}u^3 - \tfrac{1}{2}u_x^3$$

Other well-known examples are:

$$A = \frac{\partial}{\partial x}, \quad H = 1 - \cos u$$

which results in the Sine-Gordon equation:

$$u_{tx} = \sin u$$

and $$A = \frac{\partial}{\partial x} - (\frac{\partial}{\partial x})^{-1}, \quad H = \frac{1}{2}u^2 + \frac{1}{6}u^3$$

which yields the Benjamin-Bona-Mahony (BBM) equation:

$$u_t - u_{txx} + u_x + uu_x = 0 \tag{5.8}$$

In connection with this theorem the following remarks might be useful:

1. Equation (5.7) is the Euler equation of the transformed action principle:

$$\frac{\delta W}{\delta u} = 0, \; W = \int dt[\tfrac{1}{2}(u_t, Au) + H\{u\}]$$
$$= \int dt[\tfrac{1}{2}(p_t, q) - \tfrac{1}{2}(q_t, p) + H\{p,q\}]$$

2. The transformation determines A independent of H. When looking for a separation like (5.2) one therefore best begins with a simple Hamiltonian, e.g. $H = \int dx.\frac{1}{2}u^2$. The separation

$$u = q - A^{-1}p \tag{5.9}$$

will do when the function spaces p and q can be chosen such that the in-product $(q, A^{-1}p)$ always vanishes.

3. In order to exploit this hidden Hamiltonian concept it is not necessary that the transformation is known. As the general properties of Hamiltonian systems can be formulated in terms of Poisson brackets the only requirement is to find the transformation of this bracket. This turns out to be:

$$\{F,G\} = (\frac{\delta F}{\delta n}, A^{-1}\frac{\delta G}{\delta u}) \tag{5.10}$$

therefore (5.7) can be written as $u_t = \{u,H\}$

The properties can be used sometimes to relate various properties of this type of equations. We mention two results on the KdV equations (5.1)

It is easily verified that the functional

$$T = \int dx[\tfrac{1}{2}u^2 - xu] \tag{5.11}$$

is a constant of the motion for (5.1). Therefore it will be the generator of a group of invariant canonical transformations. The transformed functions u(x,t;s) satisfy the equation deduced from (5.10):

$$u_s = \{u,T\} = 1 - t\frac{\partial u}{\partial x}$$

The solution is:

$$u(x,t;s) = u(x-st,t) + s$$

which is a known invariant transformation.

The second example has to do with the well known polynominal conservation laws of (5.1), found by Miura et al [10]). These constants of the motion are

$$\mathcal{T}_n = \int dx . T_n(u, u_x, u_{xx} \ldots)$$

When the polynomia T_n are normalized in such a way that each contains a term $u^n/n!$ it can be shown [11], using (5.11), that

$$\mathcal{T}_n = \{\mathcal{T}.\mathcal{T}_{n+1}\} = \int dx \frac{\partial T}{\partial u} n+1 \qquad (5.12)$$

which is a ladder relation for the $\mathcal{T}_n$.

It is unfortunate that this ladder works in the wrong direction. Equations admitting a relation similar to (5.12) do exist, e.g.:

$$u_t + \frac{\partial}{\partial x} (\tfrac{1}{2}u^2 + Du) = 0$$

where D is any hermitian operator commutating with $\frac{\partial}{\partial x}$. If a $\mathcal{T}_n$ would exist, the $\mathcal{T}_m$'s then are readily calculated. However, apart from the case $D = -\frac{\partial^2}{\partial x^2}$, it seems not to be known whether conserved quantities with $n > 3$ do occur.

We have not been able to construct a non-trivial example of an upward-directed ladder. Intermediate cases are represented by the Sine-Gordon equation and the modified KdV equation. Here self-reproducing relations:

$$\{\mathcal{T}, \mathcal{T}_n\} = c_n, \mathcal{T}_n$$

can be obtained.

6. The relation between first- and second order equations

A formal link between equations of the Boussinesq kind and KdV, BBM, or similar equations can be laid in the following way.

Hamiltonians of the kind of (4.11) can be written as:

$$H = \int dx\ E(v,v_x\ldots,\ q,q_x\ldots)$$

where $v = p_x$.

The equations of motion then are:

$$q_t = -\left(\frac{\delta E}{\delta v}\right)_x \qquad v_t = -\left(\frac{\delta E}{\delta q}\right)_x \tag{6.1}$$

Now let T be a Hermitian operator which commutes with ∂/∂_x. Then the substitution

$$q + Tv = f,\ q - Tv = b \tag{6.2}$$

transforms (6.1) into:

$$f_t = -2T\left(\frac{\delta E}{\delta f}\right)_x \tag{6.3}$$

$$b_t = 2T\left(\frac{\delta E}{\delta b}\right)_x \tag{6.4}$$

This is still exact. Now suppose that $E\{f,b\} = F\{f\} + B\{b\} + S\{f,b\}$ and that the remainder S is a small term. Then, when the initial value of b is zero (or very small) one can hope that b will remain small and that a good approximation will be obtained by putting b = 0 in (6.3) and neglecting (6.4). It seems very hard to prove these assumptions for non-linear problems, we will proceed nevertheless, taking for granted that the equations can be chosen in such a way that, at any rate for some finite interval of time, our conditions are fulfilled.

Obviously the remain equation is of the hidden Hamiltonian type (6.5-6.7) with

$$A^{-1} = -2T\frac{\partial}{\partial x}, \qquad H\{f\} = \int dx\ \ [F\{f\} + S\{f,0\}]$$

It is seen that, upon neglecting $S\{f,b\}$ altogether, the equations (6.3) and (6.4) are an exactly separable set.

This procedure does not provide an unique solution for the connection between second and first order equations. In the first place there is some freedom in choosing T. For instance, in (4.11) we could take T = I. The resulting equation (6.3) then is the KdV equation (4.13). The separable approximation is the ordinary wave equation, that is the very long, very low wave approximation.

Another choice would be $T = R^{\frac{1}{2}}$. In that way one shifts the dispersive term in from S to the separable part. It would be complicated to do this exactly. An approximation would be $R^{-\frac{1}{2}} = T^{-1} = 1 - \frac{1}{6}\frac{\partial^2}{\partial x^2}$ which is akin to (4.13). In this way one finds a BBM equation. The corresponding separable approximation is still linear but includes the main part of the long-wave dispersion. It could be called a very low, fairly long wave approximation.

Other unidirectional approximations in the form of (5.7) could be obtained by performing a canonical transformation before starting the approximation procedure. The Hamiltonian approach therefore certainly can not solve the problem of finding the one and only best second or first order approximation for the equations given in the beginning of section 4. It can help however to systematize the discussions on these questions, which will no doubt continue for a considerable time.

References

1. Lipkin, D.M., J.Math.Phys. 5(1964), 696
2. Eckart, C., Phys. of Fluids 3(1960),421
3. Kobussen, J.A., Thesis, Eindhoven 1973
4. Broer, L.J.F. and J.A. Kobussen, Appl. Sci. Res. 29(1974) 419
5. Groesen, E.W.C. van, L.J.F. Broer and J. Lodewijk, In course of preparation
6. Broer, L.J.F., Physica 83A(1976) 471
7. Broer, L.J.F., Appl. Sci. Res. 29(1974) 430, 31(1975) 377
   Broer, L.J.F., E.W.C. van Groesen and J.M.W. Timmers, Appl. Sci. Res. 32(1976) 619
8. Broer, L.J.F. and F.W. Sluijter, Physics of Fluids 20(1977) 1458
9. Broer, L.J.F., Physica 79A(1975) 583
10. Miura, R.M., C.S. Gardner and M.D. Kruskal, J. Math. Phys. 9(1968) 1204
11. Broer, L.J.F. and S.C.M. Backerra, Appl. Sci. Res. 32(1976) 495

HOW MANY JUMPS? VARIATIONAL CHARACTERIZATION OF THE LIMIT SOLUTION OF A SINGULAR PERTURBATION PROBLEM

O. Diekmann & D. Hilhorst

Mathematisch Centrum
2e Boerhaavestraat 49
1091 AL Amsterdam
The Netherlands

1. INTRODUCTION

Consider the two-point boundary value problem

$$\text{BVP} \quad \begin{cases} \varepsilon y'' + (g-y)y' = 0, \\ y(0) = 0, \quad y(1) = 1, \end{cases}$$

where $g \in L_2 = L_2(0,1)$ is a given function and $y \in H^2$ is unknown. As we shall show, there exists for each $\varepsilon > 0$ a unique solution y_ε, which is increasing. We are interested in the limiting behaviour of y_ε as $\varepsilon \downarrow 0$.

Motivated by a physical application we previously studied a similar problem in a joint paper with L.A. Peletier [2]. Using the maximum principle as our main tool we were able to establish the existence of a unique limit solution y_0 under certain, physically reasonable, assumptions on the function g. In some cases we could characterize y_0 completely, in others, however, some ambiguity remained.

Here, inspired by the work of Grasman & Matkowsky [4], we shall resolve this ambiguity by using a variational formulation of the problem. In fact we shall present two different methods of analysis. The first one is based on the theory of maximal monotone operators, whereas the second one uses duality theory.

During our investigation of BVP we experienced that it could serve as a fairly simple, yet nontrivial, illustration of concepts and methods from abstract functional analysis. In order to demonstrate this aspect of the problem we shall spell out our arguments in some more detail than is strictly necessary.

The organization of the paper is as follows. In Section 2 we present the first method. We prove, by means of Schauder's fixed point theorem, that BVP has a solution y_ε for each $\varepsilon > 0$. Moreover, we show that BVP is equivalent to an abstract equation AE, involving a maximal monotone operator A, and to a variational problem VP, involving a convex, lower semi-continuous functional W. Subsequently we exploit these formulations in the investigation of the limiting behaviour of y_ε as $\varepsilon \downarrow 0$. (The idea of using the theory of maximal monotone operators was suggested to us by Ph. Clément.) It turns out that y_ε converges in L_2 to a limit y_0. Moreover, y_0 is abstractly characterized as the projection (in L_2) of g on $\overline{\mathcal{D}(A)}$. We conclude this section with some results about uniform convergence under restrictive assumptions.

In Section 3 we study a minimization problem P related to VP. We begin by proving that P has a unique solution. Next, we present a dual problem P^* and we deduce from the extremality relations between primal and dual problems that P and BVP are equivalent. Putting $\varepsilon = 0$ in P^* we obtain a formal limit problem P_0^*. Subsequently we associate with P_0^* a dual problem P_0^{**} and we show that the solution of P tends, as $\varepsilon \downarrow 0$, to the solution of P_0^{**}. A rewriting of P_0^{**} reveals the relation with the result of Section 2. This treatment of the problem has grown out of conversations with R. Témam who, in particular, indicated to one of us the appropriate functional analytic setting for studying the minimization problem.

In Section 4 we give concrete form to the characterization of y_0. In particular we present sufficient conditions for a function to be y_0 and we show, by means of examples, how these criteria can be used in concrete cases. The first part of the title originated from Example 4.

In Section 5 we make various remarks about generalizations and limitations of our approach.

ACKNOWLEDGEMENT.

The authors gratefully acknowledge helpful remarks and suggestions of Ph. Clément, E.W.C. van Groesen, H.J. Hilhorst, L.A. Peletier and R. Témam.

2. THE FIRST METHOD

2.1. THREE EQUIVALENT FORMULATIONS

In order to demonstrate the existence of a solution of BVP, let us first look at the auxiliary problem

$$u'' + (g-w)u' = 0,$$

$$u(0) = 0, \quad u(1) = 1,$$

where $w \in L_2$ is a given function. The solution of this linear problem is given explicitly by

$$u(x) = C(w) \int_0^x \exp\left(\int_0^\zeta (w(\xi) - g(\xi))\,d\xi\right) d\zeta$$

with

$$C(w) = \left(\int_0^1 \exp\left(\int_0^\zeta (w(\xi) - g(\xi))\,d\xi\right) d\zeta\right)^{-1}.$$

From this expression it can be concluded that $u' > 0$ and $0 \le u \le 1$. So if we write $u = Tw$, then T is a compact map of the closed convex set $\{w \in L_2 \mid 0 \le w \le 1\}$ into itself and hence, by Schauder's theorem, T must have a fixed point. Clearly this fixed point corresponds to a solution of BVP. Thus we have proved

PROPOSITION 2.1. *For each* $\varepsilon > 0$ *there exists a solution* $y_\varepsilon \in H^2$ *of* BVP. *Moreover, any solution* $y \in H^2$ *satisfies* (i) $y' > 0$ *and* (ii) $0 \le y \le 1$.

The a priori knowledge that y' is positive allows us to divide the equation by y'. In this manner we are able to reformulate the boundary value problem as an equivalent abstract equation

$$\text{AE} \qquad (I + \varepsilon A)y = g$$

where the (unbounded, nonlinear) operator $A: \mathcal{D}(A) \to L_2$ is defined by

$$(2.1) \qquad Au = -\frac{u''}{u'} = -(\ln u')'$$

with

$$(2.2) \qquad \mathcal{D}(A) = \{u \in L_2 \mid u \in H^2,\ u' > 0,\ u(0) = 0,\ u(1) = 1\}.$$

PROPOSITION 2.2. *The operator* A *is monotone. Hence the solution of* AE *(and* BVP*) is unique.*

PROOF. Let $u_i \in \mathcal{D}(A)$ for $i = 1,2$ then

$$(Au_1 - Au_2,\ u_1 - u_2) = -\int ((\ln u_1')' - (\ln u_2')')(u_1 - u_2)$$

$$= \int (\ln u_1' - \ln u_2')(u_1' - u_2') \ge 0$$

(because $z \mapsto \ln z$ is monotone on $(0,\infty)$; note that here and in the following we write $\int \phi$ to denote $\int_0^1 \phi(x)dx$.) Next, suppose $\varepsilon Ay_i = g - y_i$, $i = 1,2$, then $0 \le \varepsilon(Ay_1 - Ay_2,\ y_1 - y_2) = (g - y_1 - g + y_2,\ y_1 - y_2) = -\|y_1 - y_2\|^2$ and hence $y_1 = y_2$. □

We recall that a monotone operator A defined on a Hilbert space H is called *maximal monotone* if it admits no proper monotone extension (i.e., it is maximal in the sense of inclusion of graphs). It is well known that A is maximal monotone if and only if $\mathcal{R}(I + \varepsilon A) = H$ for each $\varepsilon > 0$ (see Brézis [1]). In our case, with $H = L_2$ and A defined in (2.1), this is just a reformulation of the existence result Proposition 2.1. Consequently we know

PROPOSITION 2.3. *A is maximal monotone.*

In search for yet another formulation let us write the equation in the form

$$-\varepsilon(\ln y')' + y - g = 0$$

Hence, for any $\phi \in H_0^1$,

$$\varepsilon \int \phi'(\ln y' + 1) + \int \phi(y - g) = 0.$$

Motivated by this calculation we define a functional $W: L_2 \to \overline{\mathbb{R}}$ by

(2.3) $$W(u) = \varepsilon\Psi(u) + \frac{1}{2}\|u - g\|^2$$

where

(2.4) $$\Psi(u) = \begin{cases} \int u' \ln u' & \text{if } u \in \mathcal{D}(\Psi), \\ +\infty & \text{otherwise,} \end{cases}$$

and

(2.5) $$\mathcal{D}(\Psi) = \{u \in L_2 \mid u \text{ is AC, } u' \geq 0,\ u' \ln u' \in L_1,\ u(0) = 0,\ u(1) = 1\}$$

(here AC means absolutely continuous). Also we define a variational problem

VP $\quad \operatorname{Inf}_{L_2} W.$

We note that the mappings $z \mapsto z \ln z$ and $z \mapsto z^2$ are (strictly) convex (on $[0,\infty)$ and $(-\infty,\infty)$ respectively) and that W inherits this property because $\mathcal{D}(\Psi)$ is convex as well. Hence VP has at most one solution. For future use we observe that the convexity of $z \mapsto z \ln z$ implies, for $z \geq 0$ and $\zeta > 0$, the inequality

$$z \ln z - \zeta \ln \zeta \geq (1 + \ln \zeta)(z - \zeta).$$

PROPOSITION 2.4. *y_ε solves* VP.

PROOF. Firstly we note that $y_\varepsilon \in \mathcal{D}(\Psi)$. So for any $u \in \mathcal{D}(\Psi)$

$$\begin{aligned} W(u) - W(y_\varepsilon) &= \varepsilon \int (u' \ln u' - y_\varepsilon' \ln y_\varepsilon') + \frac{1}{2}\|u - g\|^2 - \frac{1}{2}\|y_\varepsilon - g\|^2 \\ &\geq \varepsilon \int (1 + \ln y_\varepsilon')(u' - y_\varepsilon') + \int (y_\varepsilon - g)(u - y_\varepsilon) \\ &= \int \left(-\varepsilon \frac{y_\varepsilon''}{y_\varepsilon'} + y_\varepsilon - g\right)(u - y_\varepsilon) = 0. \qquad \square \end{aligned}$$

We recall that the *subgradient* $\partial\Psi$ of the convex functional Ψ is defined by

$$\partial\Psi(u) = \{\zeta \in L_2 \mid \Psi(v) - \Psi(u) \geq (\zeta, v-u),\ \forall v \in \mathcal{D}(\Psi)\}.$$

A calculation like the one above shows that, for $u \in \mathcal{D}(A)$ and $v \in \mathcal{D}(\Psi)$,

$$\Psi(v) - \Psi(u) \geq (Au, v-u).$$

Hence $A \subset \partial\Psi$, but, since $\partial\Psi$ is monotone and A is *maximal* monotone, we must have $A = \partial\Psi$. Likewise it follows that $\partial W = \varepsilon A + I - g$. These observations should clarify the relation between VP and AE.

One can show that Ψ (and hence W as well) is lower semicontinuous and subsequently one can use this knowledge to give a direct variational proof of the existence of a solution of VP. We refer to Theorem 3.2. below for a detailed proof of this result.

We summarize the main results of this subsection in the following theorem.

THEOREM 2.5. *The problems* BVP, AE *and* VP *are equivalent. In fact, for each* $\varepsilon > 0$, *there exists* $y_\varepsilon \in \mathcal{D}(A)$ *which solves each problem and no problem admits any other solution.*

2.2. LIMITING BEHAVIOUR AS $\varepsilon \downarrow 0$

The fact that y_ε solves AE can be expressed as

$$y_\varepsilon = (I + \varepsilon A)^{-1} g.$$

Subsequently, the observation that A is maximal monotone provides a key to describing the limiting behaviour. For, it is known from the general theory of such operators (see Brézis [1, Section II.4, in particular Th. 2.2]) that

$$\lim_{\varepsilon \downarrow 0} (I + \varepsilon A)^{-1} g = \mathrm{Proj}_{\overline{\mathcal{D}(A)}}\, g,$$

where the expression at the right-hand side denotes the projection (in the sense of the underlying Hilbert space, hence L_2 in this case) of g on the closed convex set $\overline{\mathcal{D}(A)}$, or, in other words,

$$\mathrm{Proj}_{\overline{\mathcal{D}(A)}}\, g = y_0$$

where y_0 denotes the unique solution of the variational problem

$$\mathrm{Min}_{\overline{\mathcal{D}(A)}}\ W_0$$

with

$$W_0(u) = \|u - g\|^2.$$

Below we shall give a proof of this result for this special case, using techniques as in Brézis' book, but exploiting the fact that A is the subdifferential of the functional Ψ.

THEOREM 2.6.

$$\lim_{\varepsilon \downarrow 0} \|y_\varepsilon - y_0\| = 0.$$

PROOF. First of all we note that $\|y_\varepsilon\| \le 1$. We shall split the proof into three steps.

Step 1. Take any $z \in \mathcal{D}(A)$ then from

$$\Psi(y_\varepsilon) - \Psi(z) \ge (Az, y_\varepsilon - z)$$

it follows that

$$\liminf_{\varepsilon \downarrow 0} \varepsilon(\Psi(y_\varepsilon) - \Psi(z)) \ge 0.$$

Step 2. By definition,

$$0 \ge W(y_\varepsilon) - W(z) = \varepsilon(\Psi(y_\varepsilon) - \Psi(z)) + \frac{1}{2}\|g - y_\varepsilon\|^2 - \frac{1}{2}\|g - z\|^2.$$

Hence

$$\limsup_{\varepsilon \downarrow 0} \|g - y_\varepsilon\|^2 \le \|g - z\|^2, \qquad \forall z \in \mathcal{D}(A).$$

But then, in fact, the same must hold for all $z \in \overline{\mathcal{D}(A)}$.

Step 3. Since $\|y_\varepsilon\| \le 1$, $\{y_\varepsilon\}$ is weakly precompact in L_2. Take any $\{\varepsilon_n\}$ and $\tilde{y}$ such that $y_{\varepsilon_n} \rightharpoonup \tilde{y}$ in L_2, then

$$(*) \qquad \|g-\tilde{y}\|^2 \le \liminf_{n \to \infty} \|g-y_{\varepsilon_n}\|^2 \le \limsup_{n \to \infty} \|g-y_{\varepsilon_n}\|^2 \le \|g-z\|^2, \qquad \forall z \in \overline{\mathcal{D}(A)}.$$

Consequently $\tilde{y} = y_0$, which shows that the limit does not depend on the subsequence under consideration. Hence $y_\varepsilon \rightharpoonup y_0$. Finally, by taking $z = y_0$ in (*) it follows that in fact $y_\varepsilon \to y_0$. □

We note that

$$\overline{\mathcal{D}(A)} = \{u \in L_2 \mid u \text{ is nondecreasing, } 0 \le u \le 1\}.$$

So in general y_0 need not be continuous (nor does it need to satisfy the boundary conditions). However it is possible, as our next result shows, to establish uniform convergence to a continuous limit at the price of some conditions on g.

THEOREM 2.7. *Suppose* $g \in C^1$, $g(0) < 0$ *and* $g(1) > 1$. *Then* $y_0 \in C$ *and*

$$\lim_{\varepsilon\downarrow 0} \sup_{0\le x\le 1} |y_\varepsilon(x) - y_0(x)| = 0.$$

PROOF. The idea is to derive a uniform bound for y_ε'. We know already that $y_\varepsilon' > 0$ and we are going to show that $y_\varepsilon' \le \sup g'$. To this end we first observe that $g(0)-y_\varepsilon(0)<0$, and $g(1)-y_\varepsilon(1) > 0$, which, combined with the differential equation, shows that $y_\varepsilon''(0) > 0$ and $y_\varepsilon''(1) < 0$. Hence y_ε' assumes its maximum in an interior point, say $\bar{x}$. Next, differentiation of the differential equation followed by substitution of $y_\varepsilon''(\bar{x}) = 0$, $y_\varepsilon'''(\bar{x}) \le 0$, leads to the conclusion that $y_\varepsilon'(\bar{x}) \le g'(\bar{x})$. The uniform bound for y_ε' implies, by virtue of the Arzela-Ascoli theorem, that the limit set of $\{y_\varepsilon\}$ in the space of continuous functions is nonempty. Combination of this result with Theorem 2.6 leads to the desired conclusion. □

In Section 4 we shall show that y_0 can be calculated in many concrete examples. Quite often it will turn out that y_0 is continuous (or piece-wise continuous). This motivates our next result.

THEOREM 2.8. *Suppose* y_0 *is continuous. Then* y_ε *converges to* y_0 *uniformly on compact subsets of* $(0,1)$.

PROOF. Let $I \subset (0,1)$ be a compact set. Put $\beta(\varepsilon) = \max\{y_\varepsilon(x)-y_0(x) \mid x \in I\}$ and let $\bar{x}(\varepsilon) \in I$ be such that $y_\varepsilon(\bar{x}(\varepsilon)) - y_0(\bar{x}(\varepsilon)) = \beta(\varepsilon)$. Suppose $\limsup_{\varepsilon\downarrow 0} \beta(\varepsilon) = \beta > 0$ and let $\{\varepsilon_n\}$ be such that $\beta(\varepsilon_n) \to \beta$ as $n \to \infty$. Choose $\delta \in (0,\delta_1)$, where δ_1 denotes the distance of 1 to I, such that $|y_0(x) - y_0(\xi)| \le \frac{1}{4}\beta$ if $|x-\xi| \le \delta$. Also, choose n_0 such that $\beta(\varepsilon_n) \ge \frac{3}{4}\beta$ for $n \ge n_0$. Then for $x \in [\bar{x}(\varepsilon_n), \bar{x}(\varepsilon_n)+\delta]$ and $n \ge n_0$ the following inequality holds:

$$\begin{aligned} y_{\varepsilon_n}(x) - y_0(x) &\ge y_{\varepsilon_n}(\bar{x}(\varepsilon_n)) - y_0(\bar{x}(\varepsilon_n)) + y_0(\bar{x}(\varepsilon_n)) - y_0(x) \\ &\ge \frac{3}{4}\beta - \frac{1}{4}\beta = \frac{1}{2}\beta. \end{aligned}$$

However, this leads to

$$\|y_{\varepsilon_n} - y_0\|^2 \ge \frac{1}{4}\delta\beta^2$$

which is in contradiction with Theorem 2.6. Hence our assumption $\beta > 0$ must be false and we arrive at the conclusion that $\limsup_{\varepsilon\downarrow 0} \max\{y_\varepsilon(x) - y_0(x) \mid x \in I\} \le 0$. Essentially the same argument yields that $\liminf_{\varepsilon\downarrow 0} \min\{y_\varepsilon(x) - y_0(x) \mid x \in I\} \ge 0$. Taking both statements together yields the result. □

It should be clear that appropriate analogous results can be proved if y_0 is piece-wise continuous. In Theorem 2.8 the sense of convergence is sharpened "a posteriori", that is, once the continuity of y_0 is established by other means. Note that our proof exploits the uniform one-sided bound $y'_\varepsilon > 0$.

3. THE SECOND METHOD

3.1. A VARIATIONAL EXISTENCE PROOF

In this section we study in some detail a minimization problem P which is a variant of VP. We shall use methods from convex analysis. In fact, our presentation follows closely Ekeland & Témam [3, Chapter III, Section 4] and in order to bring this out clearly we begin by introducing some notation in accordance with this reference. We define

$$(3.1) \qquad V = AC = \{v \in L_2 \mid v' \in L_1\}$$

and we consider V as a Banach space provided with the norm

$$(3.2) \qquad \|v\|_V = \|v\|_{L_2} + \|v'\|_{L_1}.$$

We denote by V^* the dual space of V. Next, we introduce $Y = L_1 \times L_2$ and a bounded linear mapping $\Lambda: V \to Y$ defined by

$$(3.3) \qquad \Lambda v = (\Lambda_1 v, \Lambda_2 v) = (v', v).$$

Moreover, we introduce functionals G_1, G_2 and F defined on L_1, L_2 and V, respectively, as follows

$$(3.4) \qquad G_1(w) = \begin{cases} \varepsilon \int w \ln w + \frac{\varepsilon}{e} & \text{if } w \geq 0 \text{ and } w \ln w \in L_1, \\ +\infty & \text{otherwise,} \end{cases}$$

$$(3.5) \qquad G_2(w) = \frac{1}{2} \int (g - w)^2,$$

$$(3.6) \qquad F(w) = \begin{cases} 0 & \text{if } w(0) = 0 \text{ and } w(1) = 1, \\ +\infty & \text{otherwise.} \end{cases}$$

Finally, we call P the minimization problem

$$(3.7) \qquad P \qquad \operatorname{Inf}_V J$$

where by definition

(3.8) $$J(v) = G_1(\Lambda_1 v) + G_2(\Lambda_2 v) + F(v).$$

Clearly G_2 is (strictly) convex and lower semicontinuous (l.s.c.); consequently it is weakly lower semicontinuous (w.l.s.c.) as well (cf. [3, p. 10]). The next result shows that the same conclusion holds for G_1.

PROPOSITION 3.1. G_1 *is convex and w.l.s.c..*

PROOF. Let the function $k: \mathbb{R} \to \overline{\mathbb{R}}$ be defined by

(3.9) $$k(y) = \begin{cases} \varepsilon y \ln y + \frac{\varepsilon}{e} & \text{if } y \geq 0, \\ +\infty & \text{otherwise.} \end{cases}$$

Then k is Borel measurable, l.s.c. and positive. Hence, in other words, it is a normal positive integrand (cf. [3, p. 216]). Rewriting G_1 as

(3.10) $$G_1(w) = \int k(w(\cdot)),$$

we observe that the l.s.c. of k and Fatou's lemma imply that G_1 is l.s.c.:

$$\int k(w(\cdot)) \leq \int \liminf_{m\to\infty} k(w_m(\cdot)) \leq \liminf_{m\to\infty} \int k(w_m(\cdot))$$

whenever $w_m \to w$ strongly in L_1. Since obviously G_1 is convex the result follows. □

THEOREM 3.2. *For each* $\varepsilon > 0$, P *has a unique solution.*

PROOF. First we note that the functional J is bounded from below on V. Let $\{u_m\}$ be a minimizing sequence. We intend to show that $\{u_m\}$ is bounded in L_2 and that $\{u'_m\}$ is bounded in L_1 and equi-integrable (cf. [3, p. 223]). Indeed, from

$$\varepsilon \int u'_m \ln u'_m + \frac{\varepsilon}{e} + \frac{1}{2}\int (g-u_m)^2 \leq C$$

we deduce that $u'_m \geq 0$, that

$$\int u_m^2 \leq C_1, \quad \int u'_m \leq C_2$$

and that

$$\int_{\Omega(M)} u'_m \leq (\ln M)^{-1} \int_{\Omega(M)} u'_m \ln u'_m \leq (\ln M)^{-1} \frac{C}{\varepsilon}$$

where $\Omega(M) = \{x \mid u'_m(x) \geq M\}$ and $M > 1$. Thus, given any constant $\delta > 0$, we have that

$$\int_{\Omega(M)} u'_m \leq \delta$$

provided $M > \exp \frac{C}{\varepsilon\delta}$.

We conclude that $\{u_m\}$ is weakly precompact in L_2 and that $\{u_m'\}$ is weakly precompact in L_1 (cf. [3, p. 223]). If $u_m \rightharpoonup u$ in L_2 and $u_m' \rightharpoonup w$ in L_1, then the usual manipulations with distributional derivatives show that $u' = w$, and consequently that $u \in V$. Moreover, from $u_m(x) = \int_0^x u_m'(\xi)d\xi$ we deduce that $u(x) = \int_0^x u'(\xi)d\xi$ and thus that $u(0) = 0$. Likewise it follows that $u(1) = 1$. So $F(u) = 0$. Since the functionals G_1 and G_2 are w.l.s.c. on L_1 and L_2 respectively, it follows that $u = u_\varepsilon$ is a solution of P. Since, furthermore, J is strictly convex the solution is unique. □

3.2. THE DUAL PROBLEM

In Subsection 2.1 we proved the equivalence of BVP and VP by showing that the solution of BVP (whose existence was proven first) also solves VP. Here we want to go the other way around, i.e., we want to show that the solution of P also solves BVP. In order to do so we shall first determine a dual problem and subsequently we shall utilize the extremality relations.

We embed P into a wider class of perturbed problems P(p) as follows:

$$\text{(3.11)} \qquad P(p) \qquad \operatorname{Inf}_V \Phi(\cdot,p)$$

where $p = (p_1,p_2) \in Y$ and where by definition

$$\text{(3.12)} \qquad \Phi(v,p) = G_1(\Lambda_1 v - p_1) + G_2(\Lambda_2 v - p_2) + F(v).$$

With respect to these perturbations the dual problem P^* is given by (cf. [3, Section III.4])

$$\text{(3.13)} \qquad P^* \qquad \operatorname{Sup}_{Y^*} -\Phi^*(0,\cdot),$$

where $Y^* = L_\infty \times L_2$ and Φ^* is the polar function of Φ, that is

$$\text{(3.14)} \qquad \Phi^*(v^*,p^*) = \sup\{\langle v^*,v\rangle_V + \langle p^*,p\rangle_Y - \Phi(v,p) \mid v \in V,\ p \in Y\}.$$

Hence,

$$\text{(3.15)} \qquad \Phi^*(0,p^*) = \sup\{\langle p^*,p\rangle_Y - \Phi(v,p) \mid v \in V,\ p \in Y\} =$$

$$= \sup_{v\in V}\sup_{p\in Y}\{\langle p^*,p\rangle_Y - G_1(\Lambda_1 v - p_1) - G_2(\Lambda_2 v - p_2) - F(v)\} =$$

$$= \sup_{v\in V}\sup_{q\in Y}\{\langle p^*,\Lambda v - q\rangle_Y - G_1(q_1) - G_2(q_2) - F(v)\} =$$

$$= \sup_{v \in V} \sup_{q \in Y} \{ \langle -p_1^*, q_1 \rangle_{L_1} - G_1(q_1) + \langle -p_2^*, q_2 \rangle_{L_2} - G_2(q_2) + \langle \Lambda^* p^*, v \rangle_V - F(v) \}$$

$$= G_1^*(-p_1^*) + G_2^*(-p_2^*) + F^*(\Lambda^* p^*),$$

where G_1^*, G_2^* and F^* denote the polar functions of G_1, G_2 and F, respectively, and $\Lambda^*: Y^* \to V^*$ denotes the adjoint of Λ. We shall determine the functionals G_1^*, G_2^* and F^* in order to arrive at an explicit expression for P^*.

Let us first consider G_1^*. We know that (cf. (3.10)) $G_1(w) = \int k(w(\cdot))$ and since k is a normal positive integrand we can interchange integration and taking the polar (cf. [3, Prop. 2.1, p. 251]):

$$G_1^*(p_1^*) = \int k^*(p_1^*(\cdot)),$$

where

$$k^*(z) = \sup\{yz - k(y) \mid y \geq 0\} = \varepsilon \exp(\frac{z}{\varepsilon} - 1) - \frac{\varepsilon}{e}.$$

In the same manner we find

$$G_2^*(p_2^*) = \int \frac{1}{2}(p_2^*)^2 + gp_2^*.$$

Next we calculate F^*:

$$F^*(\Lambda^* p^*) = \sup\{\langle p^*, \Lambda v \rangle_Y \mid v \in V,\ v(0) = 0,\ v(1) = 1\}$$

$$= \int (ip_2^* + p_1^*) + \sup\{ \int (p_1^* u' + p_2^* u) \mid u \in V,\ u(0) = u(1) = 0\}.$$

Here we made the transformation $v = u + i$, where i denotes the function $i(x) = x$, in order to arrive at homogeneous boundary conditions. Since $\mathcal{D}$ is dense in the set $\{u \in V \mid u(0) = u(1) = 0\}$ we conclude that

$$F^*(\Lambda^* p^*) = \begin{cases} \int (ip_2^* + p_1^*) & \text{if } p_2^* = (p_1^*)' \text{ in the sense of distributions,} \\ +\infty & \text{otherwise.} \end{cases}$$

Collecting all results we arrive at the following explicit formulation:

$$(3.16)\quad P^* \quad \operatorname{Sup}\{ \int (\frac{\varepsilon}{e} - \varepsilon e^{-\frac{p_1^*}{\varepsilon} - 1} - p_1^* + (g - i)p_2^* - \frac{1}{2}(p_2^*)^2) \mid p^* \in L_\infty \times L_2,\ p_2^* = (p_1^*)'\}$$

From known properties of P one can deduce that P^* has a solution. Indeed, since

(i) Φ is convex and $\inf P$ is finite,

(ii) the function $p \mapsto \Phi(i,p)$ is finite and continuous at the point $p = 0$, we are in a position to conclude from [3, Prop. 2.3, p. 51] that P^* has a solution and that $\inf P = \sup P^*$. Finally, we deduce from the strict convexity of the functional in (3.16) that the solution of P^* is unique.

3.3. THE EXTREMALITY RELATIONS

In virtue of [3, Prop. 2.4, p. 52] the following claims are equivalent:

(i) v is a solution of P, p^* is a solution of P^*

(ii) $v \in V$ and $p^* \in Y^*$ satisfy the extremality relation

$$\Phi(v,0) + \Phi^*(0,p^*) = 0. \tag{3.17}$$

In the present case (3.17) can be decoupled as follows:

$$\begin{aligned} 0 &= \Phi(v,0) + \Phi^*(0,p^*) \\ &= G_1(\Lambda_1 v) + G_1^*(-p_1^*) + G_2(\Lambda_2 v) + G_2^*(-p_2^*) + F(v) + F^*(\Lambda^* p^*) \\ &= \{G_1(\Lambda_1 v) + G_1^*(-p_1^*) - \langle -p_1^*, \Lambda_1 v\rangle_{L_1}\} + \\ &+ \{G_2(\Lambda_2 v) + G_2^*(-p_2^*) - \langle -p_2^*, \Lambda_2 v\rangle_{L_2}\} + \\ &+ \{F(v) + F^*(\Lambda^* p^*) - \langle \Lambda^* p^*, v\rangle_V\}. \end{aligned}$$

Since each of these expressions in brackets is nonnegative, actually each of them must be zero. Thus we find

$$\int (\varepsilon v' \ln v' + \varepsilon e^{-\frac{p_1^*}{\varepsilon} - 1} + v' p_1^*) = 0, \tag{3.18}$$

$$\int (\tfrac{1}{2}(g - v)^2 + \tfrac{1}{2}(p_2^*)^2 - g p_2^* + v p_2^*) = \tfrac{1}{2}\int (p_2^* - g + v)^2 = 0, \tag{3.19}$$

$$v(0) = 0, \quad v(1) = 1, \quad p_2^* = (p_1^*)'. \tag{3.20}$$

In order to draw further conclusions from (3.18), consider the function f defined by

$$f(x) = \varepsilon\lambda \ln \lambda + \varepsilon e^{-\frac{x}{\varepsilon} - 1} + \lambda x,$$

for fixed $\lambda \geq 0$. If $\lambda = 0$, then $f > 0$. If $\lambda > 0$, then the convex function f is nonnegative and it attains its minimum, zero, at the point $x = -\varepsilon(1 + \ln \lambda)$. Consequently

(3.18) implies that $v' > 0$ and that

$$p_1^* = -\varepsilon(1 + \ln v'). \tag{3.21}$$

Likewise (3.19) implies that

$$p_2^* = g - v. \tag{3.22}$$

Finally, combination of (3.20) - (3.22) leads to

$$\begin{cases} -\varepsilon(\ln v')' + v = g. \\ v(0) = 0, \quad v(1) = 1. \end{cases} \tag{3.23}$$

So if v is the solution of P then v satisfies (3.23). From the fact that $g \in L_2$ we deduce that $\ln v' \in H^1$ and consequently that $v \in H^2$. Hence v satisfies BVP.

Conversely, let v be the solution of BVP. Define p_1^* and p_2^* by (3.21) and (3.22), respectively. Then v and $p^* = (p_1^*, p_2^*)$ satisfy the extremality relation (3.17) and consequently v solves P while p^* solves P^*.

3.4. LIMITING BEHAVIOUR AS $\varepsilon \downarrow 0$

Formally we can associate with P^* the following limiting problem

$$P_0^* \qquad -\mathrm{Inf}\{\int (q + (i - g)q' + \tfrac{1}{2}(q')^2) \mid q \in C\}, \tag{3.24}$$

where by definition

$$C = \{q \in H^1 \mid q \geq 0\} \tag{3.25}$$

(note that the condition $p_2^* = (p_1^*)'$ motivates the choice of the underlying space and that we choose $q \geq 0$ because otherwise $\varepsilon \exp(-\frac{q}{\varepsilon})$ tends to $-\infty$ as $\varepsilon \downarrow 0$).

P_0^* consists of minimizing a strictly convex, continuous and coercive functional on a closed convex subset of the reflexive space H^1. Hence there exists a unique solution of P_0^*, which we shall call q_0.

Defining functionals G_3 and G_4 on L_2 and H^1 as follows:

$$G_3(w) = \int \tfrac{1}{2} w^2 + (i - g)w \tag{3.26}$$

$$G_4(w) = \begin{cases} \int w & \text{if } w \in C, \\ +\infty & \text{otherwise,} \end{cases} \tag{3.27}$$

we rewrite

P_0^* $\qquad -\text{Inf}\{G_3(\Lambda_1 q) + G_4(q) \mid q \in H^1\}$

where now Λ_1, defined by $\Lambda_1 q = q'$, is considered as a bounded linear mapping of H^1 into L_2.

Next we construct the dual problem P_0^{**} of P_0^* relative to the perturbed functional

$$G_3(\Lambda_1 q - r) + G_4(q), \qquad r \in L_2.$$

We find

$$(3.28) \qquad P_0^{**} \qquad \text{Inf}\{G_3^*(-v) + G_4^*(\Lambda_1^* v) \mid v \in L_2\},$$

where

$$(3.29) \qquad G_3^*(v) = \frac{1}{2}\int (v + g - i)^2$$

and

$$(3.30) \qquad G_4^*(v) = \sup\{\langle v,q\rangle_{H^1} - \int q \mid q \in C\}$$

$$= \sup\{\langle v-1,q\rangle_{H^1} \mid q \in C\}$$

$$= \begin{cases} 0 & \text{if } (1-v) \in C^*, \\ +\infty & \text{otherwise,} \end{cases}$$

where we have put $\int p = \langle 1,p\rangle_{H^1}$ (according to the natural identification of L_2 with a subspace of $(H^1)^*$) and where

$$(3.31) \qquad C^* = \{v \in (H^1)^* \mid \langle v,q\rangle_{H^1} \geq 0, \ \forall q \in C\}.$$

Performing the change of function $u = v + i$ we get

P_0^{**} $\qquad \text{Inf}\{\frac{1}{2}\int (g-u)^2 \mid u \in \Omega\}$,

where by definition

$$(3.32) \qquad \Omega = \{u \in L_2 \mid 1 - \Lambda_1^*(u-i) \in C^*\}.$$

<u>PROPOSITION 3.3.</u> $\Omega = \overline{\mathcal{D}(A)}$

<u>PROOF</u>. By definition

$$\mathcal{Q} = \{u \in L_2 \mid \int (q + (i-u)q') \geq 0, \ \forall q \in \mathcal{C}\}.$$

Let $u \in \mathcal{Q}$ and $q \in \mathcal{C} \cap H_0^1$ then, using $\int(q + iq') = 0$, we obtain $\int uq' \leq 0$, which shows that $u' \geq 0$ in the sense of distributions. Next, suppose that u assumes values larger than one on a set of positive measure. Since u is non-decreasing this set can be taken to be an interval with 1 as its right endpoint and, say, λ as its left endpoint. Take $q(x) = 0$ for $0 \leq x \leq \lambda$ and q strictly increasing for $x > \lambda$, then we arrive at the contradiction

$$\int (q + (i-u)q') < \int_\lambda^1 (q(x) + (x-1)q'(x))dx = 0$$

It follows that $u \leq 1$. Likewise one can show that $u \geq 0$, so we conclude that $u \in \overline{\mathcal{D}(A)}$ indeed.

Conversely, suppose $u \in \mathcal{D}(A)$ and $q \in \mathcal{C}$ then

$$\int (q + (i-u)q') = u(0)q(0) - (u(1)-1)q(1) + \int u'q \geq 0$$

and consequently $u \in \mathcal{Q}$. Finally, if $u \in \overline{\mathcal{D}(A)}$ we arrive at the same conclusion by using an approximating sequence in $\mathcal{D}(A)$ and by noting that $\int(q + (i-u)q')$ depends continuously on u. □

Thus we showed that P_0^{**} is precisely the reduced problem considered in Subsection 2.2. We recall that it has precisely one solution y_0. Expressing the extremality relation

$$G_3(\Lambda_1 q_0) + G_3^*(i - y_0) + \langle y_0 - i, \Lambda_1 q_0\rangle_{L_2} = 0$$

as

$$\int (q_0' + y_0 - g)^2 = 0$$

we obtain that

$$q_0' = g - y_0. \tag{3.33}$$

The other extremality relation

$$G_4(q_0) + G_4^*(\Lambda_1^*(y_0 - i)) - \langle \Lambda_1^*(y_0 - i), q_0\rangle_{H^1} = 0$$

yields the relation

$$q_0(1) = \int y_0 q_0'. \tag{3.34}$$

THEOREM 3.4. *For each* $\varepsilon > 0$ *let* y_ε *denote the solution of* P *and* $p^*_\varepsilon = (p^*_{\varepsilon 1}, p^*_{\varepsilon 2})$ *the solution of* P^*. *Moreover, let* y_0 *denote the solution of* P_0^{**} *and* q_0 *the solution of* P_0^*. *Then*

(i) $\lim_{\varepsilon \downarrow 0} \| p^*_{\varepsilon 1} - q_0 \|_{H^1} = 0$

(ii) $\lim_{\varepsilon \downarrow 0} \| y_\varepsilon - y_0 \|_{L_2} = 0$

PROOF. First we want to show that $p^*_{\varepsilon 1}$ is bounded in H^1 uniformly in ε. Since $0 \le y_\varepsilon \le 1$, y_ε and $(p^*_{\varepsilon 1})' = p^*_{\varepsilon 2} = g - y_\varepsilon$ are bounded in L_2 uniformly in ε. The definition of J implies

$$0 \le \operatorname{Inf} P \le \frac{1}{2} \int (g-i)^2 + \frac{\varepsilon}{e}.$$

Using $\operatorname{Sup} P^* = \operatorname{Inf} P$ we obtain

$$0 \le \int \left(\frac{\varepsilon}{e} - \varepsilon e^{-\frac{p^*_{\varepsilon 1}}{\varepsilon} - 1} - p^*_{\varepsilon 1} + (g-i)p^*_{\varepsilon 2} - \frac{1}{2}(p^*_{\varepsilon 2})^2\right) \le \frac{1}{2} \int (g-i)^2 + \frac{\varepsilon}{e}.$$

From $y'_\varepsilon = \exp(-\frac{1}{\varepsilon} p^*_{\varepsilon 1} - 1)$ and $\int y'_\varepsilon = y_\varepsilon(1) - y_\varepsilon(0) = 1$ we deduce that

$$\lim_{\varepsilon \downarrow 0} \varepsilon \int e^{-\frac{1}{\varepsilon} p^*_{\varepsilon 1} - 1} = 0 \tag{3.35}$$

Combination of these results yields a uniform bound for $|\int p^*_{\varepsilon 1}|$. Hence there exists $\theta = \theta(\varepsilon) \in [0,1]$ such that $p^*_{\varepsilon 1}(\theta)$ is uniformly bounded and, finally, we obtain

$$\int (p^*_{\varepsilon 1})^2 = \int_0^1 (p^*_{\varepsilon 1}(\theta) + \int_\theta^x (p^*_{\varepsilon 1})'(\xi)d\xi)dx \le C.$$

So there exists a sequence $\{\varepsilon_n\}$ and a function $w \in H^1$ such that $p^*_{\varepsilon_n 1}$ converges to w weakly in H^1 and strongly in L_2. Next we intend to show that $w = q_0$.

From the fact that p^*_ε solves P^* we deduce that

$$\int \left(\frac{\varepsilon}{e} - \varepsilon e^{-\frac{1}{\varepsilon} p^*_{\varepsilon 1} - 1} - p^*_{\varepsilon 1} + (g-i)(p^*_{\varepsilon 1})' - \frac{1}{2}((p^*_{\varepsilon 1})')^2\right)$$

$$\ge \int \left(\frac{\varepsilon}{e} - \varepsilon e^{-\frac{1}{\varepsilon} q_0 - 1} - q_0 + (g-i)q_0' - \frac{1}{2}(q_0')^2\right).$$

Furthermore, the functional $q \mapsto \int (q + (i-g)q' + \frac{1}{2}(q')^2)$ is convex and continuous on H^1 and thus w.l.s.c. Hence

$$(*) \qquad \int (w + (i-g)w' + \tfrac{1}{2}(w')^2)$$

$$\le \liminf_{n\to\infty} \int (p^*_{\varepsilon_n 1} + (i-g)(p^*_{\varepsilon_n 1})' + \tfrac{1}{2}((p^*_{\varepsilon_n 1})')^2)$$

$$\le \limsup_{n\to\infty} \int (p^*_{\varepsilon_n 1} + (i-g)(p^*_{\varepsilon_n 1})' + \tfrac{1}{2}((p^*_{\varepsilon_n 1})')^2)$$

$$\le \int (q_0 + (i-g)q_0' + \tfrac{1}{2}(q_0')^2).$$

We observe that $w \ge 0$ (else (3.35) could not be true). Since q_0 is the unique solution of P_0^*, necessarily $w = q_0$. Inserting this into (*) we obtain that in fact $p^*_{\varepsilon_n 1}$ converges to q_0 strongly in H^1. Moreover, since the limit does not depend on the sequence under consideration (i) follows. Finally, we arrive at (ii) by noting that $y_\varepsilon = g - (p^*_{\varepsilon 1})'$ and $y_0 = g - q_0'$. □

4. CALCULATION OF y_0

We recall that y_0 is the unique solution of the variational problem $\mathrm{Min}_{\overline{\mathcal{D}(A)}} W_0$, where $W_0(u) = \|u-g\|^2$. It is well known (for instance, see Ekeland-Témam [3, II, 2.1]) that one can equivalently characterize y_0 as the unique solution of the *variational inequality*:

$$(4.1) \qquad \text{find } y \in \overline{\mathcal{D}(A)} \text{ such that } (y-g, v-y) \ge 0,\ \forall v \in \overline{\mathcal{D}(A)}.$$

Already from the reduced differential equation $(g-y)y' = 0$, it can be guessed that y_0 is possibly composed out of pieces where it equals g and pieces where it equals a constant. Of course, if $y_0 = g$ in some open interval, g has to be nondecreasing in that interval. The characterization of y_0 by (4.1) can be used to find conditions on the "allowed" constants.

THEOREM 4.1. *Suppose* $y \in \overline{\mathcal{D}(A)}$ *has the following property: there exists a partition* $0 = x_0 < x_1 < \dots < x_{n-1} < x_n = 1$ *of* $[0,1]$ *and a subset* L *of* $\{0,1,\dots,n-1\}$ *such that:*

(i) if $i \notin L$ then $y(x) = g(x)$ for $x \in [x_i, x_{i+1}]$,

(ii) If $i \in L$ then $y(x) = C_i$ for $x \in [x_i, x_{i+1}]$ and

$$\int_x^{x_{i+1}} (C_i - g(\xi))d\xi \ge 0, \qquad \forall x \in [x_i, x_{i+1}], \text{ if } C_i \in [0,1),$$

$$\int_{x_i}^{x} (C_i - g(\xi))d\xi \le 0, \qquad \forall x \in [x_i, x_{i+1}], \text{ if } C_i \in (0,1],$$

(so in particular, if $C_i \in (0,1)$, $\int_{x_i}^{x_{i+1}} (C_i - g(\xi))d\xi = 0$).

Then $y = y_0$.

PROOF. According to (4.1) it is sufficient to check that

$$I(v) = \int (y-g)(v-y) \geq 0, \qquad \forall v \in \overline{\mathcal{D}(A)} .$$

In fact it is sufficient to check this for all $v \in \overline{\mathcal{D}(A)} \cap H^1$ (since this set is dense in $\overline{\mathcal{D}(A)}$ and I is continuous). We note that $I(v) = \Sigma_{i\in L} I_i(v)$, where

$$I_i(v) = \int_{x_i}^{x_{i+1}} (C_i - g(\xi))(v(\xi) - C_i)d\xi.$$

If $C_i = 0$ then

$$I_i(v) = -v(x_i) \int_{x_i}^{x_{i+1}} g(\xi)d\xi - \int_{x_i}^{x_{i+1}} v'(\xi) \int_{\xi}^{x_{i+1}} g(x)dxd\xi \geq 0.$$

If $C_i \in (0,1)$ then

$$I_i(v) = \int_{x_i}^{x_{i+1}} v'(\xi) \int_{\xi}^{x_{i+1}} (C_i - g(x))dxd\xi \geq 0.$$

If $C_i = 1$ then

$$I_i(v) = (v(x_{i+1}) - 1) \int_{x_i}^{x_{i+1}} (C_i - g(\xi))d\xi - \int_{x_i}^{x_{i+1}} v'(\xi) \int_{x_i}^{\xi} (C_i - g(x))dxd\xi \geq 0$$

Hence indeed $I(v) \geq 0$, $\forall v \in \overline{\mathcal{D}(A)} \cap H^1$. □

The sufficient conditions of the theorem can be used as a kind of algorithm to compute y_0 in concrete cases. We shall illustrate this idea by means of a number of examples (some of which are almost literally taken from [2]).

EXAMPLE 1. Suppose g is nondecreasing, then

$$y_0(x) = \begin{cases} 0 & \text{if } g(x) \leq 0, \\ g(x) & \text{if } 0 \leq g(x) \leq 1, \\ 1 & \text{if } g(x) \geq 1. \end{cases}$$

EXAMPLE 2. Suppose g is nonincreasing, then $y_0(x) = C$ with

$$C = \begin{cases} 0 & \text{if } \int g \le 0, \\ \int g & \text{if } 0 \le \int g \le 1, \\ 1 & \text{if } \int g \ge 1. \end{cases}$$

EXAMPLE 3. Suppose that $g \in C^1$ is such that g' vanishes at only two points b and c, b being a local maximum and c a local minimum. Assume that $0 < b < c < 1$ and $0 < g(c) < g(b) < 1$. Let g_1^{-1} denote the inverse of g on [0,b] and g_2^{-1} the inverse of g on [c,1]. Define two points a and d by

$$a = g_1^{-1}(g(c)), \qquad d = g_2^{-1}(g(b)).$$

Then $g([a,b]) = g([c,d])$. (See Figure 1).

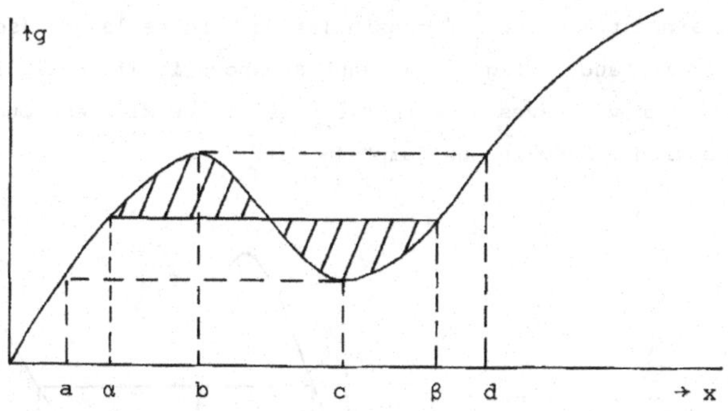

Figure 1

On [a,b] we define a mapping G by

$$G(x) = \int_x^{g_2^{-1}(g(x))} (g(x) - g(\xi))d\xi.$$

Then $G(a) < 0$, $G(b) > 0$ and on (a,b)

$$G'(x) = g'(x) \int_x^{g_2^{-1}(g(x))} d\xi > 0.$$

Consequently G has a unique zero on [a,b], say for $x = \alpha$. The function y_0 has the tendency to follow g as much as possible. However, it also has to be nondecreasing. So the inverse function of y_0 must "jump" from a point on [a,b] to a point on [c,d]. In view of Theorem 4.1 this jump can only take place between α and $\beta = g_2^{-1}(\alpha)$. We leave it to the reader to verify (by checking all requirements of Theorem 4.1) that

$$y_0(x) = \begin{cases} 0 & \text{if } x \le \alpha \text{ and } g(x) \le 0, \\ g(x) & \text{if } x \le \alpha \text{ and } g(x) \ge 0, \\ g(\alpha) & \text{if } \alpha \le x \le \beta, \\ g(x) & \text{if } x \ge \beta \text{ and } g(x) \le 1, \\ 1 & \text{if } x \ge \beta \text{ and } g(x) \ge 1. \end{cases}$$

It should be clear that the differentiability of g is not strictly necessary for our arguments to apply. In fact the monotonicity of G follows from straightforward geometrical considerations and the condition $G(\alpha) = 0$ has a corresponding interpretation (see Figure 1).

EXAMPLE 4. If g has more maxima and minima the construction of candidates for y_0 can be based on essentially the same idea as outlined in Example 3. However, it becomes more complicated since the number of possibilities becomes larger (see [2] for some more details). For instance, if g has a graph as shown in Figure 2, looking at zero's of functions like G above leaves us with two possible candidates: onc with two "jumps" (a-b,c-d) and one with a "two-in-one jump" ($\alpha - \beta$).

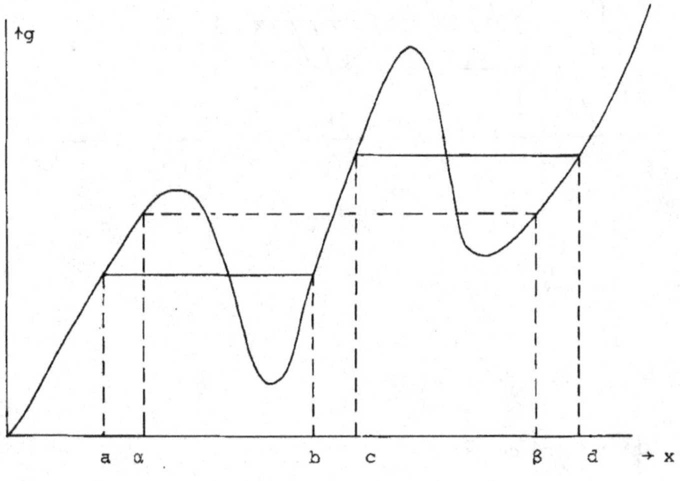

Figure 2

In [2] we were unable to decide in such a situation which was the actual limit. But now it can be read off from the picture that only the one with two "jumps" satisfies the requirements of Theorem 4.1, and hence this one must actually be y_0. (The other one corresponds to a saddle point of the functional W_0 restricted to $\overline{\mathcal{D}(A)}$.) It is in this sense that y_0 must have as many "jumps" as possible.

5. CONCLUDING REMARKS

(i) In all our examples y_0 satisfies the reduced equation $(g - y)y' = 0$. However, this equation is by no means sufficient to characterize y_0 completely. Our analysis clearly shows that the reduced variational problem $\mathrm{Min}_{\overline{\mathcal{D}(A)}} W_0$ contains much more information than the reduced differential equation.

(ii) In [2] we were actually interested in a boundary value problem of the type

(5.1) $$\varepsilon x y'' + (g-y)y' = 0, \qquad 0 < x < 1,$$

(5.2) $$y(0) = 0, \qquad y(1) = 1,$$

which arises from the assumption of radial symmetry in a two-dimensional geometry. This problem can be analysed in completely the same way as we did with BVP in this paper, by choosing as the underlying Hilbert space the weighted L_2-space corresponding to the measure $d\mu(x) = x^{-1}dx$. For instance, the operator $\tilde{A}$ defined by

$$(\tilde{A}u)(x) = -x\,\frac{u''(x)}{u'(x)}$$

with

$$\mathcal{D}(\tilde{A}) = \{u \in L^2(d\mu) \mid u' \in C(0,1],\ u' > 0,\ u(1) = 1,\ i\,\frac{u''}{u'} \in L^2(d\mu)\}$$

is clearly monotone in this space. The surjectivity of $I + \varepsilon\tilde{A}$ can be proved with the aid of an auxiliary problem and Schauder's fixed point theorem. (Note that some care is needed in checking that the functions which occur belong to the right space and that the solution operator is compact. This turns out to be all right. We refer to Martini's thesis [5] where related problems are treated in full detail.) Hence $\tilde{A}$ is maximal monotone. Subsequently it follows that, for given $g \in L_2(d\mu)$, the solution y_ε tends, as $\varepsilon \downarrow 0$, to a limit y_0 in $L_2(d\mu)$ and that y_0 is the projection in $L_2(d\mu)$ of g onto the closed convex set

$$\overline{\mathcal{D}(\tilde{A})} = \{u \in L_2(d\mu) \mid u \text{ is nondecreasing},\ 0 \le u \le 1\}.$$

The second method carries over to this situation as well.

(iii) In [2] we were also interested in the situation where the differential equation (5.1), assumed to hold for $0 < x < \infty$, is supplemented by the condition

(5.3) $$\lim_{x\to\infty} y(x) = 1.$$

Intuitively one believes that similar results should be true in this situation. However, the present approach does not carry over directly and, in fact, the noncompactness of the domain presents serious mathematical difficulties.

REFERENCES

[1] BRÉZIS, H., *Opérateurs Maximaux Monotones et Semi-groupes de Contractions dans les Espaces de Hilbert*, Math. Studies, 5, North-Holland, 1973.

[2] DIEKMANN, O., D. HILHORST & L.A. PELETIER, *A singular boundary value problem arising in a pre-breakdown gas discharge*, SIAM J. Appl. Math., in press.

[3] EKELAND, I. & R. TÉMAM, *Analyse Convexe et Problèmes Variationnels*, Dunod, Paris, 1974.

[4] GRASMAN, J. & B.J. MATKOWSKY, *A variational approach to singularly perturbed boundary value problems for ordinary and partial differential equations with turning points*, SIAM J. Appl. Math. 32, 588-597 (1977).

[5] MARTINI, R., *Differential operators degenerating at the boundary as infinitesimal generators of semi-groups*, Ph.D. thesis, Delft Technological Univ., Delft, The Netherlands, 1975.

THE CONTINUOUS NEWTON-METHOD FOR MEROMORPHIC FUNCTIONS

H.Th. Jongen, P. Jonker, F. Twilt
Twente University of Technology
Department of Applied Mathematics
P.O. Box 217, 7500 AE Enschede,
THE NETHERLANDS

1. Introduction:

Let $\mathbb{C}$ be the complex plane and f be a complex valued function defined on $\mathbb{C}$. The function f is called entire if it is analytic everywhere on $\mathbb{C}$; f is called meromorphic if all its singularities (i.e. points of $\mathbb{C}$ where f fails to be analytic) are poles.

In this paper f is always a non-constant meromorphic function. By N(f)(P(f)) we denote the set of all zeros (poles) of f. If f' stands for the usual derivative of f we call $C(f) := N(f') \setminus N(f)$ the critical set for f; its elements are called critical points for f. Note that N(f), P(f) and C(f) are discrete subsets of $\mathbb{C}$. If $f(z_0) = f'(z_0) = 0$ or if $z_0 \in P(f)$ the singularity of $\frac{f(z)}{f'(z)}$ for z_0 can be removed by defining $\frac{f(z_0)}{f'(z_0)} := 0$. Therefore, in the sequel, we shall regard $\frac{f(z)}{f'(z)}$ as an analytic function on $\mathbb{C} \setminus C(f)$. In general: A meromorphic function which has a removable singularity for z_0 will be interpreted as to be analytic at z_0. By S^2 we denote the usual one point compactification of $\mathbb{C}$ (Riemannian sphere), viewed as a differentiable manifold. In the sequel autonomous differential equations are also called flows, vectorfields or (dynamical) systems.

In this paper we study local and global properties of the phase-portrait of the dynamical system $\mathcal{N}(f)$ described by:

$$\frac{dz}{dt} = -\frac{f(z(t))}{f'(z(t))}, \quad z \in \mathbb{C} \setminus C(f). \qquad (\mathcal{N}(f)) \tag{1.1}$$

Since $\mathcal{N}(f)$ is the infinitesimal version of the well-known Newton iteration formula for finding the zeros for f we refer to $\mathcal{N}(f)$ as the continuous "Newton-method", its flows are called Newton-flows (w.r.t. f).

The paper is organized as follows:

In Section 2 we extend the system $\mathcal{N}(f)$ to a real analytical system $\bar{\mathcal{N}}(f)$ (in the sense that the components of the vectorfield $\bar{\mathcal{N}}(f)$ depend analytically on x and y; z = x + iy) defined on the whole $\mathbb{C}$. In the case f is a rational function, or an entire function of finite order with finitely many zeros we extend $\mathcal{N}(f)$ to a real analytical system $\bar{\bar{\mathcal{N}}}(f)$ defined on S^2. Applying the theory of real analytical dynamical systems defined on $\mathbb{R}^2(S^2)$ to $\bar{\mathcal{N}}(f)(\bar{\bar{\mathcal{N}}}(f))$ we give a quite complete description of the phase-portrait of $\mathcal{N}(f)$.

It is easily to be seen that $\mathcal{N}(f)$ is the differential equation for the streamlines of a steady stream with complex potential - log f(z), (cf.[17]). So we may expect in the case that, extended to S^2, the stream has only finitely many "sources" and "sinks" (this corresponds to: f is a rational function), in general the phase-portrait of $\bar{\bar{\mathcal{N}}}(f)$ behaves extremely regular (w.r.t. small perturbations of the coëfficients of f). In fact this is the gist of the main result in Section 3 which states that "generically" the systems $\bar{\bar{\mathcal{N}}}(f)$ - f rational - are structurally stable.

In section 4 we give some applications:

1. The way in which we extend $\mathcal{N}(f)$ to $\bar{\mathcal{N}}(f)$ is similar to the method proposed by F.H. Branin (cf.[6]) in order to "desingularize" the Newton differential equation for functions F: $\mathbb{R}^2 \to \mathbb{R}^2$. This leads to a counterexample of a conjecture due to Branin on the global convergence of the "Branin-method" for finding the zeros for F (cf. [6], [7], [8]).

2. Using the concept of structural stability we prove a conjecture due to D. Braess (cf. [5]) on the phase-portrait of $\mathcal{N}(f)$ for the case that f is a polynomial of degree three.

3. For a certain class of rational functions (the so called non-degenerate rational functions) the systems $\bar{\bar{\mathcal{N}}}(f)$ are examples of Morse-Smale systems. Using the results of Peixoto (cf. [21]) on the latter systems, we estimate the number of different phase-portraits of the system $\mathcal{N}(f)$ (up to their topological type), if f is a non-degenerate polynomial.

4. There is a strong relationship between the theory on (complex) Newton-flows and the theory on "the distribution of functional values" (cf.[19]). This will be made clear by giving some results on the asymptotic behaviour of entire functions of finite order with a Picard-exceptional value, which are easy consequences of the theory developed in Section 2.

Finally, in Section 4, we illustrate the theory by some numerical examples.

2. Newton-flows for meromorphic functions:

By $\gamma(z_0)$ we denote the maximal trajectory of $\mathcal{N}(f)$ through $z_0 (\notin C(f))$. If z_0 is not an equilibrium state for $\mathcal{N}(f)$ (i.e. at z_0 the r.h.s. of (1.1) does not vanish) we may describe $\gamma(z_0)$ by the solution:

$$z(t) \ , \ t \in \ ]\, a,b\, [\ , \ z_0 = z(0); \text{ eventually } a = -\infty, \quad b = +\infty.$$

Note that the set of equilibrium states for $\mathcal{N}(f)$ is $N(f) \cup P(f)$, while $\mathcal{N}(f)$ is not defined on $C(f)$. If $z_0 \in N(f) \cup P(f)$ then $\gamma(z_0) = \{z_0\}$. If $z_0 \notin N(f) \cup P(f) \cup C(f)$ we find by direct integration:

$$(2.1) \qquad f(z(t)) = e^{-t} f(z_0); \quad t \in \ ]\, a,b\, [.$$

Thus on $\gamma(z_0)$ we have $\arg f(z(t)) = \text{constant}$ ($= \arg f(z(0))$). (Compare also [22]). In view of (2.1) in a neighborhood of $z_0 \in \mathbb{C}$ the trajectories of $\mathcal{N}(f)$ are the inverse images under f of the lines $\arg f(z) = \text{constant}$. So from the theory of functions of a complex variable one immediately concludes that the local phase-portraits around z_0 are of the following four types (cf. [17]):

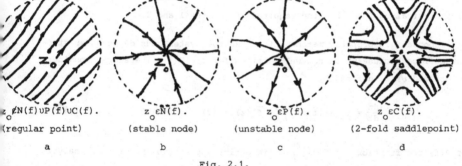

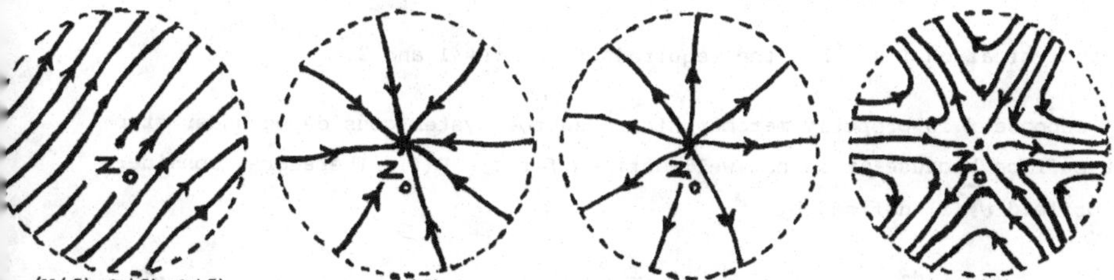

$z_0 \notin N(f) \cup P(f) \cup C(f)$.	$z_0 \in N(f)$.	$z_0 \in P(f)$.	$z_0 \in C(f)$.
(regular point)	(stable node)	(unstable node)	(2-fold saddlepoint)
a	b	c	d

Fig. 2.1.

In Fig.2.1d, let k = multiplicity of z_0 as a zero for $f'(z)$. Then the angle between two subsequent trajectories, one of which tends to z_0, the other leaves from z_0, equals $\frac{\pi}{k+1}$. We call z_0 in this case a critical point of order k (or k-fold saddlepoint) for f. This situation is well-known from the theory on steady 2 dim. streams (cf. [17]).

If we turn over from local to global properties of the phase-portrait of $\mathcal{N}(f)$ we encounter the problem that $\mathcal{N}(f)$ is not defined on the whole $\mathbb{C}$, so we cannot exploit the theory of 2 dim. dynamical systems at full strength. By means of the following lemma we overcome this problem.

Definition 2.1: $\mathcal{E}$ is the set of all transcendental entire functions of finite order with finitely many zeros. $\mathcal{R}$ is the set of all rational functions $f = \frac{p}{q}$ with p and q polynomials which are relatively prime.

Lemma 2.1: (Extension lemma)

- For each (meromorphic) function f there exists a real analytical system $\overline{\mathcal{N}}(f)$ defined on the whole $\mathbb{C}$ with the properties:
  1. Trajectories of $\mathcal{N}(f)$ are also trajectories of $\overline{\mathcal{N}}(f)$.
  2. A critical point for f is an equilibrium state for $\overline{\mathcal{N}}(f)$.
- For each $f \in \mathcal{E} \cup \mathcal{R}$ the system $\overline{\mathcal{N}}(f)$ can be extended to a real analytical system $\overline{\overline{\mathcal{N}}}(f)$ defined on S^2.

Proof: The proof will be given in four steps:

Step 1: Suppose f is an entire function. For each $z \notin C(f)$ the vector $-\overline{f}'(z).\ f(z)$ is a real positive multiple of $-\frac{f(z)}{f'(z)}$. Moreover, for $z_0 \in C(f)$: $-\overline{f}'(z_0).f(z_0) = 0$. The real and imaginary parts of the analytic function f depend analytically on x and y ($z = x+iy$), so the real and imaginary parts of $\overline{f}'(z).f(z)$ do. Consequently, the system:

$$\frac{dz}{dt} = -\overline{f}'(z(t)).f(z(t)) \qquad (\mathcal{N}^*(f))$$

is real analytical on $\mathbb{C}$ and has the required properties 1 and 2.

Step 2: Suppose f is properly meromorphic, then the system considered under step 1 does not suffice because it is not well-defined for $z \in P(f)$. Therefore, consider the vectorfield $\overline{\mathcal{N}}(f)$ defined by:

$$\frac{dz}{dt} = -\left(1+|f(z)|^4\right)^{-1}.\ \overline{f}'(z).\ f(z). \qquad (\overline{\mathcal{N}}(f))$$

For $z \notin P(f)$ the function $\left(1+|f(z)|^4\right)^{-1}$ is strictly positive and depends analytically on x and y. We find, using the results of step 1, that we are done if we prove that for $z \in P(f)$ the vectorfield $\overline{\mathcal{N}}(f)$ vanishes and depends analytically on x and y. This follows from the useful property $\overline{\mathcal{N}}(f) = -\overline{\mathcal{N}}(\frac{1}{f})$ which can be verified by inspection.

Step 3: Suppose $f \in \mathcal{R}$; $f = \frac{p_n}{q_m}$, $p_n(q_m)$ a polynomial of degree $n(m)$. Denote by $a_k(b_1)$ the coëfficients of $z^k(z^1)$ in $p_n(z)(q_m(z))$; $0 \leq k \leq n(0 \leq 1 \leq m)$.
Firstly we treat the case $m > n$:
Consider the system $\overline{\overline{\mathcal{N}}}(f)$:

$$\frac{dz}{dt} = -(1+|z|^2)^{m-n+1}.(1+|f(z)|^4)^{-1}.\ \overline{f}'(z).\ f(z). \qquad (\overline{\overline{\mathcal{N}}}(f))$$

Since $\overline{\overline{\mathcal{N}}}(f) = (1+|z|^2)^{m-n+1} \cdot \overline{\mathcal{N}}(f)$ it follows that $\overline{\overline{\mathcal{N}}}(f)$ is real analytical on $\mathbb{C}$ and has the required properties 1 and 2. We extend $\overline{\overline{\mathcal{N}}}(f)$ to S^2 by using the transformation $z = \frac{1}{w}$. In the neighborhood of $z = \infty$ we find for the extension of $\overline{\overline{\mathcal{N}}}(f)$ w.r.t. the local coördinate w:

$$\frac{dw}{dt} = -(1+|w|^2)^{m-n+1} \cdot (1+|f(\tfrac{1}{w})|^4)^{-1} \cdot |w|^{-2m+2n+2} \cdot f(\tfrac{1}{w}) \cdot \overline{\frac{df(\frac{1}{w})}{dw}} . \qquad (\overline{\overline{\mathcal{N}}}_w(f))$$

If we denote the r.h.s. of this expression by $\Theta(w)$ it is easily seen that $\lim_{w\to 0} \frac{\Theta(w)}{w} = -(m-n) \cdot |\frac{a_n}{b_m}|^2$. Consequently $\overline{\overline{\mathcal{N}}}_w(f)$ has a non-degenerate equilibrium state at $w = 0$. The proof that $\overline{\overline{\mathcal{N}}}_w(f)$ is real analytical in $w = 0$ runs along the same lines as in step 1 and 2.

We now treat the case $m < n$:

Consider the system $\overline{\overline{\mathcal{N}}}(f)$:

$$\frac{dz}{dt} = -(1+|z|^2)^{-m+n+1} \cdot (1+|f(z)|^4)^{-1} \cdot \overline{f}'(z) \cdot f(z) . \qquad (\overline{\overline{\mathcal{N}}}(f))$$

Using the relation $\overline{\mathcal{N}}(f) = -\overline{\mathcal{N}}(\frac{1}{f})$ a direct calculation shows that $\overline{\overline{\mathcal{N}}}(\frac{1}{f}) \left(= \overline{\overline{\mathcal{N}}}(\frac{q_m}{p_n})\right) = -\overline{\overline{\mathcal{N}}}(f)$ which yields the desired result.

Special attention must be given to the case $m = n$.

We now define $\overline{\overline{\mathcal{N}}}(f)$ by:

$$\frac{dz}{dt} = -(1+|z|^2)^2 \cdot (1+|f(z)|^4)^{-1} \cdot \overline{f}'(z) \cdot f(z) . \qquad (\overline{\overline{\mathcal{N}}}(f))$$

Similarly to the case $m > n$ one proves that $\overline{\overline{\mathcal{N}}}(f)$ fulfils all the required conditions. In a neighborhood of $z = \infty$ one finds for $\overline{\overline{\mathcal{N}}}(f)$ w.r.t. the w-coördinate:

$$\frac{dw}{dt} = -(1+|w|^2)^2 \cdot (1+|f(\tfrac{1}{w})|^4)^{-1} \cdot \overline{\frac{d}{dw}(f(\tfrac{1}{w}))} \cdot f(\tfrac{1}{w}) \; (=\Theta(w)) . \qquad (\overline{\overline{\mathcal{N}}}_w(f))$$

By direct calculation we find that $\lim_{w\to 0} \Theta(w)$ exists (in $\mathbb{C}$) and that $\lim_{w\to 0} \Theta(w) \neq 0$ iff $a_n \cdot b_{n-1} \neq a_{n-1} \cdot b_n$. For the case that $a_n \cdot b_{n-1} = a_{n-1} \cdot b_n$ we have $\lim_{w\to 0} \frac{\Theta(w)}{\overline{w}} = \alpha$, with $\alpha \neq 0$ iff $a_n \cdot b_{n-2} \neq a_{n-2} \cdot b_n$ (Note that if $n = 1$ we always have: $\lim_{w\to 0} \Theta(w) \neq 0$). Consequently the system $\overline{\overline{\mathcal{N}}}_w(f)$ is regular at $w = 0$ iff $a_n \cdot b_{n-1} \neq a_{n-1} \cdot b_n$ and it has a non-degenerate equilibrium state at $w = 0$ iff $a_n \cdot b_{n-1} = a_{n-1} \cdot b_n$ and $a_n \cdot b_{n-2} \neq a_{n-2} \cdot b_n$. Using the same techniques as in the case $m > n$ one proves that $\bar{\bar{\mathcal{N}}}_w(f)$ is real analytical at $w = 0$.

Step 4: If $f \in \mathcal{E}$, then $f(z)$ has the form: $f(z) = p_n(z)e^{q_m(z)}$, $p_n(q_m)$ a polynomial of degree $n(m)$; $m \geq 1$, m = order (f), (cf. [17]). For $\mathcal{N}(f)$ we find:

$$\frac{dz}{dt} = -(p_n'(z) + q_m'(z)p_n(z))^{-1} \cdot p_n(z). \qquad (\mathcal{N}(f))$$

Playing the same game as we did in step 1 we find a new system $\mathcal{N}^*(f)$ given by:

$$\frac{dz}{dt} = -\overline{(p_n'(z)} + \overline{q_m'(z)} \cdot \overline{p_n(z))} \cdot p_n(z) \qquad (\mathcal{N}^*(f))$$

which obviously fulfils all the conditions required for $\bar{\mathcal{N}}(f)$.

Now consider the system $\bar{\bar{\mathcal{N}}}(f)$ defined by:

$$\frac{dz}{dt} = -(1+|z|^2)^{-m-n+1} \cdot \overline{(p_n'(z)} + \overline{q_m'(z)} \cdot \overline{p_n(z))} \cdot p_n(z). \qquad (\bar{\bar{\mathcal{N}}}(f))$$

Note that $\bar{\bar{\mathcal{N}}}(f) = (1+|z|^2)^{-m-n+1} \cdot \mathcal{N}^*(f)$ and that $(1+|z|^2)^{-m-n+1}$ depends analytically on x and y. So the system $\bar{\bar{\mathcal{N}}}(f)$ suffices as well. If we extend $\bar{\bar{\mathcal{N}}}(f)$ (by use of the transformation $w = \frac{1}{z}$) to S^2 we find in a neighborhood of $z = \infty$ - w.r.t. the w-coördinate - the system $\bar{\bar{\mathcal{N}}}_w(f)$ given by:

$$\frac{dw}{dt} = -(1+|w|^2)^{-m-n+1} \cdot |w|^{2m+2n+2} \cdot \left[\frac{\overline{dp_n(\frac{1}{w})}}{dw} + \frac{\overline{dq_m(\frac{1}{w})}}{dw} \cdot \overline{p}_n(\tfrac{1}{w})\right] \cdot p_n(\tfrac{1}{w}) \; (=\Theta(w))$$

A straight-forward calculation shows that $\lim_{w \to 0} w^{-m-1} \cdot \Theta(w) = \beta(\neq 0)$.

Consequently the system $\bar{\bar{\mathcal{N}}}_w(f)$ has at $w = 0$ an isolated degenerate equilibrium-state. Arguing as above we find that $\bar{\bar{\mathcal{N}}}_w(f)$ is real analytical at $w = 0$. □

Remark 2.1: In the sequel we always regard systems $\bar{\bar{\mathcal{N}}}(f)$ as dynamical systems on S^2 (strictly speaking we consider the pair $(\bar{\bar{\mathcal{N}}}(f), \bar{\bar{\mathcal{N}}}_w(f)$ and tacitly assume that it is defined for functions $f \in \mathcal{E} \cup \mathcal{R}$ only.

Remark 2.2: Since there exist meromorphic functions for which the zeros accumulate at $z = \infty$ (e.g. tan z) it is impossible to extend Newton-flows for arbitrary meromorphic functions to S^2 as global dynamical systems with only isolated equilibrium-states.

Remark 2.3: From Lemma 2.1 it follows: z is equilibrium-state for $\bar{\mathcal{N}}(f)$ iff z is either equilibrium-state for $\mathcal{N}(f)$ or $z \in C(f)$. This easily implies that the maximal trajectories (which are not reduced to a point) of both systems are the same.

Lemma 2.2: Let the maximal trajectory of $\mathcal{N}(f)$ through $z_0 \notin N(f) \cup P(f) \cup C(f)$ be given by $z(t)$, $t \in]a,b[$, $z_0 = z(0)$. Then:

1. Either $\lim_{t \uparrow b} z(t) = \infty$ or $\lim_{t \uparrow b} z(t) = l \in \mathbb{C}$. In the latter case: $l \in N(f)\,(l \in C(f))$ if $b = +\infty$ $(b < +\infty)$.

2. Either $\lim_{t \downarrow a} z(t) = \infty$ or $\lim_{t \downarrow a} z(t) = l' \in \mathbb{C}$. In the latter case: $l' \in P(f)\,(l' \in C(f))$ if $a = -\infty\,(a > -\infty)$.

Proof: Since the proofs of 1 and 2 are similar, we only prove 1. Let (t_n) be an arbitrary strictly increasing sequence with $\lim_{n\to\infty} t_n = b$. We define $z_n := z(t_n)$. Suppose (z_n) has no finite accumulation points. Since $|f(z(t))|$ is a strictly decreasing function of t all the elements of (z_n) must be different. So $\lim_{n\to\infty} z(t_n) = \infty$. In the case that this holds for every such sequence (t_n) one concludes that $\lim_{t \uparrow b} f(z(t)) = \infty$. Otherwise we may suppose that there is a sequence (t_n)-defined as above- such that $(z_n) = (z(t_n))$ has a finite accumulation point, say z_*. If $z_* \notin N(f) \cup P(f) \cup C(f)$, then f is conformal in z_* and so there exists a neighborhood Ω of z_* such that $f \mid \Omega \to f(\Omega)$ is a diffeomorphism. Since $\arg f(z_n) = \arg f(z_0)$, for all $n \in \mathbb{N}$, the points z_n for which $z_n \in \Omega$ (such points exist) must lie on the trajectory $\gamma(z_*)$ through z_*. Consequently: $\gamma(z_*) = \gamma(z_0)$. If we choose Ω sufficiently small, then $\gamma(z_*) \cap \Omega$ can be described by $z(t)$, $t \in \mathcal{J} =]\tau - \tau_1, \tau + \tau_2[$, $\tau_{1,2} > 0$, $z(\tau) = z_*$. Obviously $\tau + \tau_2 < b$. Since z_* is an accumulation point of (z_n), there exists $t_{n_1} \in]\tau + \tau_2, b[$ such that $z(t_{n_1}) \in \Omega$. Thus $z(t_{n_1}) = z(\tau')$ for some $\tau' \in \mathcal{J}$. This contradicts the fact that $|f(z(t)|$ is a strictly decreasing function of t. It follows that $z_* \in N(f) \cup P(f) \cup C(f)$.

Firstly suppose: $z_* \in N(f)$. Since z_* is a stable node for N(f) there is a neighborhood of z_* such that the trajectory of any point of this neighborhood tends to z_* for increasing t. From this it follows that z_* is the only accumulation point for (z_n) and $\lim_{t\uparrow b} z(t) = z_*$. So $\lim_{t\uparrow b} f(z(t)) = f(z_*) = 0$; in view of (2.1) this is only possible if $b = +\infty$. In view of (2.1) it obviously is impossible that $z_* \in P(f)$. Finally suppose $z_* \in C(f)$. There exists a subsequence (z_{n_k})

of (z_n) such that $\lim_{k\to\infty} z_{n_k} = z_*$. Consequently: $\lim_{k\to\infty} f(z_{n_k}) = f(z_*)$ and $\arg f(z_*) = \arg f(z_0)$ $(= \arg f(z_{n_k}))$. From this it follows that $\gamma(z_0)$ is one of the trajectories tending (for increasing t) to the "saddlepoint" z_*. So z_* is the only accumulation point for (z_n) and $\lim_{t\uparrow b} z(t) = z_*$. This only can be the case if $b < \infty$, again in view of (2.1). □

<u>Remark 2.4</u>: For the case $f \in \mathcal{E} \cup \mathcal{R}$ Lemma 2.2 is a direct consequence of the celebrated theorem of Poincaré-Bendixon-Schwartz on the limiting set of trajectories of dynamical systems defined on S^2, Lemma 2.1 and the fact that in view of (2.1) periodic trajectories and a so called "path-polygon" [16] with trajectories spiralling to it cannot occur (cf. [10]).

<u>Definition 2.2</u>: The basin of a zero resp. pole z^* for f is the set

$$B(z^*) = \{z^*\} \cup \{z_0 \in \mathbb{C} \mid \lim_{t\to+\infty} z(t) = z^*;\ z(0) = z_0\} \quad \text{resp.}$$

$$B(z^*) = \{z^*\} \cup \{z_0 \in \mathbb{C} \mid \lim_{t\to-\infty} z(t) = z^*;\ z(0) = z_0\}.$$

Note that the basin of a zero (pole) for f is the stable (unstable) manifold of this point w.r.t. $\mathcal{N}(f)$. Obviously, the basin of a zero (pole) for f is an open, path-connected subset of $\mathbb{C}$.

The following lemma gives some information on the possible structure of the boundary $\partial B(z^*)$ of the basin $B(z^*)$. In the sequel, a maximal trajectory which not reduces to a single point, is called a regular trajectory.

<u>Lemma 2.3</u>: For $z^* \in N(f)$ $(z^* \in P(f))$ there are two mutually exclusive possibilities:

1. $\partial B(z^*) = \emptyset$ and in this case f has the form: $f(z) = \alpha(z-z^*)^n$ $(f(z) = \alpha(z-z^*)^{-n})$, $\alpha \in \mathbb{C}$, $n \in \mathbb{N}$.
2. $\partial B(z^*)$ = union of the (topological) closures of regular trajectories w.r.t. $\mathcal{N}(f)$. Moreover, if $B(z^*)$ is bounded there is a pole (zero) for f on $\partial B(z^*)$.

<u>Proof</u>: Since the proofs for $z^* \in N(f)$ and $z^* \in P(f)$ are similar we only prove the case $z^* \in N(f)$.

1. Suppose $\partial B(z^*) = \emptyset$ then $B(z^*) = \overline{B}(z^*)$ (= topological closure of $B(z^*)$). Thus ($\mathbb{C}$ is connected) it follows that $B(z^*) = \mathbb{C}$. Consequently: $P(f) = C(f) = \emptyset$; $N(f) = \{z^*\}$. Suppose f(z) is a transcendental entire function. Since z^* is a stable node, there exists a circle C_r around z^* with radius r such that every trajectory tending to z^* must cross C_r once. Let $\beta := \min_{z\in C(r)} |f(z)|$. Obviously

$\beta > 0$. Since $B(z^*) = \mathbb{C}$, for each z_0 outside C_r there exists $t_0 > 0$ such that $z(t_0) \in C_r$, where $z(t)$ is the solution of $\mathcal{N}(f)$ with $z(0) = z_0$. In view of (2.1) it follows that $|f(z_0)| > \beta$ for every z_0 outside C_r. This conclusion however is violated by the Casorati-Weierstrass theorem which states that in the case of a transcendental entire function f, the image of a neighborhood of $z = \infty$ under f is dense in $\mathbb{C}$. From this it follows: $f(z) = \alpha(z-z^*)^n$.

2. Suppose: $\partial B(z^*) \neq \emptyset$, then we have for $z \in \partial B(z^*)$ the following possibilities:

(i) <u>$z \notin N(f) \cup P(f) \cup C(f)$</u>. Since the solution of the system $\mathcal{N}(f)$ depends continuously on the initial conditions (cf. [2]) one proves that $\gamma(z) \subset \partial B(z^*)$, and thus $\overline{\gamma(z)} \subset \partial B(z^*)$.

(ii) <u>$z \in N(f)$</u>. This would imply $B(z) \cap B(z^*) \neq \emptyset$, which obviously is impossible.

(iii) <u>$z \in C(f)$</u>. In at least one hyperbolic sector H (cf. [2], [10]) at z there exists a sequence (z_n), $z_n \in B(z^*)$ with $\lim_{n\to\infty} z_n = z$.
From this it follows that the separatrix of H which tends to z for increasing t belongs to $\partial B(z^*)$.

(iv) <u>$z \in P(f)$</u>. Choose a reduced ε-neighborhood Ω of z (i.e. $z \notin \Omega$) such that $\Omega \subset B(z)$. If $\Omega \cap \partial B(z^*) \neq \emptyset$, then we are done since we may use the result of (i). Suppose $\Omega \cap \partial B(z^*) = \emptyset$. We prove that this leads to a contradiction. A connectedness argument shows: $\Omega \cap B(z^*) = \Omega$. Choose in Ω a circle C_1 around z such that every trajectory tending to z for decreasing t crosses C_1 once. Choose C_r as in 1. and define $g: C_1 \to C_r$ by $z_0 \in C_1 \to \gamma(z_0) \cap C_r$. Using standard arguments one proves that g is a homeomorphism of C_1 onto C_r. A homotopy argument shows that this is impossible.

Suppose $B(z^*)$ is bounded. Since $C(f)$ is at most countable, there exists $z' \in B(z^*)$ such that $\arg f(z')$ does not equal $\arg f(z)$ for any $z \in C(f)$. Consequently the trajectory $\gamma(z')$ can tend (for decreasing t) neither to a critical point, nor to $z = \infty$. In view of Lemma 2.2 the trajectory $\gamma(z')$ must tend (for decreasing t) to a pole, which necessarily lies on $\partial B(z^*)$. □

<u>Corollary 2.1</u>: If f is an entire function then the basins of the zeros for f are unbounded (cf. [5]). □

In the case we are able to extend $\mathcal{N}(f)$ to S^2 (see Lemma 2.1) we describe the phase-portrait around $z = \infty$ in full detail:

<u>Lemma 2.4</u>: (i) Let $f \in \mathcal{R}$, $f = \frac{p_n}{q_m}$, $p_n(q_m)$ a polynomial of degree n (m). Let s = total number of critical points for f, each counted a number of times equal to its order. Then:

a. If $n < m$ $(n > m)$, $z = \infty$ is a stable (unstable) non-degenerate node for $\overline{\overline{\mathcal{N}}}(f)$.

b. If $n = m$, $z = \infty$ is a k-fold saddlepoint for $\overline{\overline{\mathcal{N}}}(f)$,

$k = \# N(f) + \# P(f) - s - 2 = 2n - \deg(p_n'.q_n - p_n.q_n') - 2$; $k = 0$: regular.

(ii) Let $f \in \mathcal{E}$, order $(f) = m$. Then $z = \infty$ is an isolated equilibrium-state for $\overline{\overline{\mathcal{N}}}(f)$ with - apart from the (eventual) parabolic sectors - exactly 2m elliptic sectors and no hyperbolic sectors. (For the definitions of parabolic, elliptic and hyperbolic sectors see e.g. [2]).

Proof: Part (i)-a follows directly from the proof of Lemma 2.1, Step 3.

(i)-b. Adopting the notation of the proof of Lemma 2.1 (Step 3) we have: $\lim_{z\to\infty} f(z) = \frac{a_n}{b_n}$ $(\neq 0)$. In view of (2.1) it is impossible for a trajectory to tend (for increasing and decreasing t) to $z = \infty$. Consequently $z = \infty$ has no elliptic sectors. Since on each trajectory tending to (or emanating from) $z = \infty$ we have $\arg f(z) = \text{constant}$ $(= \arg(\frac{a_n}{b_n}))$ there are no parabolic sectors at $z = \infty$. If $\Theta(w)$ is defined as in Lemma 2.1, Step 3, a direct calculation shows: $\lim_{w\to 0} \frac{\Theta(w)}{w^k}$ exists and does not vanish if $k = 2n - \deg.(p_n'.q_n - p_n.q_n') - 2$. Consequently the Poincaré-index of $z = \infty$ w.r.t. $\overline{\overline{\mathcal{N}}}(f)$ equals $(-k)$. Using the classification theorem on isolated equilibrium-states for real analytical dynamical systems (cf. [2], [10]) one finds that $z = \infty$ is a k-fold saddlepoint. Applying the Poincaré-Hopf index-theorem to $\overline{\overline{\mathcal{N}}}(f)$ we find: $k = \# N(f) + \# P(f) - s - 2$.

(ii). We adopt the notations of the proof of Lemma 2.1, Step 4. If $w(t)$ stands for the solution of $\overline{\overline{\mathcal{N}}}_w(f)$, with $w(0) = w_0 \neq 0$, the system $\overline{\overline{\mathcal{N}}}_w(f)$ is described by:

$$\frac{dw}{dt} = \beta . w^{m+1}(1 + h(w,\bar{w})) \qquad (*)$$

where $h(w,\bar{w})$ is rational in $w,\bar{w}$ with $h(0,0) = 0$. A direct calculation shows:

$$\frac{d}{dt}|w(t)|^2 = \frac{d}{dt}(w(t)\,,\bar{w}(t)) = 2|w(t)|^2 . \operatorname{Re}\{\beta w^m(t)(1+h(w(t),\bar{w}(t))\}.$$

Thus (write $w(t) = w$):

$$\frac{d}{dt}(\log|w|) = \operatorname{Re}\,[\beta w^m(1+h(w,\bar{w}))] = |w|^m[\operatorname{Re}(\beta.(\frac{w}{|w|})^m) + O(|w|)].$$

Let for $t = t_0$ the function $|w(t)|$ be stationary, then:

$$\operatorname{Re}\{\beta(\frac{w}{|w|})^m\} = O(|w|) \qquad \text{for } t = t_0.$$

Thus:

$$\beta\left(\frac{w(t_0)}{|w(t_0)|}\right)^m = \pm i|\beta| + O(|w|). \qquad (**)$$

For such t_0 we have:

$$\frac{d^2}{dt^2}(\log|w|) = \text{Re}[m\beta^2 w^{2m}(1+h(w,\bar{w}))^2 + \beta w^m . \frac{d}{dt}h(w,\bar{w})]$$

$$= m.|w|^{2m}[\text{Re}\{\beta^2(\frac{w}{|w|})^{2m}\} + O(|w|)]$$

$$\overset{(**)}{=} |w|^{2m}[-m|\beta|^2 + O(|w|)] < 0$$

for $|w|$ sufficiently small.

Conclusions:

1. For $t = t_0$ there is a strict maximum value of the function $|w(t)|$;
2. In a (sufficiently) small neighborhood of $w = 0$ the function $|w(t)|$ has at most one stationary value;
3. No trajectory spirals to $w = 0$ (since for $|w|$ sufficiently small no stationary points of $|w(t)|$ exist, consequently $\frac{d}{dt}\log|w|$ has constant sign, whereas $\text{Re}(\beta(\frac{w}{|w|})^m)$ takes both values $\pm|\beta|$ in case of a spiral);
4. There are no hyperbolic sectors at $w = 0$ (since $|w(t)|$ has no minimum).

In view of (*) the Poincaré-index of $z = \infty$ equals m+1. We now are in the position to use the formula: Index $(z=\infty) = 1 + (\frac{\theta-\sigma}{2})$, where σ (θ) stands for the number of hyperbolic (elliptic) sectors at $z = \infty$ (cf. [10]. Conclusion 4 above yields to the desired result. □

<u>Lemma 2.5:</u> In the situation of Lemma 2.4-(ii) we denote the elliptic sectors at $z = \infty$ by ε_k, $k = 1, \ldots, 2m$. . Then: For each k there exist two rays, say $L_{k,1}$, $L_{k,2}$ emanating from $w = 0$, determining the angle $\frac{\pi}{m}$ and with the property that each trajectory in ε_k emanates from $w = 0$ tangent to $L_{k,1}$ and tends to $w = 0$ tangent to $L_{k,2}$.

<u>Proof:</u> (Notations as in Lemma 2.4) Since on a trajectory of $\mathcal{N}(f)$ or $\overline{\mathcal{N}}(f)$ we have arg f(z) = constant, it follows:

$$\arg[p_n(\frac{1}{w(t)}), \exp(q_m(\frac{1}{w(t)})] = \text{constant}. \qquad (*)$$

We suppose that w(t) describes a trajectory in ε_k, consequently $\lim_{t\to\pm\infty}|w(t)| = 0$.

Let arg $w(t) = \Theta(t)$. Note that since $\bar{\bar{\mathcal{W}}}(f)$ is complete, $w(t)$ is defined for all $t \in \mathbb{R}$. Since $\bar{\bar{\mathcal{W}}}(f)$ is analytical, and $w(t)$ has a non-spiralling character, it is well-known that $\lim_{t\to\infty} \Theta(t)$ exists, say Θ^* (cf. [2]).

If we denote the coefficients of $p_n(q_m)$ by $a_l(b_j)$, $l = 0, \ldots, n$, $j = 0, \ldots, m$ we find for (*):

$$\arg[a_n|w(t)|^{-n}\exp(-ni\Theta(t)) + \ldots + a_0] + \mathrm{Im}\left[b_m|w(t)|^{-m}\exp(-mi\Theta(t)) + \ldots + b_0\right] = \text{constant.} \qquad (**)$$

Denote the first term of the l.h.s. by $\psi(t)$. Since $\lim_{t\to\infty} \psi(t)$ exists, for t sufficiently large, $\psi(t)$ is in a fixed interval of length 2π and the r.h.s. of (**) is a fixed constant.

From (**) it follows:

$$|w(t)|^m\psi(t) + \mathrm{Im}[b_m \exp(-mi\Theta(t)) + \ldots + |w(t)|^m b_0] = \text{const. } |w(t)|^m.$$

Since $\lim_{t\to\infty} |w(t)| = 0$ it follows:

$$\lim_{t\to\infty} \mathrm{Im}(b_m \exp(-mi\Theta(t))) = 0 \quad \text{or} \quad \sin(\arg b_m - m.\Theta^*) = 0.$$

Consequently: $\Theta^* = \frac{\pi}{m} k + \frac{1}{m} \arg b_m$, $\quad k = 1, \ldots, 2m.$

Essentially the same calculations yield to the same possible values for $\lim_{t\to-\infty} \Theta(t) = \Theta^{**}$. From the relation (*) in the proof of Lemma 2.4 it follows that:

$$-\frac{1}{m}\frac{d}{dt}(|w|^{-m}) = |\beta|\cos(m\,\Theta(t) - \arg b_m) + O(|w|).$$

In view of conclusions 1,2 in the proof of Lemma 2.4 the function $\mathbb{R} \to \mathbb{R}$; $t \to |w(t)|$ attains exactly one strict minimum, if $w(t)$ describes a trajectory which is in an ε-neighborhood of $w = 0$, ε sufficiently small.

Consequently we have: $|\Theta^* - \Theta^{**}| = \frac{\pi}{m}$. The assertion follows from that fact there are exactly 2m elliptic sectors. □

For the case m = 2 the result of the preceding two lemmas are summarized in Fig. 2.2.

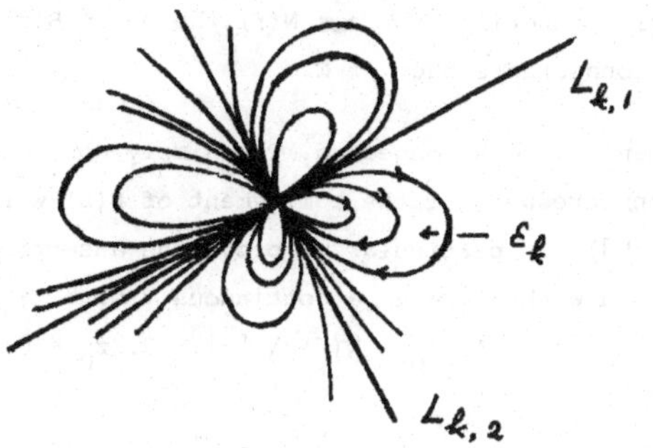

Fig. 2.2.

We proceed by giving some results on the structure of the set of points which "move" to (from) $z = \infty$ along a trajectory of $\mathcal{N}(f)$.

Definition 2.3: The basin of attraction ($A(\infty)$) resp. the basin of repulsion ($R(\infty)$) of $z = \infty$ for f is defined to be the interior of the set of all $z \in \mathbb{C}$ with the property that $\gamma(z)$ tends for increasing resp. decreasing t to $z = \infty$.

In case that $f \in \mathcal{R}$, $f = \frac{p_n}{q_m}$ there are the following possibilities:

If $m > n$ ($m<n$), $A(\infty)$ ($R(\infty)$) is the basin of $z = \infty$ considered as a stable (unstable) node for $\bar{\mathcal{N}}(f)$ and $R(\infty) = \emptyset$ ($A(\infty) = \emptyset$).

If $m = n$, $A(\infty) = R(\infty) = \emptyset$.

In case that $f \in \mathcal{E}$ it is possible that $A(\infty) \cap R(\infty) \neq \emptyset$; moreover $A(\infty)$ need not to be a connected set. (see Lemma 2.4 and Example 2.5).

Lemma 2.6: Let $f \in \mathcal{E} \cup \mathcal{R}$ and $z \in C(f)$. Then:

1. There exists $z^* \in N(f)$ such that $z \in \partial B(z^*)$ or $z \in \partial A(\infty)$.
2. There exists $z^{**} \in P(f)$ such that $z \in \partial B(z^{**})$ or $z \in \partial R(\infty)$.

Proof: We only prove 1., since the proof of 2. is similar.

The set $V := A(\infty) \cup \bigcup_{z^* \in N(f)} B(z^*)$ is dense in $\mathbb{C}$. Suppose this is not true. Then there exists an open set, say Ω, which is contained in $\mathbb{C} \setminus V$. Since $\# C(f) < \infty$ there exists an open subset $\Omega' \subset \Omega$ such that for any $z \in \Omega'$: $\arg f(z) \notin \{\arg f(z') \mid z' \in C(f)\}$. But from this one concludes (in view of Lemma 2.2) that for any $z \in \Omega'$ the trajectory $\gamma(z)$ tends for increasing t to $z = \infty$. Consequently $\Omega' \subset V$. Contradiction.

It follows, since $\# N(f) < \infty$ that $\bar{V} = \bar{A}(\infty) \cup \bigcup_{z^* \in N(f)} \bar{B}(z^*) = \mathbb{C}$ and thus: For each $z \in C(f)$ we have $z \in \partial B(z^*)$, some $z^* \in N(f)$ or $z \in \partial A(\infty)$. □

Lemma 2.7: Let f be an entire function. For $z \in N(f)$ the basin $B(z)$ is an open, simply connected, path-connected subset of $\mathbb{C}$.

Proof: Obviously $B(z)$ is open and path-connected. The simply connectedness of $B(z)$ then follows from the connectedness of the complement of $B(z)$ w.r.t. the extended complex plane (cf. [1]). In particular we prove the assertion that every $z_0 \in \mathbb{C} \setminus B(z)$ is connected with $z = \infty$ by a continuous γ with $\gamma \subset \mathbb{C} \setminus B(z)$. Distinguish the following three cases: 1. $z_0 \in N(f) \setminus \{z\}$; 2. $z_0 \in C(f)$; 3. $z_0 \notin N(f) \cup C(f)$.

In case 1, the assertion follows from the facts that $B(z_0)$ is unbounded (Corollary 2.1) and $B(z_0)$ is path-connected.

In case 2 at least two trajectories (not intersecting $B(z)$) tend for increasing t to z_0. In view of Lemma 2.2 - for decreasing t - at least one of these trajectories tends to $z = \infty$ or all of them tend to a critical point for f. In the first case we are done, in the second case we are back in the initial situation of case 2. It follows that there exists a continuous curve $\gamma \subset \mathbb{C} \setminus B(z)$ consisting either of finitely many trajectories, the last one tending to $z = \infty$, or infinitely many (countable) trajectories each of them connecting two points of $C(f)$. In the latter case γ also tends to $z = \infty$ since the critical points for f cannot accumulate in a bounded subset of $\mathbb{C}$.

In case 3, the unique trajectory $\gamma(z_0)$ through z_0 which does not intersect $B(z)$ tends for decreasing t to $z = \infty$ (then we are done) or to a point of $C(f)$ (then we are in case 2). □

Lemma 2.8: Let f be a polynomial, then the following properties hold:

(i) $\partial R(\infty) \supset N(f) \cup C(f)$.

(ii) $\partial R(\infty)$ does not contain a closed Jordan curve.

(iii) $\partial R(\infty)$ is compact and path-connected.

Proof: (i) This follows immediately from the fact that (since $P(f) = \emptyset$) $\overline{R}(\infty) = \mathbb{C}$ (cf. Lemma 2.6, proof of Part 2).

(ii) As in Lemma 2.3 (Part 2) one may prove that - considering $z = \infty$ as a pole for f - $\partial R(\infty)$ is the union of the closures of regular trajectories of $\mathcal{N}(f)$. Since a trajectory of $\mathcal{N}(f)$ does not connect two zeros for f and there are finitely many critical points it follows that $\partial R(\infty)$ consists of finitely many trajectories. Suppose there exists a closed Jordan curve J, $J \subset \partial R(\infty)$. Then J is built up by the closures of regular trajectories. Since $\overline{R}(\infty) = \mathbb{C}$, in the "interior" of J there exists a point z^* such that $\gamma(z^*)$ tends to $z = \infty$, which obviously is impossible.

(iii) Obviously $\partial R(\infty)$ is compact. We prove that $\partial R(\infty)$ is connected, path-connectedness being a trivial consequence in this case.

Suppose that $\partial R(\infty)$ is not connected and denote its components by $\Gamma_1, \ldots, \Gamma_m$. Since $\partial R(\infty)$ is bounded and $z = \infty$ is an unstable node for $\bar{\bar{\mathcal{N}}}(f)$ there exists an $r > 0$ such that the circle C_r around $z = 0$ with radius r has the properties:

1. For each $z \in C_r$ the vectorfield $\bar{\bar{\mathcal{N}}}(f)$ points inwards C_r.
2. $\partial R(\infty)$ is contained in the circular disc determined by C_r.

Consider the map $\phi: C_r \to \{1,2,\ldots,m\}$ defined as follows:

$z \mapsto i$, $i = 1, \ldots, m$ if $\bar{\gamma}(z) \cap \partial R(\infty) \in \Gamma_i$.

We are done if we prove that ϕ is continuous, since in that case from the connectedness of C_r it follows that $m = 1$. For $z_* \in C_r$ such that $\bar{\gamma}(z_*) \cap \partial R(\infty) \in N(f)$ the continuity of ϕ at z_* is an easy consequence of the fact that the basin of a zero for f is an open subset of $\mathbb{C}$. Consider the case $\bar{\gamma}(z_*) \cap \partial R(\infty) \in C(f)$. Since $\# N(f) < \infty$ and $\# C(f) < \infty$ a reduced ε-neighborhood Ω of $z_* \in C_r$ exists such that: If C_1, C_2 are the two connected components of $C_r \cap \Omega$, then for all $z \in C_1(C_2)$ $\gamma(z)$ tends - for increasing t - to a $z_1(z_2) \in N(f)$. The map ϕ is continuous at z_* if the following assertion is true: If $\phi(z_*) = j$, then $z_1, z_2 \in \Gamma_j$. Suppose the assertion is not true, so suppose $z_1 \in \Gamma_i$, $i \neq j$. Then $\gamma(z_*) \subset \partial B(z_1)$. There exists a trajectory, say γ', issuing from $\bar{\gamma}(z_*) \cap \partial R(\infty)$ $(\in C(f))$, such that $\bar{\gamma}' \subset \partial B(z_1) \cap \partial R(\infty)$. Since $N(f) \cap \partial B(z_1) = \emptyset$, γ' cannot tend (for increasing t) to a zero for f, and since γ' cannot cross C_r it cannot tend to $z = \infty$ either. Consequently - for increasing t - γ' tends to a critical point for f, say z_3. There exists a trajectory issuing from z_3 which also belongs to $\partial B(z_1) \cap \partial R(\infty)$ and also tends - for increasing t - to a critical point. In view of (ii) from this it follows that there are infinitely many critical points in the closed disc determined by C_r which obviously is impossible.

Since either $\bar{\gamma}(z_*) \cap \partial R(\infty) \in N(f)$ or $\bar{\gamma}(z_*) \cap \partial R(\infty) \in C(f)$ we proved the continuity of ϕ and hence the assertion (iii) of the lemma. □

For later use (see Section 4) we now introduce a class of dynamical systems which are closely related to the systems we studied up till now:

<u>Definition 2.4</u>: The conjugate $\mathcal{N}^{\perp}(f)$ $(\bar{\mathcal{N}}^{\perp}(f))$ of the system $\mathcal{N}(f)$ $(\bar{\mathcal{N}}(f))$ is defined by:

$$\frac{dz}{dt} = -i\,\frac{f(z(t))}{f'(z(t))} \quad \left(\frac{dz}{dt} = -i.\,\frac{1}{(1+|f(z(t))|^4)}\,.\,\overline{f'}(z(t)).\,f(z(t))\right).$$

In case $\bar{\bar{\mathcal{N}}}(f)$ is defined, one defines $\bar{\bar{\mathcal{N}}}^{\perp}(f)$ in a similar way.

Again $\overline{\mathcal{N}}^{\perp}(f)$ and $\overline{\overline{\mathcal{N}}}^{\perp}(f)$ are extensions of $\mathcal{N}^{\perp}(f)$ (which only is defined on $\mathbb{C} \setminus C(f)$) to $\mathbb{C}(S^2)$.

By $\gamma^{\perp}(z_0)$ we denote the maximal trajectory of $\mathcal{N}^{\perp}(f)$ through $z_0 \notin C(f)$. This trajectory $\gamma^{\perp}(z_0)$ is given by a solution denoted by $z^{\perp}(t)$, $z^{\perp}(0) = z_0$, $t \in]a,b[$, eventually $a = -\infty$, $b = +\infty$. Note that the trajectories of $\mathcal{N}^{\perp}(f)$ are the orthogonal trajectories of $\mathcal{N}(f)$.

Lemma 2.9: Situation as above.

1. $f(z^{\perp}(t)) = e^{-it}.f(z_0)$ (consequently: $|f(z^{\perp}(t))| = \text{constant}$).

2. If $z_0 \in N(f)$ ($\in P(f)$) then z_0 is a centre for $\mathcal{N}^{\perp}(f)$.

3. If z_0 is a critical point for f of order k, then z_0 is a k-fold saddle point for $\overline{\mathcal{N}}^{\perp}(f)$.

Proof: 1. Trivial, see also (2.1).

2. Let $z_0 \in N(f)$, k = multiplicity (z_0). In a neighborhood of z_0, the system $\mathcal{N}^{\perp}(f)$ can be linearly approximated by: $\frac{dz}{dt} = -i\,\frac{(z-z_0)}{k}$. Consequently z_0 is a non-degenerate equilibrium-state for $\mathcal{N}^{\perp}(f)$ with characteristic roots $\pm \frac{1}{k}i$. In view of 1. and since $f(z) \neq$ constant z_0 cannot be either a focus or a centro-focus for $\mathcal{N}^{\perp}(f)$ (cf. [2]). From this it follows (cf. [2]) that at z_0 the system $\mathcal{N}^{\perp}(f)$ has a centre.

For $z_0 \in P(f)$ the same argument holds.

3. Since in a neighborhood of z_0 $\mathcal{N}^{\perp}(f)$ may be considered as a Newton-flow w.r.t. $\exp[-i \log f(z)]$ the assertion follows directly from the results already obtained for systems $\mathcal{N}(f)$. □

In the cases we have extended the conjugate Newton-flow to S^2 we may give a similar characterization of the local phase-portrait at $z = \infty$ as was given in Lemmas 2.4,5. The proof, running essentially along the same lines, will be omitted.

Lemma 2.10: (i) Adopting the notations of Lemma 2.4 (Part i):

a. If $n \neq m$, $z = \infty$ is a centre for $\overline{\overline{\mathcal{N}}}^{\perp}(f)$.

b. If $n = m$, $z = \infty$ is a k-fold saddlepoint,

$$k = \#N(f) + \#P(f) - s - 2 = 2n - \deg.(p_n'.q_n - p_n.q_n') - 2.$$

(ii) Adopting the notations of Lemma 2.4 (Part ii)):

At $z = \infty$ the system $\overline{\overline{\mathcal{N}}}^{\perp}(f)$ has an isolated equilibrium-state with - apart from the (eventual) parabolic sectors - exactly 2m elliptic sectors and no hyperbolic sectors.

<u>Remark 2.5</u>: An assertion similar to the result of Lemma 2.5 holds for the elliptic sectors of $\mathcal{N}^{\perp}(f)$ at $w = 0$.

<u>Remark 2.6</u>: An assertion similar to the result of Lemma 2.2 holds:

Let the maximal trajectory of $\mathcal{N}^{\perp}(f)$ through $z_0 \notin N(f) \cup P(f) \cup C(f)$ be given by $z(t)$, $t \in \,]a,b[$, $z_0 = z(0)$. Then:

Either $z(t)$ is a periodic trajectory (and $a = -\infty$; $b = +\infty$) or
$a < t_1 < t_2 < b \Longrightarrow z(t_1) \neq z(t_2)$.
Moreover in the latter case we have:

1. Either $\lim\limits_{t \uparrow b} z(t) = \infty$ or $\lim\limits_{t \uparrow b} z(t) \in C(f)$ (and $b < \infty$).

2. Either $\lim\limits_{t \downarrow a} z(t) = \infty$ or $\lim\limits_{t \downarrow a} z(t) \in C(f)$ (and $a > -\infty$).

We conclude this section by giving some examples, illustrating the preceding theory:

<u>Example 2.1</u>: (See Fig. 2.3)

$f(z) = z^n$; $\mathcal{N}(f)$: $\dfrac{dz}{dt} = -\dfrac{1}{n}\, z(t)$

$N(f) = \{0\}$; $P(f) = C(f) = \emptyset$

Note $\partial B(0) = \emptyset$, compare Lemma 2.3.

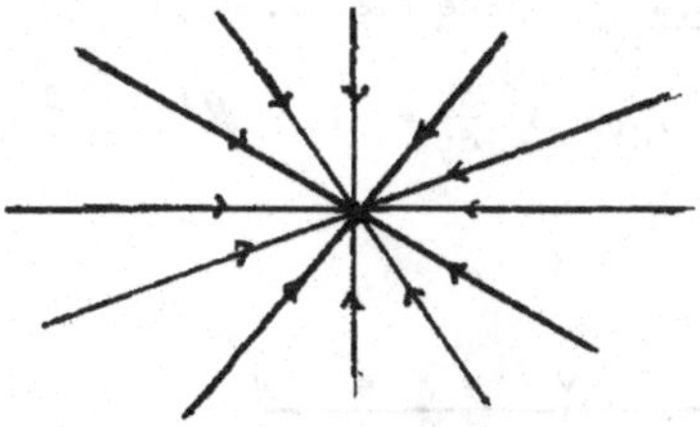

Fig. 2.3.

<u>Example 2.2</u>: (See Fig. 2.4 a,b)

$f(z) = z^3 - 1$; $\mathcal{N}(f)$: $\dfrac{dz}{dt} = -\dfrac{1}{3}(z - z^{-2})$; $\bar{\bar{\mathcal{N}}}(f)$: $\dfrac{dz}{dt} = -(1+|z|^2)^{-2}(1+|z^3-1|^4)^{-1}\cdot(3\bar{z}^2)\cdot(z^3-1)$.

$N(f) = \{z_k = \exp(\frac{2k\pi i}{3})$; $k = 0,1,2\}$, $P(f) = \emptyset$, $C(f) = \{0\}$, order $(0) = 2$.

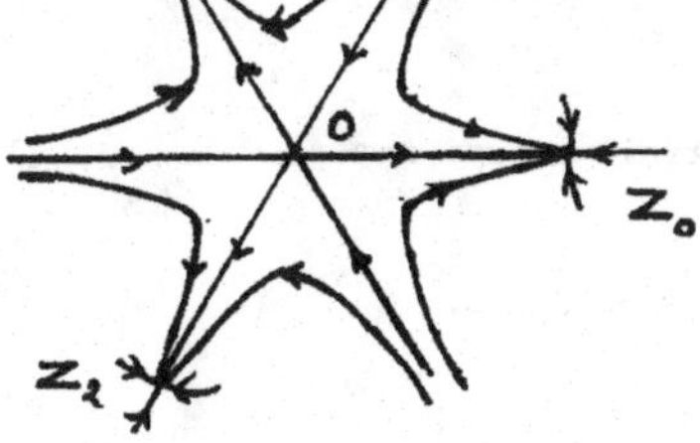

Fig. 2.4.a.

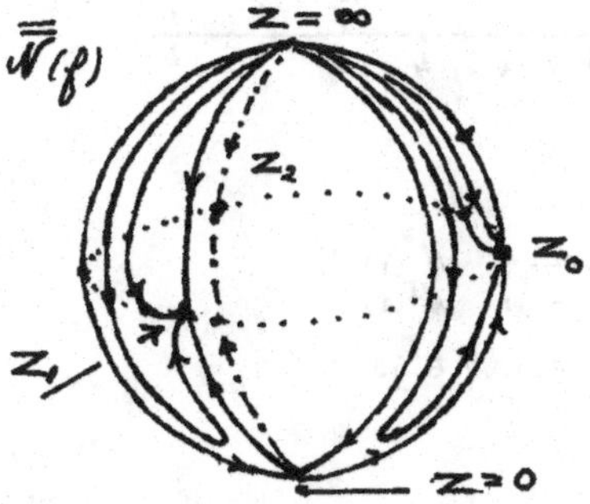

Fig. 2.4.b.

<u>Example 2.3:</u> (See Fig. 2.5)

$$f(z) = \frac{z^2+1}{z} \quad ; \quad \mathcal{N}(f): \quad \frac{dz}{dt} = -z. \left(\frac{z^2+1}{z^2-1}\right)$$

$N(f) = \{i, -i\}$; $P(f) = \{0\}$; $C(f) = \{1, -1\}$

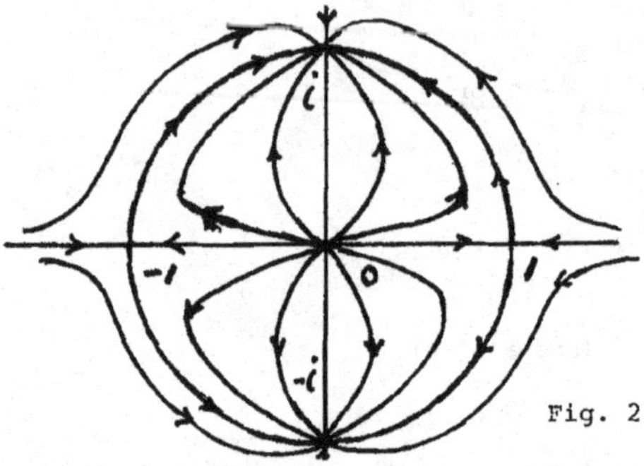

Fig. 2.5.

Note that: $B(0)$ is bounded; $A(\infty) = \emptyset$; and $\partial R(\infty)$ contains a closed Jordan curve (compare Lemma 2.8, case (ii)).

<u>Example 2.4</u> (See Fig. 2.6 a,b)

$f(z) = e^z$; $\mathcal{N}(f): \frac{dz}{dt} = -1$; $\mathcal{N}^{\perp}(f): \frac{dz}{dt} = -i$; $N(f) = P(f) = C(f) = \emptyset$

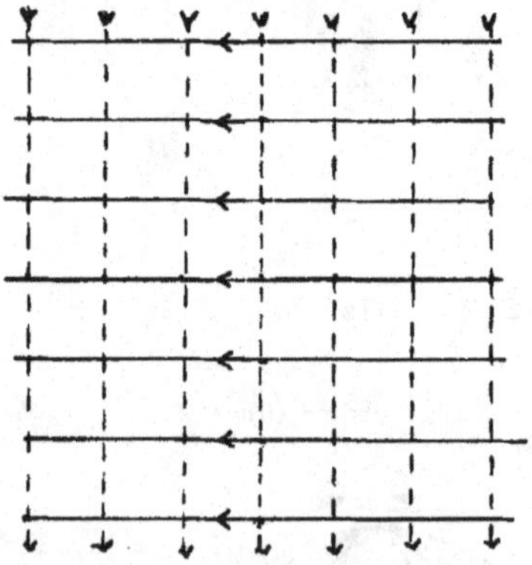

——— : $\mathcal{N}(f)$
- - - - : $\mathcal{N}^{\perp}(f)$

Fig. 2.6.a.

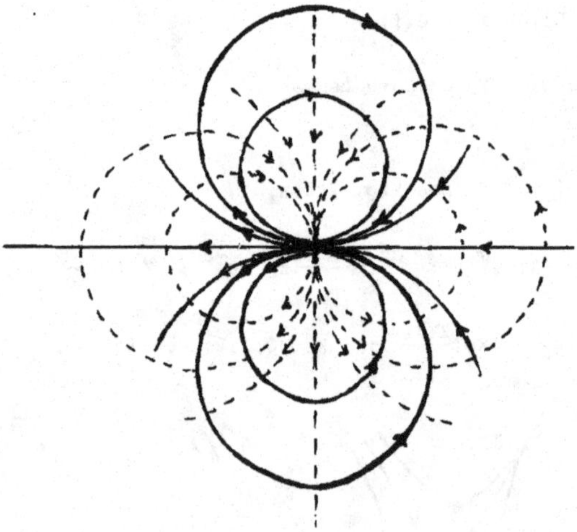

Local phase-portrait around $z=\infty$

Fig. 2.6.b.

Example 2.5 (See Fig. 2.7 a,b)

$f(z) = z.e^{(z+8)^2}$; $N(f) = \{0\}$; $C(f) = \{z_1 = -4 - \frac{1}{2}\sqrt{62}\ \ z_2 = -4 + \frac{1}{2}\sqrt{62}\}$;

$P(f) = \emptyset$; order (f) = 2. Note that $\bar{\bar{\mathcal{N}}}(f)$ as well as $\bar{\bar{\mathcal{N}}}^{\perp}(f)$ has exactly 4 elliptic sectors at $z = \infty$

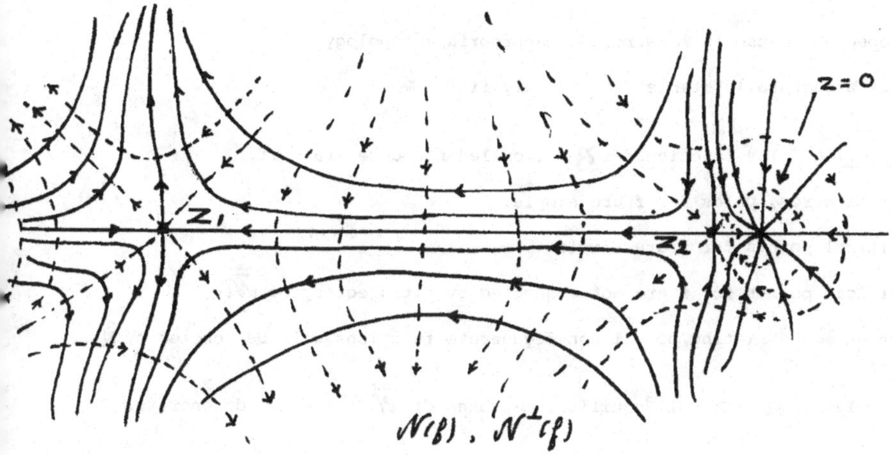

——— : = $\mathcal{N}(f)$
- - - - : = $\mathcal{N}^{\perp}(f)$

Fig. 2.7.a.

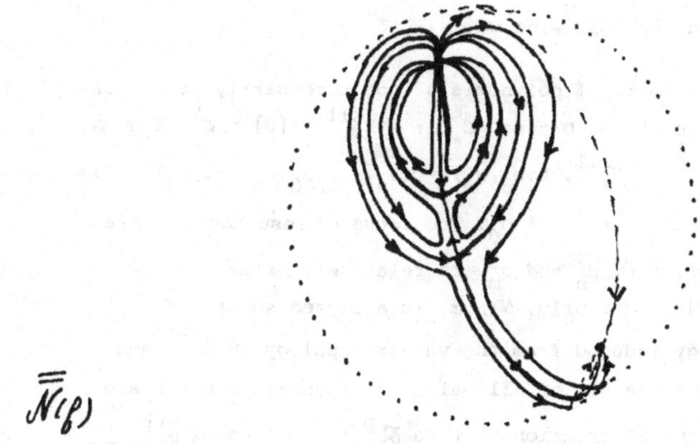

Fig. 2.7.b.

3. Newton-flows for rational functions; structural stability aspects.

In this section we restrict ourselves to functions $f \in \mathcal{R}$. Such functions can be extended (in the usual way) to meromorphic functions on S^2; the set of all these extensions will also be denoted by $\mathcal{R}$. The point $z = \infty$ is called a zero, pole, critical point for f if $w = 0$ is a zero, pole, critical point for $f(\frac{1}{w})$, $w = \frac{1}{z}$. As the main result of this section we shall define a subset $\tilde{\mathcal{R}}$ of $\mathcal{R}$ with the following properties:

1. $\tilde{\mathcal{R}}$ is open and dense in $\mathcal{R}$ w.r.t. an appropriate topology τ.
2. $\bar{\bar{\mathcal{N}}}(f)$ is "structurally stable" if and only if $f \in \tilde{\mathcal{R}}$.

Definition 3.1: The function $f \in \mathcal{R}$ is called non-degenerate if:

1. All finite zeros, poles for f are simple.
2. All critical points for f are simple.
3. Two critical points for f are not connected by a trajectory of $\bar{\bar{\mathcal{N}}}(f)$.

The subset of $\mathcal{R}$ consisting of all non-degenerate functions will be denoted by $\tilde{\mathcal{R}}$.

Lemma 3.1: If $f \in \tilde{\mathcal{R}} \Longrightarrow$ All equilibrium-states of $\bar{\bar{\mathcal{N}}}(f)$ are non-degenerate.

The proof of this lemma follows directly from the proof of Lemma 2.1 (Step 3).

Remark 3.1: Let $f \in \mathcal{R}$, $f = \frac{p_n}{q_n}$, $p_n(z) = a_n z^n + \ldots + a_0$, $q_n(z) = b_n z^n + \ldots + b_0$, $a_n \neq 0$, $b_n \neq 0$. Then: f has a regular or a simple critical point at $z = \infty$ iff:

$$|a_n . b_{n-1} - a_{n-1} . b_n| + |a_n . b_{n-2} - a_{n-2} . b_n| \neq 0.$$

We introduce a topology on $\mathcal{R}$ in the following way:

Denote by $\mathcal{R}(n,m)$ the set of all pairs of polynomials (not necessarily prime) of degree (n), (m) respectively; n,m fixed. Define $\mathbb{C}_*^{k+1} := \mathbb{C}^{k+1} \setminus \{0\} \times \mathbb{C}^k$, $k \in \mathbb{N}$. Consider the map $t: \mathcal{R}(n,m) \to \mathbb{C}_*^{n+1} \times \mathbb{C}_*^{m+1}$, $t(p_n, q_m) = (a_n, \ldots, a_0, b_m, \ldots, b_0)$ where $p_n(z) = a_n z^n + \ldots + a_0$, $q_m(z) = b_m z^m + \ldots + b_0$. The locus of resultant for the pair (n,m) is denoted by $\rho(n,m)$, i.e. p_n and q_m are relatively prime iff $t(p_n, q_m) \notin \rho(n,m)$ (cf. [15]). Obviously, $\rho(n,m)$ is a closed subset of $\mathbb{C}_*^{n+1} \times \mathbb{C}_*^{m+1}$ w.r.t. the topology induced from the vector topology on $\mathbb{C}^{n+1}$ and $\mathbb{C}^{m+1}$. Let $\mathcal{R}^p_{(n,m)} \subset \mathcal{R}_{(n,m)}$ be the set of all pairs of polynomials which are relatively prime. Define $\tilde{t}$ as the restriction of t to $\mathcal{R}^p_{(n,m)}$. We endow $\mathcal{R}^p_{(n,m)}$

with the weakest topology making $\tilde{t}$ continuous. The pairs (p_n, q_m) and $(\hat{p}_n, \hat{q}_m)$ in $\mathcal{R}^p_{(n,m)}$ are said to be equivalent ($\sim$) iff $\frac{p_n}{q_m} = \frac{\hat{p}_n}{\hat{q}_m}$, i.e. they represent the same function in $\mathcal{R}$. We endow $\mathcal{R}^p_{(n,m)} / \sim$ with the quotient topology. A standard argument (cf. [4]) shows that the canonical map $\kappa : \mathcal{R}^p_{(n,m)} \longrightarrow \mathcal{R}^p_{(n,m)} / \sim$ is open w.r.t. the topologies involved. The topology τ will be the strongest topology making all canonical injections $\mathcal{R}^p_{(n,m)} \longrightarrow \mathcal{R}$ continuous.

Remark 3.2: The topology τ is a natural one in the following sense: Let $f \in \mathcal{R}$ and choose $\frac{p_n}{q_m}$ as its representant. Given $\varepsilon > 0$ sufficiently small, there exists a τ-neighborhood $\mathcal{O}$ of f such that for $g \in \mathcal{O}$ the function g can be represented by $\frac{\tilde{p}_n}{\tilde{q}_m}$ such that the coefficients of $\tilde{p}_n, \tilde{q}_m$ are in ε-neighborhoods of the corresponding coefficients of p_n, q_m.

Lemma 3.2: Let $\mathcal{X}(S^2)$ be the set of analytical vectorfields on S^2 endowed with C^1-topology c. (cf. [12]) Consider the map $T : \mathcal{R} \longrightarrow \mathcal{X}(S^2)$ given by $T(f) = \bar{\mathcal{N}}(f)$. Then: T is (τ, c)-continuous.

Proof: The set $\mathcal{X}(S^2)$ is a vectorspace and can be made into a Banach space in the following way: Choose a finite atlas $\{U_\alpha\}$ with the property that each closure $\bar{U}_\alpha$ is itself contained in a coordinate neighborhood Ω_α. Let $X \in \mathcal{X}(S^2)$ and X_α correspond to X w.r.t. the coordinates valid in Ω_α; let DX_α correspond to the first derivative of X (w.r.t. Ω_α). Define:

$$\|X\| := \max_\alpha \{ \| X_\alpha \| + \| DX_\alpha \| \}, \text{ with } \|X_\alpha\| := \sup_{U_\alpha} \| X_\alpha \| , \ \|DX_\alpha\| := \sup_{U_\alpha} \|DX_\alpha\|.$$

Let Ω_1, Ω_2 be the charts induced by the stereographic projection of S^2 (considered as the subspace $x_1^2 + x_2^2 + x_3^2 = 1$ of $\mathbb{R}^3$) w.r.t. the north, south pole. Let $r \in]0,1[$. We define U_1, U_2 as follows:

$$U_1 := \{(x_1, x_2, x_3) \in S^2 \mid -1 \leqslant x_3 < r\} \qquad \text{(z-chart)}.$$

$$U_2 := \{(x_1, x_2, x_3) \in S^2 \mid 1 \geqslant x_3 > -r\} \qquad \text{(w-chart)}.$$

In $\mathbb{C} (= \mathbb{R}^2)$ these charts are given by $|z| < R$ resp. $|w| < R$ for some $R > 1$. Note that the projection of $U_1 \cap U_2$ is given by: $\frac{1}{R} < |z| < R$ or $\frac{1}{R} < |w| < R$.

We denote by D_R the disc given w.r.t. the z-chart (w-chart) by $|z| \leq R$ ($|w| \leq R$).

Let $f \in \mathcal{R}$, $f = \dfrac{p_n}{q_m}$ and let $\tilde{O}$ be an ε-neighborhood (w.r.t. the topology c) of $\bar{\bar{\mathcal{N}}}(f) \in \mathcal{X}(S^2)$.

Firstly we assume that $m > n$.

Adopting the notations of Lemma 2.1 (Step 3):

$$\bar{\bar{\mathcal{N}}}(f) \equiv - \frac{(1+|z|^2)^{m-n+1}}{1+|f(z)|^4} \; . \; \bar{f}'(z) \, . \, f(z) \qquad (*)$$

Substituting $f = \dfrac{p_n}{q_m}$ in (*) we obtain:

$$\bar{\bar{\mathcal{N}}}(f) \equiv - (1+|z|^2)^{m-n+1} . \frac{|q_m|^2 . p_n . \bar{p}_n' - |p_n|^2 . q_m . \bar{q}_m'}{|p_n|^4 + |q_m|^4} \qquad (**)$$

The real and imaginary parts of the r.h.s of (**) may be considered as rational functions F_1, F_2 in the 2m + 2n + 6 real parameters:

{Re a_i; Im a_i; Re b_j; Im b_j; x and y ($z = x + iy$)}. Obviously F_1, F_2 are continuous at the points $(\tilde{t}(f), x, y) \in \Omega \times D_R$, where $\Omega = \mathbb{C}_*^{n+1} \times \mathbb{C}_*^{m+1} \setminus \rho(n,m)$. Here we adopted the notations used by introducing τ. Note that $|p_n(z)|^4 + |q_m(z)|^4$ is strictly positive on $\Omega \times D_R$.

The same observations can be made on the first order partial derivatives w.r.t. x, y for F_1, F_2. Since D_R is compact, there exists a neighborhood O_1 of $\tilde{t}(f) \in \Omega$ such that $\| \bar{\bar{\mathcal{N}}}(f) - \bar{\bar{\mathcal{N}}}(g) \|_{\bar{U}_1} < \varepsilon$ as soon as $\tilde{t}(g) \in O_1$.

Now we consider $\bar{\bar{\mathcal{N}}}(f)$ w.r.t. the w-chart. Obviously $\bar{\bar{\mathcal{N}}}_w(f)$ can be written in the following form:

$$(1+|w|^2)^{m-n+1} . w . \frac{|w^n . p_n(\frac{1}{w})|^2 . (w^m . q_m(\frac{1}{w})) . (\overline{w^{-m+1}} . \bar{q}_m'(\frac{1}{w})) - |w^m . q_m(\frac{1}{w})|^2 . (w^n . p_n(\frac{1}{w})) . (\overline{w^{-n+1}} . \bar{p}_n'(\frac{1}{w}))}{|w^m . q_m(\frac{1}{w})|^4 + |w^m . p_n(\frac{1}{w})|^4} \qquad (***)$$

where $p_n'(\frac{1}{w})$, $q_m'(\frac{1}{w})$ stand for $\frac{d}{dw}[p_n(\frac{1}{w})]$, $\frac{d}{dw}[q_m(\frac{1}{w})]$.

Again, the components of this real analytical vectorfield are rational functions G_1, G_2 in 2m + 2n + 6 real parameters, being continuous in $(\tilde{t}(f),x,y)$, with $w = x + iy$, $w \in D_R$. Note that the denominator in (***) does not vanish in D_R. In particular for $x = y = 0$ its value equals $|b_m|^4$. Again, a similar observation can be made on the first order partial derivatives of G_1, G_2 w.r.t. x and y. As in the previous case there exists a neighborhood O_2 of $\tilde{t}(f) \in \Omega$ such that $\|\bar{\bar{\mathcal{N}}}(f) - \bar{\bar{\mathcal{N}}}(g)\|_{\bar{U}_2} < \varepsilon$ as soon as $\tilde{t}(g) \in O_2$.

Define $O := O_1 \cap O_2$ then: For $g \in \mathcal{R}$ with $\tilde{t}(g) \in O$ we have the property that $\bar{\bar{\mathcal{N}}}(g) \in \tilde{O}$. This completes the proof for the case $m > n$.

In case $m < n$, $m = n$ the proof runs essentially along the same lines. □

Before stating the main result of this section, we need one more definition:

Definition 3.2: The system $\bar{\bar{\mathcal{N}}}(f)$ is called structurally stable if there exists a τ-neighborhood Ω of f such that for each $g \in \Omega$ the phase-portraits of $\bar{\bar{\mathcal{N}}}(f)$ and $\bar{\bar{\mathcal{N}}}(g)$ are topologically equivalent, i.e. there exists a homeomorphism $h : S^2 \to S^2$ such that h maps trajectories of $\bar{\bar{\mathcal{N}}}(f)$ onto trajectories of $\bar{\bar{\mathcal{N}}}(g)$.

Theorem 3.1:

(i) $\tilde{\mathcal{R}}$ is τ-open and τ-dense in $\mathcal{R}$.

(ii) $\bar{\bar{\mathcal{N}}}(f)$ is structurally stable if and only if $f \in \tilde{\mathcal{R}}$.

Proof: Firstly we prove that $\tilde{\mathcal{R}}$ is τ-dense in $\mathcal{R}$. This is a direct consequence of the following three observations: (see Fig. 3.1).

a. If not all critical points or finite zeros, poles for $f \in \mathcal{R}$ are simple, then, given an arbitrary τ-neighborhood Ω of f, there exists an $\tilde{f} \in \Omega$ such that all critical points and finite zeros, poles for $\tilde{f}$ are simple.

b. For $\tilde{f} \in \Omega$ (as in a) there exists a τ-neighborhood $\tilde{\Omega}(\subset\Omega)$ of $\tilde{f}$ such that for all $g \in \tilde{\Omega}$ the critical points, finite zeros, poles are simple.

c. For an arbitrary τ-neighborhood $\tilde{\tilde{\Omega}}$ of $\tilde{f}$ with $\tilde{\tilde{\Omega}} \subset \tilde{\Omega}$ there exists an $\tilde{\tilde{f}} \in \tilde{\tilde{\Omega}}$ such that $\tilde{\tilde{f}} \in \tilde{\mathcal{R}}$.

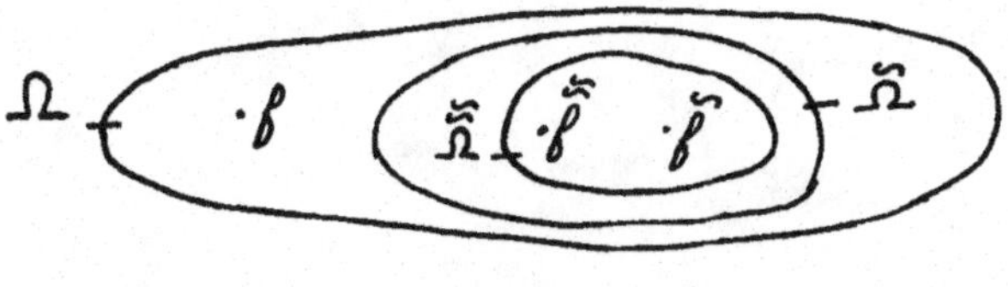

Fig. 3.1.

ad a,b: Let $f \in \mathcal{R}$, $f(z) = \frac{p_n(z)}{q_m(z)} = \frac{a_n z^n + \ldots + a_0}{b_m z^m + \ldots + b_0}$, $a_n \neq 0$, $b_m \neq 0$.

We say that f has the regularity property (R.P.) if the following two conditions are fulfilled:

1. $(\forall \varepsilon > 0)(\exists\, \hat{f}(z) = \frac{\hat{p}_n(z)}{\hat{q}_m(z)} = \frac{\hat{a}_n z^n + \ldots + \hat{a}_0}{\hat{b}_m z^m + \ldots + \hat{b}_0} \in \mathcal{R})$ with

   $|a_i - \hat{a}_i| < \varepsilon$, $|b_j - \hat{b}_j| < \varepsilon$, $i = 0, \ldots, n$; $j = 0, \ldots, m$ such that

   $\hat{f}$ has only simple finite zeros, poles.

2. Let $\hat{f}$ be as in 1, then:

   $(\exists \varepsilon_1 > 0)(\forall h(z) = \frac{\alpha_n z^n + \ldots + \alpha_0}{\beta_m z^m + \ldots + \beta_0} \in \mathcal{R}$ with $|\alpha_i - \hat{a}_i| < \varepsilon_1$, $|\beta_j - \hat{b}_j| < \varepsilon_1$; $i = 0, \ldots, n$;

   $j = 0, \ldots, m)$ (h has only simple finite zeros, poles).

From the very construction of the topology τ it follows: If $f \in \mathcal{R}$ has the (R.P.) and Ω is a τ-neighborhood of f, then there exists an $\hat{f} \in \Omega$ and a τ-neighborhood Ω_1 of $\hat{f}$ such that all $h \in \Omega_1$ have only simple finite zeros, poles. Using the properties of the discriminants of p_n and q_m (cf. [15]) one shows that $f = \frac{p_n}{q_m}$ $(\in \mathcal{R})$ has the (R.P.). Let h as in 2., then a simple calculation shows that: $h' \in \mathcal{R}$ and $C(h) = N(h')$. Since $h' \in \mathcal{R}$ has the (R.P.) as well and the coefficients of h' depend continuously on the coefficients of h the assertions a. and b. are justified. (Strictly speaking, for the case m = n special attention should be given to $z = \infty$ which can be either a regular or a critical point for f. However, the behaviour of f at $z = \infty$ is governed by algebraic equations on the coefficients of f (compare Remark 3.1). Consequently, for this case, we tackle the problem in a similar way as we did for finite critical points).

ad c: Let $f \in \mathcal{R}$, $f = \frac{p_n}{q_m}$, $\tilde{f}$, $\tilde{\Omega}$ as in assertions a. and b.

Firstly we consider the case $n \geq m$.

For $\varepsilon > 0$ sufficiently small we have $\tilde{f}(z) + \alpha \in \tilde{\Omega}$ as well as $C(\tilde{f}+\alpha) = C(\tilde{f})$ for all $\alpha \in \mathbb{C}$ with $|\alpha| < \varepsilon$.

Necessary conditions for the pair $(z_1, z_2) \in C(\tilde{f}) \times C(\tilde{f})$ to be connected by a trajectory of $\bar{\bar{\mathcal{N}}}(\tilde{f})$ are:

- $\tilde{f}(z_1) \neq \tilde{f}(z_2)$.
- There is a ray in the $\tilde{f}(z)$-plane issuing from the origin and connecting $\tilde{f}(z_1)$ and $\tilde{f}(z_2)$.

Note that this holds also if (for the case $m = n$) two finite critical points are connected by a trajectory passing through $z = \infty$ or if a finite critical point is connected with a critical point at $z = \infty$.

It is always possible to choose an α_0, with $|\alpha_0| < \varepsilon$ such that: All straight lines through any pair of different critical values $(\tilde{f}(z_i) + \alpha_0, \tilde{f}(z_j) + \alpha_0)$, $z_i, z_j \in C(\tilde{f})$ do not contain the origin of the $f(z)$-plane. (Note that if for $z_1, z_2 \in C(\tilde{f})$ we have $\tilde{f}(z_1) = \tilde{f}(z_2)$, in view of (2.1), they cannot be connected by a trajectory of $\bar{\bar{\mathcal{N}}}(\tilde{f}+\alpha)$). From this it follows that (in case $n \geq m$) assertion c is justified. The case that $n < m$ can be reduced to the case $m < n$ by making the following observations: $C(\tilde{f}) = C(\frac{1}{\tilde{f}})$; if $\tilde{f}$ is τ-near to $\tilde{g}$ then $\frac{1}{\tilde{f}}$ is τ-near to $\frac{1}{\tilde{g}}$; $\bar{\bar{\mathcal{N}}}(f) = -\bar{\bar{\mathcal{N}}}(\frac{1}{f})$.

We proceed by proving (ii):

Let $f \in \tilde{\mathcal{R}}$. In view of Lemma 3.1 all equilibrium-states of $\bar{\bar{\mathcal{N}}}(f)$ are non-degenerate nodes or saddlepoints, no saddlepoints being connected by a trajectory of $\bar{\bar{\mathcal{N}}}(f)$. In view of Lemma 2.2 the only possible limiting sets for trajectories of $\bar{\bar{\mathcal{N}}}(f)$ are isolated points of S^2. Applying the well-known characterization theorem of de Baggis-Peixoto (cf. [3], [20]) on structural stability for dynamical systems on a compact 2 dim. manifold we conclude in view of Lemma 3.2: $\bar{\bar{\mathcal{N}}}(f)$ is structurally stable.

Let $\bar{\bar{\mathcal{N}}}(f)$ be structurally stable and suppose $f \notin \tilde{\mathcal{R}}$. We already proved that since $\tilde{\mathcal{R}}$ is τ-dense in $\mathcal{R}$ the function f can be τ-approximated by elements of $\tilde{\mathcal{R}}$. This leads to the conclusion that the phase-portrait of $\bar{\bar{\mathcal{N}}}(f)$, $f \notin \tilde{\mathcal{R}}$, $f = \frac{p_n}{q_m}$ is topologically equivalent to the phase-portrait of $\bar{\bar{\mathcal{N}}}(\hat{f})$, $\hat{f} \in \tilde{\mathcal{R}}$, $\hat{f} = \frac{\hat{p}_n}{\hat{q}_m}$. A careful analysis of definitions 3.1,2 shows that this is impossible.

Finally we prove the τ-openess of $\tilde{\mathcal{R}}$ in $\mathcal{R}$. This is direct consequence of the (already proved) assertion (ii). □

Remark 3.3: For the proof of Theorem 3.1 we used the theorem of de Baggis-Peixoto. In our special case it might be possible to give a proof which is independent from this theorem. The advantage of our strategy however is that we may interpret the systems $\overline{\overline{\mathcal{N}}}(f)$, $f \in \tilde{\mathcal{R}}$ as examples of the so called Morse-Smale systems (cf. [21]). This point of view will be used in Section 4.

Remark 3.4: Although we restricted ourselves here to systems $\overline{\overline{\mathcal{N}}}(f)$, $f \in \mathcal{R}$ it might be possible to make some structural stability statements on systems $\overline{\overline{\mathcal{N}}}(f)$, $f \in \mathcal{E}$. Such statements however cannot be based on the de Baggis-Peixoto theorem, since for $f \in \mathcal{E}$ we always encounter an equilibrium-state at $z = \infty$ (see Lemma 2.4) which is degenerate. Moreover these systems obviously cannot be Morse-Smale.

Remark 3.5: Let $f \in \tilde{\mathcal{R}}$, $f = \dfrac{p_n}{q_m}$, $m > n$ ($m < n$). Then there exists a τ-neighborhood Ω of f such that: 1. $\Omega \subset \tilde{\mathcal{R}}$, 2. All $\tilde{f} \in \Omega$ have at $z = \infty$ a stable (unstable) node. Using the theorem of de Baggis-Peixoto as well as Lemma 3.2, Remark 2.3 it follows: The system $\mathcal{N}(f)$ is structurally stable in the following sense: If $\tilde{f} \in \Omega$, a homeomorphism ϕ: $\mathbb{C} \to \mathbb{C}$ exists such that ϕ maps the trajectories of $\mathcal{N}(f)$ onto the trajectories of $\mathcal{N}(\tilde{f})$ and the critical points for f on the critical points for $\tilde{f}$.

If $m = n$ and $z = \infty$ is a critical point for f, the saddlepoint $z = \infty$ for $\overline{\overline{\mathcal{N}}}(f)$ can "move" into a finite saddlepoint for $\overline{\overline{\mathcal{N}}}(\tilde{f})$, where $\tilde{f}$ is obtained from f by an arbitrary small perturbation of the coefficients of f. It follows that although $\overline{\overline{\mathcal{N}}}(f)$ is structurally stable, the system $\mathcal{N}(f)$ is not structurally stable (in the sense as defined above).

4. Applications

4.1. Justification of a conjecture of D. Braess.

Let $f \in \mathcal{R}$, $f(z) = \alpha(z-z_1)(z-z_2)(z-z_3)$ - z_1, z_2, z_3 as points of $\mathbb{R}^2$ non-collinear - be a polynomial of degree 3 which is non-degenerate in the sense of Definition 3.1. Then the triangle Δ with z_1, z_2, z_3 as its vertices has exactly one longest side. In fact - z_1, z_2, z_3 being distinct - in the case Δ is fully symmetric, the barycenter of Δ is a degenerate critical point for f and if there are exactly two longest sides, then the two distinct critical points for f are connected by a trajectory of $\mathcal{N}(f)$. The converse is also true, cf. [5]. In view of the Lemmas 2.2, 2.3, 2.6 and the non-degeneracy of f it is easily seen that $\bigcup_{i=1,2,3} \partial B(z_i)$ consists exactly of two components, each being the union of a critical point for f and the two trajectories tending to it for increasing t. Following Braess we call these components critical lines for f (they separate the basins $B(z_i)$, $i = 1,2,3$).

We denote the side of Δ opposite to z_i by l_i, $i = 1,2,3$.

For $z \in l_1 (z \neq z_2, z_3)$ we have:

$$(z_2-z_3) \cdot \frac{f'(z)}{f(z)} = (z_2-z_3) \left[\frac{1}{z-z_1} + \frac{1}{z-z_2} + \frac{1}{z-z_3} \right] .$$

Thus: $$\mathrm{Im}\left[(z_2-z_3) \cdot \frac{f'(z)}{f(z)}\right] = \mathrm{Im}\left(\frac{z_2-z_3}{z-z_1}\right) \neq 0.$$

Consequently in z the vectorfield $\overline{\mathcal{N}}(f)$ is transversal to l_1 and so a critical line may intersect l_1 transversally and only once. Obviously the same results hold for l_2, l_3. (See Fig. 4.1.1; note that by Lucas theorem [13] the critical points of f are situated in the convex hull of z_1, z_2, z_3).

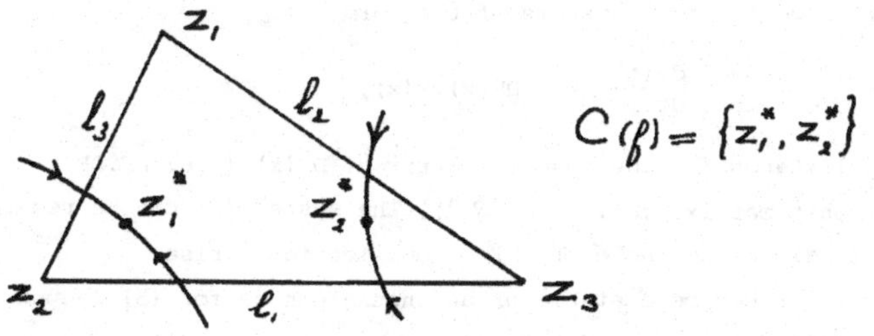

Fig. 4.4.1.

In [5] Braess states the following conjecture:

C_1: Let l_1 be the longest side of Δ. Then both critical lines for f intersect l_1. The sides l_2, l_3 are intersected by only one critical line.

In the special case that l_2 and l_3 are of equal length, Braess proved C_1 by elementary means (cf. [5]). We shall justify C_1 for the general case by using the concept of structural stability as developed in Section 3. Let z_2, z_3 be fixed and consider:

$$\{z \in \mathbb{C} \mid |z-z_2| < |z_2-z_3|, \ |z-z_3| < |z_2-z_3|; \ z, z_2, z_3 \ \text{non-collinear}\}.$$

Define one of its components to be G.

We are done if we prove that for all $z \in G$ the conjecture C_1 holds. Define $G_1 = \{z \in G \mid C_1 \text{ holds}\}$, $G_2 = \{z \in G \mid C_1 \text{ does not hold}\}$. Note that for $z \in G$ we have $f \in \tilde{\mathcal{R}}$ and in view of Theorem 3.1 we find that $\overline{\overline{\mathcal{N}}}(f)$ is structurally stable if $z \in G$. Structural stability of dynamical systems on S^2 is equivalent with ε-structural stability (cf. [3]) (i.e. in the situation of Definition 3.2, given an arbitrary $\varepsilon > 0$ and any compact subset D of the z-chart, the homeomorphism h may be constructed such that on D: $|h(z) - z| < \varepsilon$ if g is sufficiently τ-near to f; in fact we implicitely use Lemma 3.2 again). From the preceding observations it follows that G_1, G_2 are open subsets of G. Since all points $z \in G$ with $|z - z_2| = |z - z_3|$ belong to G_1 we have $G_1 \neq \emptyset$. From the connectedness of G it follows that $G = G_1$. □

4.2. A counterexample for a conjecture of F.H. Branin

Let $F: \mathbb{R}^2 \to \mathbb{R}^2$ be of class C^∞, $DF(x)$ the Jacobian matrix of F at x and suppose that F is generic in the sense that $\|F(x)\| + |\det DF(x)| > 0$ on $\mathbb{R}^2$.

Consider the system:

$$\frac{dx(t)}{dt} = -DF^{-1}(x).F(x), \quad x \text{ such that } \det DF(x) \neq 0. \quad (\alpha)$$

The zeros for F are the only equilibrium-states of (α) and all of them are stable nodes (cf. [7]). In general (α) is not defined on the whole $\mathbb{R}^2$ and in [6] Branin proposed to study systems of the form:

$$\frac{dx(t)}{dt} = -\widetilde{DF}(x).F(x), \quad (\beta)$$

where $\widetilde{DF}(x)$ stands for the adjoint matrix of $DF(x)$ (i.e. $\widetilde{DF}.DF = I.\det DF$, I being the 2×2-unit matrix) (cf. also [7]). The system (β) can be regarded as an extension of (α) to the whole $\mathbb{R}^2$. Two complications arise:

1. A zero for F can be a stable or an unstable node for (β) (depending on the sign of det DF).

2. In general (β) will have (apart from the zeros for F) other equilibrium-states, namely those $x \in \mathbb{R}^2$ at which $\widetilde{DF}(x).F(x) = 0$, $F(x) \neq 0$. Following Branin we call these equilibrium-states extraneous singularities for (β).

On "numerical evidence" Branin stated the following conjecture:

C_2: If there are no extraneous singularities for (β) one can find all zeros for F by following (for in-or de-creasing t) a trajectory of (β) until one reaches a zero for F and following another trajectory issuing from this zero until one reaches another zero, and so on. In other words: (β) provides a globally convergent method for finding all zeros for f in absence of extraneous singularities (cf. [6], [7], [8]).

We shall reject this conjecture by giving a counterexample.
Put $F(x,y) = (u(x,y), v(x,y))^T$ and suppose that u(x,y) and v(x,y) are continuously differentiable and fulfil the Cauchy-Riemann equations on $\mathbb{R}^2$. Then F can be considered as an entire function $f: \mathbb{C} \to \mathbb{C}$. The Newton-flow $\mathcal{N}(f)$ is extended to $\mathbb{C}$ by means of the system: $\frac{dz}{dt} = -\overline{f'(z)}.f(z)$ (cf. Lemma 2.2, Step 1).
Separating this equation in real and imaginary parts and using the Cauchy-Riemann equations we find:

$$\frac{d}{dt}\begin{pmatrix} x(t) \\ y(t) \end{pmatrix} = -\begin{pmatrix} v_y & -u_y \\ -v_x & u_x \end{pmatrix}\begin{pmatrix} u \\ v \end{pmatrix}, \qquad (\gamma)$$

where u_x, u_y, v_x, v_y denote the partial derivatives w.r.t. x,y. Obviously

$\widetilde{DF} = \begin{pmatrix} v_y & -u_y \\ -v_x & u_x \end{pmatrix}$, so (γ) is a special case of the Branin-system (β).

Using again the Cauchy-Riemann equations we find:

(x,y) is an extraneous singularity iff $z = x + iy \in C(f)$.

Let us consider $f(z) = e^z - 1$. Since $C(f) = \emptyset$ the corresponding Branin system has no extraneous singularities. Moreover, the zeros for f are all stable nodes (cf. Section 2). Consequently one cannot go from one zero to another in following trajectories of (γ). So the system

$$\begin{cases} \dot{x} = -e^{2x} + e^x.\cos y \\ \dot{y} = -e^x.\sin y \end{cases}$$

provides a counterexample for C_2. In fact, it is easily seen that the phase-portrait of $\mathcal{N}(f)$ has the form as depicted in Fig. 4.2.1; the basins of the zeros for f are separated by the trajectories $\{z \mid \text{Im } z = (2k+1)\pi\}$, $k \in \mathbb{Z}$.
For a detailed study on Branin-systems w.r.t. F = grad g we refer to [14].

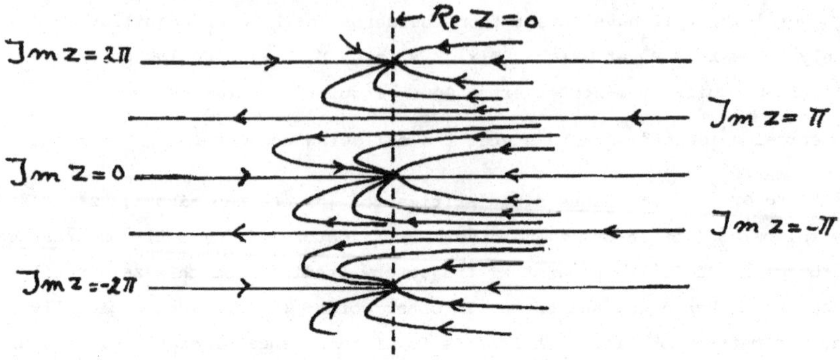

Fig. 4.2.1.

4.3 On the number of different phase-portraits of $\mathcal{N}(f)$, up to their topological type, if $f \in \tilde{\mathcal{R}}$ is a polynomial of degree n.

In this subsection f is always a non-degenerate polynomial (in the sense of Definition 3.1) of degree n.
By the unstable (stable) manifold $\gamma_-(z)$ ($\gamma_+(z)$) of $z \in C(f)$ we denote the union of $\{z\}$ and the two trajectories of $\mathcal{N}(f)$ emanating from (tending to) z.

Lemma 4.3.1. In the situation as above:

1. $\partial R(\infty) = \bigcup_{z \in C(f)} \overline{\gamma_-}(z)$.
2. Each element of C(f) is connected - by trajectories of $\mathcal{N}(f)$ - to two different elements of N(f).

Proof: 1. Since $\overline{R}(\infty) = \mathbb{C}$ (cf. Lemma 2.6, proof of Part 2) a trajectory which emanates from $z \in C(f)$ must be contained in $\partial R(\infty)$. It follows from Lemma 2.8 (i) that $\partial R(\infty) \supset \bigcup_{z \in C(f)} \overline{\gamma_-}(z)$. Since $\partial R(\infty)$ is compact (Lemma 2.8) the trajectory through regular points of $\partial R(\infty)$ cannot tend -for decreasing t- to $z = \infty$ and consequently tends to an element of C(f) (cf. Lemma 2.2). From this it follows:
$\partial R(\infty) \subset \bigcup_{z \in C(f)} \overline{\gamma_-}(z)$.

2. The two separatrices contained in the unstable manifold of $z \in C(f)$ tend to zeros for f. These zeros are different as a consequence of Lemma 2.8 (ii). □

Definition 4.3.1. The graph $\mathcal{G}(f)$ is defined as follows:

- The vertices of $\mathcal{G}(f)$ are the elements of N(f).
- The edges of $\mathcal{G}(f)$ are the unstable manifolds $\gamma_-(z)$, $z \in C(f)$.
- The vertex $z_1 \in N(f)$ and the edge $\gamma_-(z_2)$, $z_2 \in C(f)$ are said to be incident iff $z_1 \in \overline{\gamma_-}(z_2)$.

Remark 4.3.1. From Lemma 4.3.1 it follows that $\partial R(\infty)$ may be viewed as a realization of the graph $\mathcal{G}(f)$.

Lemma 4.3.2. $\mathcal{G}(f)$ is a (connected) tree.

Proof: This is a direct consequence of Lemma 2.8.

Remark 4.3.2. Using Lemmas 2.2, 4.3.1 (2) it is not difficult to prove the following assertion: For $z_{1,2} \in N(f)$ we have: $\partial B(z_1) \cap \partial B(z_2) \neq \emptyset$ iff z_1 and z_2 are incident with the same edge of $\mathcal{G}(f)$. This shows that $\mathcal{G}(f)$ and the graph introduced by Braess in [5] are essentially the same.

Remark 4.3.3. The graph $\mathcal{G}(f)$ is closely related to the graph $\mathcal{P}(f)$ - introduced by Peixoto in [21] - which is defined as follows:

- The vertices of $\mathcal{P}(f)$ are the zeros, critical points for f and $z = \infty$.
- The edges of $\mathcal{P}(f)$ are the (regular) trajectories w.r.t. $\bar{\bar{\mathcal{N}}}(f)$ tending to or emanating from a critical point for f.
- A vertex z_0 of $\mathcal{P}(f)$ is said to be incident with an edge $\gamma(z_1)$ iff $z_0 \in \bar{\gamma}(z_1)$.

The set $\bigcup_{z \in C(f)} (\bar{\gamma}_+(z) \cup \bar{\gamma}_-(z))$ in S^2 may be viewed as the realization of $\mathcal{P}(f)$ and will also be denoted by $\mathcal{P}(f)$, if no confusion is possible.

Lemma 4.3.3. Let z(t) be the solution of $\mathcal{N}(f)$ describing a regular trajectory in the stable manifold of $z^* \in C(f)$; $s = \#C(f)$ (= n-1). Then we have:

1. $\lim_{t \to -\infty} \arg z(t)$ exists.

2. No two limits as defined in 1. are equal. (Consequently the stable manifolds $\gamma_+(z)$, $z \in C(f)$ define 2s different "limit-directions").

Proof: Consider the system $\bar{\bar{\mathcal{N}}}(f)$ defined on S^2, then we know that $z = \infty$ is a non-degenerate unstable node and moreover the linear approximation of $\bar{\bar{\mathcal{N}}}(f)$ at $z = \infty$ is - w.r.t. the w-chart - of the form $+n.|a_n|^2.w$ (See Lemma 2.1, Step 3; case $m = 0$, $b_0 = 1$). Consequently, each trajectory emanates from $z = \infty$ in a definite direction, different trajectories corresponding to different directions (cf.[2]).

□

Note that $\bar{\bar{\mathcal{N}}}(f)$ is a so called Morse-Smale system defined on S^2 (cf.[21]). In order to make use of the classification-theorem on these systems we introduce the following lemma (Lemma 4.3.4).

Let C_R be a circle around $z = 0$, with radius R such that $\bar{\bar{\mathcal{N}}}(f)$ is pointing inwards C_R at every point of C_R; and moreover $N(f) \cup C(f)$ is contained in the interior of C_R. Note that by choosing R sufficiently large this is always possible. Let $C(f) = \{z_1, \ldots, z_s\}$. For each z_j, $j = 1, \ldots, s$, the stable manifold $\gamma_+(z_j)$ intersects C_R in exactly two points, say j_1, j_2. Let j_α, $j'_{\alpha'}$ be two such points with the property that at least one of the two open arcs of C_R determined by them is not intersected by any $\gamma_+(z_k)$, $z_k \in C(f)$. Denote such an arc by: $\mathrm{arc}(j_\alpha, j'_{\alpha'})$.

In the sequel the pair $(j_\alpha, j'_{\alpha'})$ will be called consecutive.

Then we have, see also Figure 4.3.1^{a,b}:

Lemma 4.3.4.

1. For every $z_0 \in \mathrm{arc}(j_\alpha, j'_{\alpha'})$ the trajectory $\gamma(z_0)$ tends (for $t \to \infty$) to the same $z^* \in N(f)$; in fact - in terms of the graph $\mathcal{G}(f)$ - z^* is the only vertex incident with both edges $\gamma_-(z_j)$, $\gamma_-(z_{j'})$ if $\gamma_-(z_j) \neq \gamma_-(z_{j'})$, and a vertex of degree 1 if $\gamma_-(z_j) = \gamma_-(z_{j'})$.

2. The union of all trajectories $\gamma(z_0)$, $z_0 \in \mathrm{arc}\,(j_\alpha, j'_{\alpha'})$ is a component of $S^2 \setminus \mathcal{P}(f)$, being a distinghuished set of type 1 if $j \neq j'$, and of type 3 if $j = j'$. (For a definition of distinguished sets, see [21]).

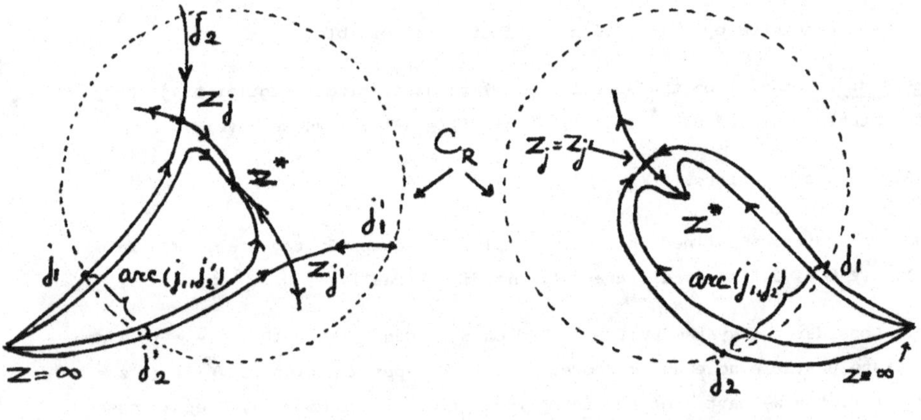

Fig. 4.3.1.a. Fig. 4.3.1.b.

Proof: 1. Let V be the set of zeros for f which are reached by a trajectory passing through $\mathrm{arc}(j_\alpha, j'_{\alpha'})$. Then $\{B(z) \cap \mathrm{arc}(j_\alpha, j'_{\alpha'}) \mid z \in V\}$ provides an open covering of $\mathrm{arc}(j_\alpha, j'_{\alpha'})$. From the connectedness of $\mathrm{arc}(j_\alpha, j'_{\alpha'})$ it follows that V consists of one single element, say z^*. It follows that $j_\alpha, j'_{\alpha'} \in \partial B(z^*)$ and consequently $z_j, z_{j'} \in \partial B(z^*)$ (use Lemma 2.3(2)). In view of the proof of Step (iii) of this latter lemma, one of the two separatrices in $\gamma_-(z_j)$ tends to z^*; the same property holds for $\gamma_-(z_{j'})$.

2. One easily shows that the union of all trajectories $\gamma(z_0)$, $z_0 \in \mathrm{arc}(j_\alpha, j'_{\alpha'})$ is a path-connected, open subset of $S^2 \setminus \mathcal{P}(f)$. Different arcs between consecutive points define disjoint subsets. Moreover, let $z \in S^2 \setminus \mathcal{P}(f)$. Since $\overline{R}(\infty) = \mathbb{C}$ the (regular) trajectory through z must intersect C_R in one of these arcs, i.e. all subsets as defined above provide an open covering of $S^2 \setminus \mathcal{P}(f)$. Consequently these subsets are the connected components of $S^2 \setminus \mathcal{P}(f)$. In case $j = j'$ only one critical point $z_j = z_{j'}$ is involved; apparently - in terms of the graph $\mathcal{G}(f)$ - z^* is a vertex incident with just one edge, namely $\gamma_-(z_j)$. The remaining part of the assertion is a direct consequence of the definition of distinguished sets. □

From Lemma 4.3.1(2) it follows that the so called distinguished sets of type 2 (cf. [21]) do not occur in our case.

Remark 4.3.4. In case $j = j'$, let the "limit-directions" determined by $\gamma_+(z_j)$ be denoted by Θ_1, Θ_2. Then $|\Theta_1 - \Theta_2| = \frac{2\pi}{n}$. This follows from the facts that z^* is a simple zero and $z = \infty$ is a pole of order n for f.

We now turn to the converse statement of Lemma 4.3.2.

Theorem 4.3.1. Let Γ be a (connected) tree with n vertices. Then there exists a non-degenerate polynomial f, with degree n such that Γ and $\mathcal{G}(f)$ are isomorphic.

Proof: The proof will be given by induction on n. We only pay attention to the induction-step since the assertion is proved by inspection up to n = 5 by Braess (cf. [5]). To be precise we assume the existence of a polynomial $f \in \tilde{\mathcal{R}}$ (degree n) and we shall construct a polynomial $g \in \tilde{\mathcal{R}}$ (degree n+1) such that $\mathcal{G}(g)$ is obtained from $\mathcal{G}(f)$ by attaching an edge to one of the n vertices (which may be arbitrarily chosen) of $\mathcal{G}(f)$.

Without loss of generality we assume that the coefficient of z^n in f equals 1. By a_{n-1} we denote the coefficient of z^{n-1}.

According to Lemma 4.3.3 there are 2s (s = n-1) different directions in which the stable manifolds $\gamma_+(z)$, $z \in C(f)$, "approach" $z = \infty$. The set of (positive) angles determined by all pairs of consecutive directions has a minimum, say Θ.

I.{Let $\delta_1 > 0$ be such that $\arg(1+z) \ll \Theta$ for all those $z \in \mathbb{C}$ with $|z| < \delta_1$. Here and in the sequel the expression "$\ll\Theta$" will be used to denote a quantity which is negligibly small in comparison with Θ.

Now we choose R > 1 such that for the circle C_R around z = 0 with radius R the following assertions hold:

II.
1. $N(f) \cup C(f)$ is contained in the interior of the compact disc D_R, determined by C_R, while $\mathcal{N}(f)$ points inwards D_R along C_R.
2. $|\frac{f(z)}{z^n} -1| < \delta_1$ for $|z| \geq R$.
3. Using the notation of Lemmas 4.3.3, 4.3.4:

$$|\arg j_\alpha - \lim_{t\to-\infty} \arg z_{j_\alpha}(t)| \ll \Theta, \qquad j = 1,\ldots,s;\ \alpha = 1,2,$$

where $z_{j_\alpha}(t)$ stands for the solution of $\mathcal{N}(f)$ with $z_{j_\alpha}(0) = j_\alpha$.

This R will be kept fixed for the sequel.

We introduce the function $g_\varepsilon(z) := (\varepsilon z - e^{i\psi}).f(z)$, where $\varepsilon \in]0,\frac{1}{R}[$ and the angle ψ will be fixed later on.

Note that in the disc D_R the function g_ε has n zeros which coincide with the zeros for f.

One easily proves that on D_R the following assertion holds:

$$\mathcal{N}^*(g_\varepsilon) \ (= -\overline{g}_\varepsilon'(z).g_\varepsilon(z)) = \mathcal{N}^*(f) + \varepsilon\lambda,$$

where λ stands for a polynomial expression in ε, z and $\overline{z}$. (For $\mathcal{N}^*(f)$ also see Lemma 2.1 (Step 1)).

Since D_R is compact the vectorfields $\mathcal{N}^*(f)$ and $\mathcal{N}^*(g_\varepsilon)$ can be made arbitrarily near in the usual metric on $\mathcal{X}(D_R)$ by choosing ε sufficiently small; compare also Lemma 3.2 (proof).

Since $f \in \tilde{\mathcal{R}}$ we know that $\mathcal{N}^*(f)$ is "ε-structurally stable" on D_R (cf. [3], Section 4.1).

We now fix a number $\eta_0 > 0$ such that:

III.
1. $\eta_0 \ll \Theta$
2. The disc $\{\zeta \mid |z-\zeta| \leq \eta_0\}$ is contained in the basin B(z) for every $z \in N(f)$.

In view of the ε-structural stability of $\mathcal{N}^*(f)$ on D_R one may fix a number $\delta_2 \in]0,\frac{1}{R}[$ such that:

For all $\varepsilon \in]0,\delta_2[$ a homeomorphism $h: D_R \to D_R$ exists such that:

IV.
1. For all $z \in D_R$: $|h(z) - z| < \eta_0$.
2. $\mathcal{N}^*(g_\varepsilon)$ points inwards to D_R everywhere on C_R.
3. h maps (maximal) trajectories of $\mathcal{N}^*(f)$ onto (maximal) trajectories of $\mathcal{N}^*(g_\varepsilon)$; h preserves the orientation.

Note that although the zeros for f coincide with zeros for g_ε their critical points may be different. However, in view of IV all but one of the critical points for g_ε are contained in $\overset{\circ}{D}_R$.

Let $z^* \in N(f)$ and $z_j \in C(f)$ such that $z^* \in \overline{\gamma}_-(z_j)$. Since $h(\gamma_-(z_j))$ is the unstable manifold of $h(z_j)$ w.r.t. $\mathcal{N}^*(g_\varepsilon)$ and moreover $h(z^*) \in N(g_\varepsilon) \cap D_R$ it follows that $h(z^*) \in N(f)$. In view of III and IV one concludes that $h(z^*) = z^*$ and $h(z_j)$ is "connected" (by a trajectory of $\mathcal{N}^*(g_\varepsilon)$) with z^* (so h maps "connected singularities" of $\mathcal{N}^*(f)$ to "connected singularities" of $\mathcal{N}^*(g_\varepsilon)$ (as far as they are contained in D_R)). In terms of the graphs $\mathcal{G}(f)$ and $\mathcal{G}(g_\varepsilon)$ we have proved: If two vertices of $\mathcal{G}(f)$ are incident with an edge of this graph, these vertices (considered as vertices of $\mathcal{G}(g_\varepsilon)$) are also incident with an edge of $\mathcal{G}(g_\varepsilon)$, i.e. deleting $\frac{1}{\varepsilon} e^{i\psi}$ from $\mathcal{G}(g_\varepsilon)$ just gives $\mathcal{G}(f)$.

Let $z_\varepsilon = \frac{1}{\varepsilon} e^{i\psi}$ be the only zero for g_ε outside D_R and $\tilde{z}_\varepsilon$ the only critical point for g_ε outside D_R.

Since:

$$g'_\varepsilon(z) = (n+1)\varepsilon z^n + n(\varepsilon a_{n-1} - e^{i\psi})z^{n-1} + \dots ,$$

we find:

$$\tilde{z}_\varepsilon = \frac{n}{n+1} \cdot \frac{1}{\varepsilon} e^{i\psi} - \frac{n}{n+1} \cdot a_{n-1} - \Sigma_\varepsilon ,$$

where Σ_ε stands for Σz taken over all $z \in C(g_\varepsilon) \cap D_R$. Obviously: $|a_{n-1}| < nR$ and $|\Sigma_\varepsilon| < nR$.

From now on we take a <u>fixed</u> ε, such that $\varepsilon < \min\left\{\delta_2, \frac{\delta_1}{3n^2R}\right\}$.

It follows that:

$$\text{V.}\quad \begin{cases} \tilde{z}_\varepsilon = \frac{n}{n+1} z_\varepsilon (1+u), & \text{with } |u| < \delta_1; \\ (\varepsilon \tilde{z}_\varepsilon - e^{i\psi}) = -\frac{1}{n+1} e^{i\psi}(1+v), & \text{with } |v| < \delta_1. \end{cases}$$

Using the conditions I, II(2) and V one verifies:

$$\text{VI.}\quad \begin{cases} 1.\ \arg g_\varepsilon(z) \approx \arg(-z^n . e^{i\psi}) & \text{for } z \in C_R; \\ 2.\ \arg(g_\varepsilon(z)) \approx \arg(-z_\varepsilon^n . e^{i\psi}) & \text{for } z \text{ in the open segment }]Re^{i\psi}, z_\varepsilon[\\ \quad \text{determined by } Re^{i\psi} \text{ and } z_\varepsilon; & \\ 3.\ \arg(g_\varepsilon(\tilde{z}_\varepsilon)) \approx \arg(-z_\varepsilon^n . e^{i\psi}). & \end{cases}$$

where $\alpha \approx \beta$ means $|\alpha - \beta| << \Theta$.

Adopting the notations of Lemma 4.3.4 we consider two consecutive intersections of the stable manifolds $\gamma_+(z_j)$, $\gamma_+(z_{j'})$ with C_R, which will be denoted by $j_\alpha (j'_{\alpha'})$. By Lemma 4.3.4 all trajectories of $\mathcal{N}(f)$ through $\mathrm{arc}(j_\alpha, j'_{\alpha'})$ tend for $t \to \infty$ to the same $z^* \in N(f)$; in terms of $\mathcal{G}(f)$: Vertex z^* is incident with the edges $\gamma_-(z_j)$, $\gamma_-(z_{j'})$. Since h is a homeomorphism of D_R onto itself we have: $h(j_\alpha), h(j'_{\alpha'}) \in C_R$. We already proved that $h(z^*) = z^*$. From IV it now follows that $h(j_\alpha)$ and $h(j'_{\alpha'})$ are two consecutive intersection points of stable manifolds $h(\gamma_+(z_j))$ and $h(\gamma_+(z_{j'}))$ with C_R. Moreover trajectories (w.r.t. $\mathcal{N}^*(g_\varepsilon)$) through points of arc $(h(j_\alpha), h(j'_{\alpha'}))$ tend to z^*.

Using II(3) and IV we find:

$$(*) \qquad \arg h(j_\alpha) \approx \lim_{t \to -\infty} \arg z_{j_\alpha}(t), \qquad j = 1,\dots,s, \quad \alpha = 1,2.$$

Finally we are going to fix ψ:

We want to choose ψ such that $Re^{i\psi} \in \mathrm{arc}\,(h(j_\alpha), h(j'_{\alpha'}))$. Consequently, in view of (*), the possible values for ψ are restricted to an interval I of minimum length $\approx \Theta$. Using the condition VI (1), for $z = h(j_\alpha)$ and $z = h(j'_{\alpha'})$ as well as the conditions VI (2), (3) one proves that it is always possible to choose $\psi \in I$ such that:

$$\text{VII.} \quad \begin{cases} 1. \quad Re^{i\psi} \in \mathrm{arc}\,(h(j_\alpha), h(j'_{\alpha'})). \\ 2. \quad \arg g_\varepsilon(z) \neq \arg g_\varepsilon(h(j_\alpha)) \text{ and } \neq \arg g_\varepsilon(h(j'_{\alpha'})) \\ \qquad \text{for all } z \in\,]Re^{i\psi}, z_\varepsilon[. \\ 3. \quad \arg g_\varepsilon(\tilde{z}_\varepsilon) \neq \arg g_\varepsilon\,(h(j_\alpha)) \text{ and } \neq \arg g_\varepsilon(h(j'_{\alpha'})). \end{cases}$$

The part of the trajectory - w.r.t. $\bar{\bar{\mathcal{N}}}(g_\varepsilon)$- through $h(j_\alpha)$ $(h(j'_{\alpha'}))$ outside D_R is denoted by $\kappa(\kappa')$. Obviously $z_\varepsilon \notin \bar{\kappa} \cup \bar{\kappa}'$; by (2.1) and VII(3) also $\tilde{z}_\varepsilon \notin \bar{\kappa} \cup \bar{\kappa}'$.

Define the Jordan-curve J in S^2 by:

$$J := \kappa \cup \kappa' \cup \mathrm{arc}\,(h(j_\alpha), (h(j'_{\alpha'})) \cup \{z = \infty\}$$

Obviously J is closed and piece-wise differentiable with the property that every trajectory of $\bar{\bar{\mathcal{N}}}(g_\varepsilon)$ - apart from $\gamma(h(j_\alpha))$ and $\gamma(h(j'_{\alpha'}))$ - crosses J at most once in a point of arc $(h(j_\alpha), h(j'_{\alpha'}))$.
Clearly: $z_\varepsilon \notin J$ and $\tilde{z}_\varepsilon \notin J$.

It is well-known that $S^2 \setminus J$ consists of two (connected) components, say A and B (cf. [23]). Suppose $z_\varepsilon \in A$.

Since for $z \in\,]Re^{i\psi}, z_\varepsilon[$ $\arg g_\varepsilon(z)$ is different from the arguments of g_ε on κ and κ' (see VII(2)) we have $(R+\delta)e^{i\psi} \in A$ for sufficiently small $\delta > 0$. Consequently $(R-\delta)e^{i\psi} \in B$, for $\delta > 0$ sufficiently small (cf. [9], [23]). Since $\mathring{D}_R \cap J = \emptyset$ it follows that $\mathring{D}_R \subset B$.

Now consider z_ε. Every trajectory of $\overline{\overline{\mathcal{N}}}(g_\varepsilon)$ tending to z_ε (for $t \to \infty$) must emanate from $z = \infty$ or from an element $\hat{z} \in C(g_\varepsilon)$. Assume $\hat{z} \in \mathring{D}_R (\subset B)$ then this trajectory has to cross J in an element of C_R. This is impossible by IV (2). Thus $\hat{z} = \tilde{z}_\varepsilon$. Since at least one trajectory tending to z_ε must emanate from a point of $C(g_\varepsilon)$ (cf. Lemma 2.8 (iii)) we conclude that exactly one trajectory connects z_ε with $\tilde{z}_\varepsilon$ (cf. Lemma 4.3.1) and all other trajectories tending to z_ε must emanate from $z = \infty$. This proves that $\tilde{z}_\varepsilon \in A$ and that the other separatrix in $\gamma_-(\tilde{z}_\varepsilon)$ must tend to an element $\tilde{z} \in N(g_\varepsilon) \cup C(g_\varepsilon)$, which consequently is contained in $\mathring{D}_R$. It follows that this separatrix has to cross J in arc $(h(j_\alpha), h(j'_{\alpha'}))$ and thus this $\tilde{z}$ must be z^*. We now conclude:

1. g_ε is a non-degenerate polynomial of degree $(n+1)$.
2. In terms of $\mathcal{G}(g_\varepsilon)$: Vertex z_ε is incident with only one edge, namely $\gamma_-(\tilde{z}_\varepsilon)$; z^* also incident with $\gamma_-(\tilde{z}_\varepsilon)$.
3. The union of all trajectories tending to z_ε is a distinguished set of type 3 (since $\gamma_+(\tilde{z}_\varepsilon) \subset A$).

This concludes the proof of Theorem 4.3.1. □

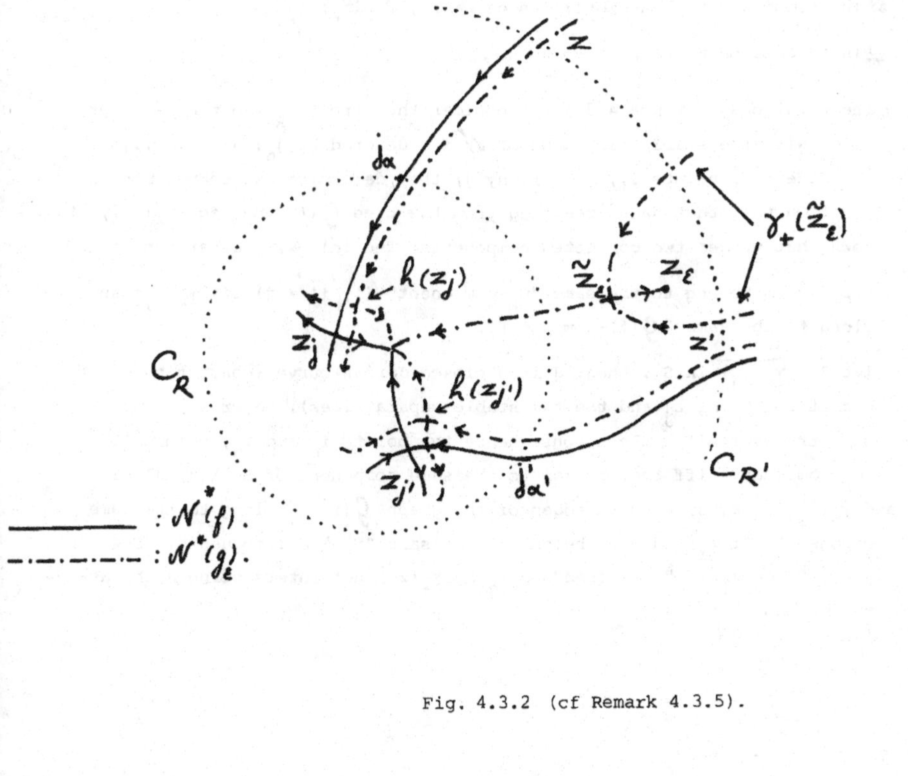

Fig. 4.3.2 (cf Remark 4.3.5).

Remark 4.3.5.

For later use we emphasize that actually much more than the assertion of Theorem 4.3.1 has been proved:

Taking R' sufficiently large, in fact $>> \frac{1}{\varepsilon}$, the patterns of intersections of stable manifolds (w.r.t. $\mathcal{N}^*(f)$, $\mathcal{N}^*(g_\varepsilon)$) with $C_{R'}$ are the same, with the following exception: If z, z' $\in C_{R'}$ are "connected" to $h(j_\alpha)$, $h(j'_{\alpha'})$ the points $\gamma_+(\tilde{z}_\varepsilon) \cap C_{R'}$ are to be found on arc (z,z'), forming a consecutive pair w.r.t. g_ε. (Note that $\gamma_+(\tilde{z}_\varepsilon)$ stands for the stable manifold of $\tilde{z}_\varepsilon$ w.r.t. $\mathcal{N}^*(g_\varepsilon)$). See also Figure 4.3.2.

Remark 4.3.6.

If z_ε moves to infinity (by taking $\varepsilon \to 0$) and g_ε degenerates to f, the limit-directions of the stable manifolds change discontinuously. Compare also Remark 4.3.4.

Let $\mathcal{P}_n$ be the set of all non-degenerate polynomials of degree n. Theorem 4.3.1 implies the following corollary:

Corollary 4.3.1

$$\#(\text{non-equivalent phase-portraits of } \mathcal{N}(f),\ f \in \mathcal{P}_n) \geq$$
$$\#(\text{non-isomorphic trees of order } n).$$

Adopting the notations of Lemma 4.3.4 we consider the circle C_R and the 2s intersections of C_R with the stable manifolds of $\mathcal{N}(f)$, denoted by j_α; $j = 1, \dots, s$; $\alpha = 1,2$. Consider the points j_1, j_2 for any j; they determine two components on the circle C_R. On the other hand, removing from the tree $\mathcal{G}(f)$ the edge $\gamma_-(z_j)$, it is well-known that we get two connected components. The following assertion holds:

The points i_α, $i'_{\alpha'}$ belong to the same "C_R-component" (i, i' $\neq$ j) iff $\gamma_-(z_i)$ and $\gamma_-(z_{i'})$ belong to the same "$\mathcal{G}$(f)-component". (*)

In fact, let $J = \overline{\gamma_+(z_j)}$ in S^2, then: J is a closed Jordan-curve (consisting of $z = \infty$, the critical point z_j and the two stable separatrices). J crosses $\gamma_-(z_j)$ as well as C_R transversally. As a consequence the points i_α and $i'_{\alpha'}$ belong to the same "C_R-component" iff they belong to the same component of $S^2 \setminus J$; also: $\gamma_-(z_i)$ and $\gamma_-(z_{i'})$ - considered as edges of the graph $\mathcal{G}$(f) - belong to the same "$\mathcal{G}$(f)-component" iff z_i and $z_{i'}$ belong to the same "$S^2 \setminus J$-component". (See Figure 4.3.3). Since z_i is "connected" to i_α by $\gamma_+(z_i)$ not intersecting J the assertion (*) follows.

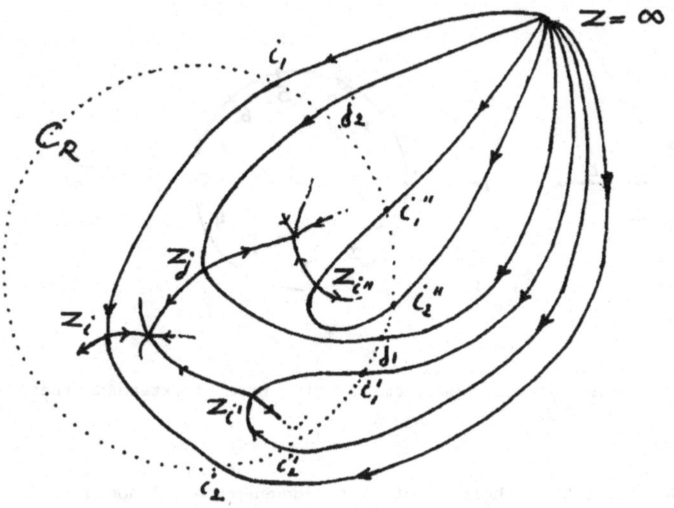

Fig. 4.3.3.

Note that there is no canonical way for labeling the two stable separatrices in any $\gamma_+(z_i)$ (whereas the instable separatrices in $\gamma_-(z_i)$ may be labeled by the zeros for f). Thus, although we may choose a definite labeling for separatrices in the stable manifolds $\gamma_+(z_i)$ - this is only necessary to label the edges of $\mathcal{P}(f)$ (cf. Remark 4.3.3) - the essential information is given by the configuration of the 2s "intersection"-points j_α on C_R. This leads to the following definition:

Definition 4.3.2. Let Γ be a tree with labeled edges (from 1 to s). A configuration of 2s points on a circle C is called compatible with Γ if: To each i ($1 \leq i \leq s$) correspond two different points on C (denoted by (i,i)) in such a way that for any j ($1 \leq j \leq s$) the configurations induced on the two components of C determined by the pair of points (j,j) correspond - in the sense of (*) - to the two (tree-)components of $\Gamma \setminus \{j\}$.

A tree Γ together with a compatible configuration will be called an extended tree Γ^e.

Example 4.3.1:

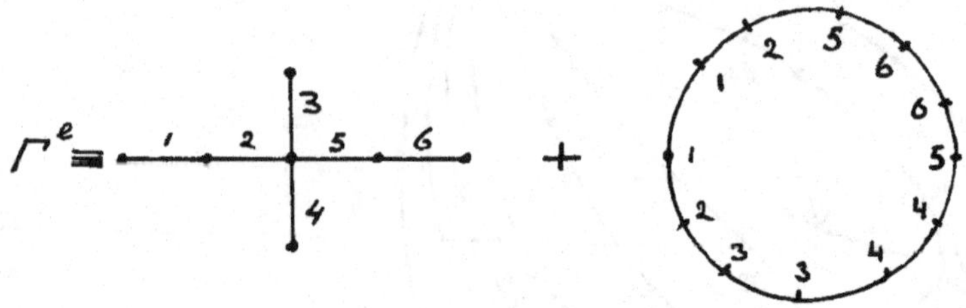

We have already seen that non-degenerate polynomials give rise to extended trees.

Theorem 4.3.2.

1. Let Γ^e be an extended tree. Then there exists a non-degenerate polynomial such that its extended tree is isomorphic to Γ^e.

2. There exists a bijective correspondence between the set of extended trees (up to isomorphism) and the set of phase-portraits of $\mathcal{N}(f)$ (up to equivalence), f a non-degenerate polynomial.

Proof: 1. (By induction on the number of vertices of Γ).

Since for n=3 the assertion may be proved by inspection (cf. [5]), we only pay attention to the induction-step. Let Γ^e be an extended tree with (n+1) vertices. Since Γ is a tree at least two vertices are of degree 1; let i be the label of an edge incident with such a vertex $(1 \leq i \leq n)$. It is easily seen that the two points on C (of Γ^e) corresponding to i form a consecutive pair (in the sense of Lemma 4.3.4). Let the labels j, j' (both $\neq$ i) be such that at least one of the pair (j,j) and one of the pair (j',j') forms a consecutive pair with one of the two points (i,i); e.g. in Example 4.3.1 for i = 3, j, j' turn out to be 2 resp. 4.

By deleting edge i from Γ we get a new tree Γ' (the labeling of its edges being induced by the labeling of Γ); similarly by deleting the pair (i,i) from C we get a new configuration. It is not hard to see that this new configuration is compatible with Γ', i.e. we get a new extended tree Γ'^e. By induction we may assume that Γ'^e - up to isomorphism - is realized by a non-degenerate polynomial (of degree n). In this realization let j_α and $j'_{\alpha'}$ (as points of C_R; cf. proof of Theorem 4.3.1) be such that j and j' are the labels as introduced above. According to Remark 4.3.5 it is possible to construct a non-degenerate polynomial $g_\varepsilon(z)$ - of degree (n+1) - such that the configuration corresponding to g_ε is obtained from the configuration corresponding to f by "implanting" a consecutive pair (i,i) within arc$(j_\alpha, j'_{\alpha'})$. The zero $\frac{1}{\varepsilon} e^{i\psi}$ - considered as a vertex of $\mathcal{G}(g_\varepsilon)$ - (cf.

Theorem 4.3.1 (Proof)) has to be attached (via an edge to be denoted by i) to the (only) vertex β of Γ' which is incident with j and j' (in case $j \neq j'$) or the (only) vertex β of Γ' of degree 1 incident with j (in case $j = j'$). However, by the very choice of j and j', the edge i (of Γ) has to be incident with a vertex common to i, j and j' (in case $j \neq j'$), i.e. incident with β. (See also Remark 4.3.7). The case $j = j'$ yields to the same conclusion. Consequently the extended tree corresponding to g_ε just is Γ^e. This concludes the induction step.

2. Let Γ^e be an extended tree then $\mathcal{P}(f)$ is derived from Γ by introducing a labeling j_α, $\alpha = 1,2$, - non-canonical w.r.t. the sub-index - of all points on C (corresponding to labeling the stable separatrices) and a (canonical) labeling i_{z^*} of the unstable separatrices - where z^* is a vertex of Γ incident with edge i and i_{z^*} the only separatrix connected to z^* - compare Lemma 4.3.1 (2).

The distinguished graph corresponding to $\mathcal{P}(f)$ (cf. [21]) is derived from Γ^e (in a canonical way) as follows:

a. A consecutive pair of type (i_1, i_2) defines a distinguished set of type 3: $[i_1, i_2; i_{z^*}]$, where z^* is the vertex of Γ of degree 1 incident with edge i.

b. A consecutive pair of type $(j_\alpha, j'_{\alpha'})$, where $j \neq j'$ defines a distinguished set of type 1: $[j_\alpha, j'_{\alpha'}; j_{z^*}, j'_{z^*}]$, where z^* is the (only) vertex of Γ incident with the edges j and j'.

(Compare also Lemma 4.3.4).

From the observations a. and b. the assertion follows as a consequence of the results of Peixoto on distinguished graphs [21]. □

Remark 4.3.7. From Definition 4.3.2 it follows that a consecutive pair (i,j), with $i \neq j$ has the property that the edges i and j of Γ are incident with a common vertex z^*.

For the proof of this assertion we recall that a walk on Γ from edge i to edge j is defined as a sequence $k_0, \ldots, k_m$ of different edges such that $k_0 = i$, $k_m = j$ and consecutive edges have a vertex in common. Assume $m > 1$, then $\Gamma \setminus \{k_1\}$ consists of two components such that i and j are in different ones. On C however i and j are in the same component w.r.t. (k_1, k_1). Contradiction. Consequently $m = 1$, i.e. a vertex z^* exists, incident with i and j. □

Remark 4.3.8. Let i, j, j' be labels ($i \neq j$, $i \neq j'$) such that both (i,j) and (i,j') are consecutive pairs on C:

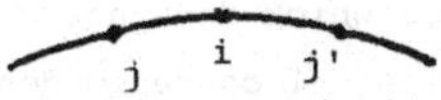

In Example 4.3.1: $i = 5$, $j = 2$, $j' = 6$ will do. If $z^*(z^{**})$ is related to (i,j) ((i,j')) - in the sense of Remark 4.3.7 - then we have $z^* \neq z^{**}$. In fact $j \neq j'$ since otherwise both points labeled with i would belong to different components determined by the pair (j,j) on C. Furthermore the edges j and j' are in different components of $\Gamma \setminus \{i\}$. Consequently j and j' cannot have a vertex in common. We find that z^* and z^{**} are just the two vertices, incident with edge i.

The result still holds if $i = j \neq j'$, z^* being of degree 1 in this case. (The result just derived is in fact Peixoto's Condition 4.2.3 in Definition 4.2 [21]). □

Given a tree Γ. A Reduced-Configuration derived from Γ (R.C.Γ) is given by an arrangement of consecutive pairs on a circle C corresponding to edges of Γ incident with a vertex of degree 1.

W.r.t. the tree Γ in Example 4.3.1: (1,1), (6,6), (4,4), (3,3) (cyclic) is an R.C.Γ. But also (1,1), (4,4), (6,6), (3,3) (cyclic) is an R.C.Γ.

Given an extended tree Γ^e, by deleting all non-consecutive points i corresponding to edge i of Γ we obtain a special R.C.Γ., called Reduced-Configuration-Compatible with Γ^e (R.C.C.Γ^e).

In Example 4.3.1 (1,1), (6,6), (4,4), (3,3) (cyclic) is an R.C.C.Γ^e.

Given a tree Γ and an R.C.Γ. By a special-walk we denote a walk "connecting" two vertices of degree 1 corresponding to pairs (i,i), (j,j) on C, which are adjacent w.r.t. R.C.Γ.

Let f be a non-degenerate polynomial. A special-walk sequence will be a sequence of the lengths of the special-walks on $\Gamma = \mathcal{G}(f)$ ordered in accordance to a cyclic ordering of the pairs of points in R.C.C.Γ^e (obtained from Γ^e induced by f). We define $\rho(f)$ as the equivalence class of all such special-walk sequences w.r.t. orientation of C and cyclic permutations.

In Example 4.3.1 the sequence (2,3,4,3) represents $\rho(f)$.

In the following lemma we will investigate the relation between the concepts introduced above and phase-portraits of $\mathcal{N}(f)$; we denote points on C corresponding to labeled edges of Γ simply by points.

Lemma 4.3.5.

a. For $z^* \in N(f)$ the number of distinguished sets having z^* as a boundary point equals the degree of z^* as a vertex in $\Gamma = \mathcal{G}(f)$.

b. Let Γ^e be an extended tree and (i,i), (j,j) adjacent in R.C.C.Γ^e. The component of $C \setminus (\text{arc}(i,i) \cup \text{arc}(j,j))$ which does not contain consecutive points is denoted by arc[i,j]. Then there is an 1-1 correspondence between the edges on the walk (on Γ) from i to j and the points on arc[i,j].

c. An R.C.Γ. is an R.C.C.Γ^e iff the sum of the lengths of all special walks equals 2s, s = order (Γ) -1.

d. If the phase-portraits of $\mathcal{N}(f)$ and $\mathcal{N}(g)$ - f,g nondegenerate polynomials - are equivalent, then $\rho(f) = \rho(g)$.

Proof. a. If z^* is of degree 1 as vertex of Γ, we know that only one distinguished set is involved, namely a set of type 3. Now assume: (degree of z^*) ≥ 2. Then the only distinguished sets with z^* as a boundary point are of type 1. Obviously their number equals the number of separatrices tending to z^*. From Lemma 4.3.1 (2) it follows that - w.r.t. Γ - this number equals degree z^*.

b. Assume that on arc[i,j] there are two points corresponding to the same edge on Γ and that - restricted to arc[i,j] - between these points there are no other such points. We denote this pair of points by (k,k). Since (k,k) is not consecutive, there must be a point corresponding to an edge k' between (k,k) - restricted to arc[i,j]. The component of $C \setminus (k,k)$ which contains k' must contain another point k' as well (cf. Definition 4.3.2). This contradicts the assumptions on (k,k). Consequently there is a path from edge i to edge j with different edges only, i.e. this path is the unique walk from i to j (on Γ).

c. The assertion $\Longrightarrow$) is a direct consequence of result b. Thus we only have to prove $\Longleftarrow$). Let (i,i) and (j,j) be two adjacent pairs in R.C.Γ. Since i and j are incident with vertices of degree 1, the R.C.Γ induces two paths on Γ, both connecting i to j (see Figure 4.3.4), namely:

1. The special-walk from i to j;

2. A chain of walks composed by the walks from (i,i) to (i_1,i_1), from (i_1,i_1) to (i_2,i_2), ..., from (i_m,i_m) to (j,j).

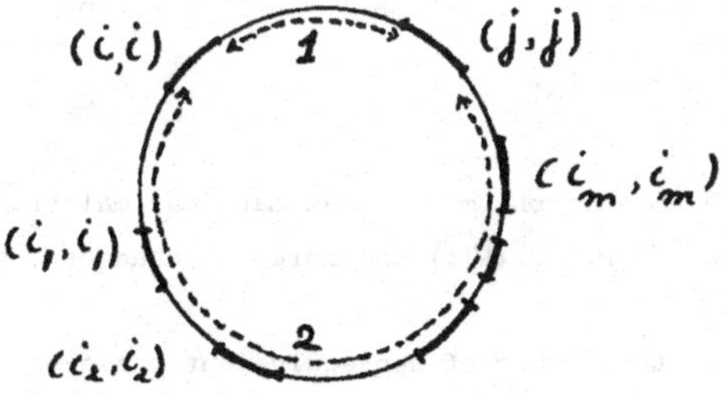

Fig. 4.3.4.

Unless Γ has only two vertices of degree 1 (i.e. Γ is "linear") - and the assertion becomes trivial - the second path cannot be a walk. Path 2 may be "thinned out" until we get a walk from i to j, being equal to path 1. This proves, that, if one induces on the arcs of C - determined by two adjacent pairs of R.C.Γ.- points labeled in accordance with the edges of the corresponding special-walks, each edge occuring in some special-walk appears at least twice on C.

Let k be an edge different from those represented in R.C.Γ, then $\Gamma\setminus\{k\}$ consists of two components, each of which contains a vertex of degree 1 w.r.t. Γ. Thus: not all edges (corresponding to R.C.Γ) can be in the same component of $\Gamma \setminus \{k\}$. Consequently, k must be an element of one of the special-walks. In fact, if not, all vertices of degree 1 could be connected by a path not using edge k, which is impossible. Combining these results we conclude: each k $(1<k<s)$ is used exactly twice as an element of the special-walks.

We now shall prove that the set of points of R.C.Γ together with the points induced (see above) by R.C.Γ forms a configuration compatible with Γ. It is sufficient to inspect the condition on compatibility for those edges k which are not incident with a vertex of degree 1.

Two points which are in the same component of C w.r.t. the pair (k,k) can be connected by an arc, not "passing" a point k, consequently they must correspond to edges in the same component of $\Gamma\setminus\{k\}$. On the other hand, k is element of some special walk with two adjacent edges k', k" $(k'\neq k")$, and thus the unique walk from k' to k" contains k, i.e. k' and k" belong to different components of $\Gamma\setminus\{k\}$. This concludes the proof of c.

d. This is a direct consequence of Theorem 4.3.2 since the extended trees w.r.t. f and g are isomorphic.

Remark 4.3.9. Two configurations (eventually reduced) on C which are compatible with the same tree Γ and which differ only w.r.t. the "orientation" on C, give rise to isomorphic extended trees and therefore represent "equivalent" phase-portraits. This can be made explicit as follows:

For $f(z)= a_n z^n + a_{n-1} z^{n-1} + \ldots + a_o$, $f \in \mathcal{P}_n$, we define $\tilde{f} \in \mathcal{P}_n$ by :

$$\tilde{f}(z) = \bar{a}_n z^n + \bar{a}_{n-1} z^{n-1} + \ldots + \bar{a}_o .$$

It is easily seen that the homeomorphism $\mathbb{C}\to\mathbb{C}, z\mapsto\bar{z}$ maps maximal trajectories w.r.t. $\mathcal{N}(f)$ onto maximal trajectories w.r.t. $\mathcal{N}(\tilde{f})$ and moreover changes the orientation of the induced configuration. □

According to Lemma 4.3.5 the number of non-equivalent phase-portraits corresponding to the same tree Γ is determined by the non-isomorphic R.C.C.Γ^e's. So, for the determination of this number, the number of vertices of degree 1, denoted by [Γ], is essential. In this paper we restrict ourselves to trees Γ with $[\Gamma] \leq 4$, but of course this restriction is not principal.

Lemma 4.3.6. Let Γ be a tree, then for the number of non-isomorphic extended trees, denoted by {Γ}, we have:

$\{\Gamma\} = 1$, if $2 \leqslant [\Gamma] \leqslant 3$.

$\{\Gamma\} \leqslant 2$, if $[\Gamma] = 4$ and Γ has no vertices of degree 4.

$\{\Gamma\} \leqslant 3$, if $[\Gamma] = 4$ and Γ has a vertex of degree 4.

Proof: For $[\Gamma]=2$ there is just one R.C.Γ. For $[\Gamma]=3$ there are two possible R.C.Γ's, but they only differ w.r.t. the orientation on C (cf. Remark 4.3.9.). Let $[\Gamma]=4$, and (as usual) # (edges) = s, # (vertices) = s+1. The number of vertices of degree $\geqslant 2$ equals (s-3). Consequently the sum of the degrees of all vertices has to be $\geqslant 4+2(s-3)=2s-2$. It is well-known (cf.[11]) that this sum equals (2s). Thus there are two possibilities:

1. There are two vertices of degree3, say α, β, the others being of degree 1 or 2.

2. There is one vertex of degree 4, say γ, the others being of degree 1 or 2.

ad.1. The tree Γ may be realized as in Figure 4.3.5.

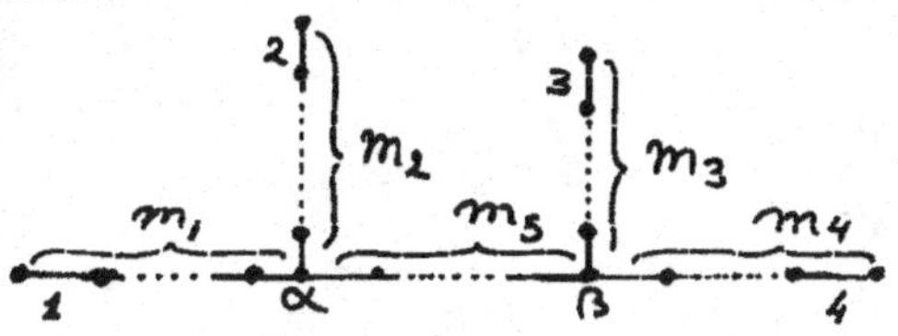

Fig. 4.3.5.

m_i = # edges in the assigned branch.

$(\sum_{i=1}^{5} m_i = s)$

The possible R.C.Γ's are (up to orientation (=reflection in this case))depicted in Figure 4.3.6.

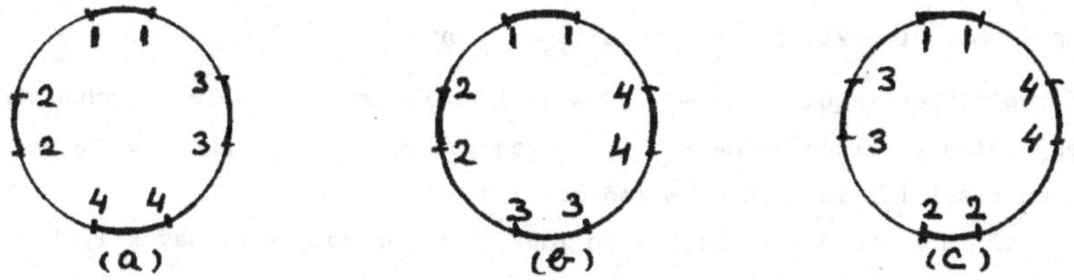

Fig. 4.3.6.

Let $\rho(\Gamma^e)$ denote some special walk sequence.

In case a : $\rho(\Gamma^e)=(m_1+m_2,\ m_2+m_4+m_5,\ m_3+m_4,\ m_1+m_3+m_5)$.

In case b : $\rho(\Gamma^e)=(m_1+m_2,\ m_2+m_3+m_5,\ m_3+m_4,\ m_1+m_4+m_5)$.

In case c, the configuration is not compatible with Γ because the sum of the lengths of the four special walks equals $(2s+2m_5) \neq 2s$. Obviously in cases a and b $\rho(\Gamma^e)$ is the same, if $m_3=m_4$. In this case 3 and 4 may be exchanged, keeping 1 and 2 fixed and we see that there is just one equivalence class. Note that $m_3 \neq m_4$ is not sufficient for the extended trees to be nonisomorphic; in fact if $m_1=m_2$, $m_3 \neq m_4$ cases a. and b. provide two isomorphic extended trees.

In order to obtain two non-isomorphic extended trees we assume the existence of an unique walk with maximal length on Γ; let this be the walk from 1 to 4. Under this assumption it is obvious that the sequences $\rho(\Gamma^e)$ in case a. en b. are strictly different.

ad.2. The tree Γ may be realized as in Figure 4.3.7.

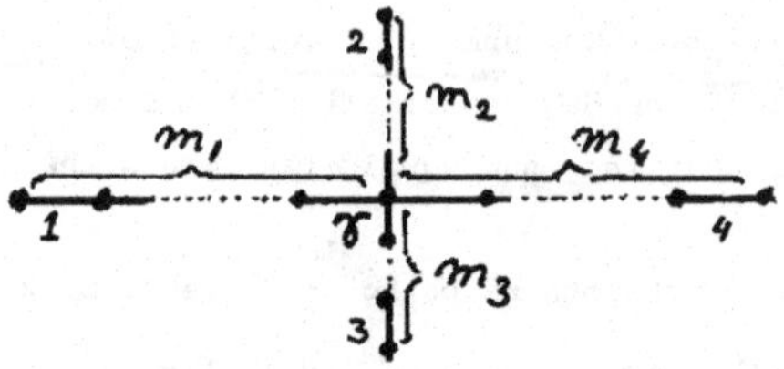

Fig. 4.3.7.

The possible R.C.Γ's are the same as in the preceding case, but now the corresponding sequences $\rho(\Gamma^e)$ are:

In case a : $\rho(\Gamma^e)=(m_1+m_2,\ m_2+m_4,\ m_3+m_4,\ m_1+m_3)$.

In case b : $\rho(\Gamma^e)=(m_1+m_2,\ m_2+m_3,\ m_3+m_4,\ m_1+m_4)$.

In case c : $\rho(\Gamma^e)=(m_1+m_3,\ m_2+m_3,\ m_2+m_4,\ m_1+m_4)$.

All three sequences obviously fulfil the condition that the sum of their elements equals 2s.

As in case 1. there are at least two non-isomorphic extended trees if there exists a unique walk with maximal length. A necessary and sufficient condition for Γ to have three non-isomorphic extended trees is: $m_k \neq m_l$ for $k\neq l$.

Necessity: If two m's are equal, there exists an isomorphism of Γ which "exchanges" the two corresponding vertices of degree 1. Consequently, an isomorphism between two of the three possibilities a,b,c is induced.

Sufficiency: If all m's are different, it follows that for some k,l, say k=1, l=4, m_k+m_l is maximal w.r.t. all other sums of two different m's. Consequently in case a, the extended tree Γ^e cannot be isomorphic to Γ^e in either case b or case c. Since in case b (case c) in the sequence $\rho(\Gamma^e)$ (m_1+m_4) is "adjacent" to (m_3+m_4) and (m_1+m_2) (to (m_2+m_4) and (m_1+m_3)) and since $m_3+m_4\neq m_2+m_4$; $m_3+m_4\neq m_1+m_3$ also in cases b and c there is no isomorphism.

As a corollary of Lemma 4.3.6. we obtain:

Corollary 4.3.2.

$\#$ (non-equivalent phase-portraits of $\mathcal{N}(f)$, $f\in\mathcal{P}_n$) $\geq$

$\#$ (non-isomorphic trees of order n)

The equality holds iff $n<6$.

Proof: For $n \leq 6$ (thus $s \leq 5$) the only trees Γ to be considered are those for which $[\Gamma] \geq 4$. These trees are depicted in Figure 4.3.8:

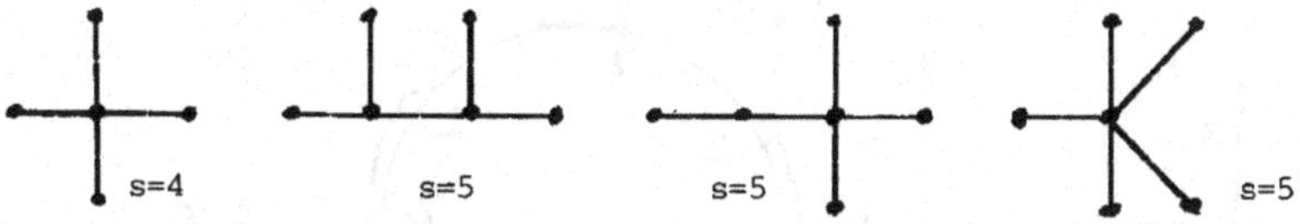

(1) (2) (3) (4)

Fig. 4.3.8.

For (1) resp.(4) there is essentially one R.C.Γ which obviously is an R.C.C.Γ^e. For (2) - in view of Lemma 4.3.6(ad.1) - there is only one equivalence class. For (3) - in view of Lemma 4.3.6(ad.2)-there are at most two non-isomorphic extended trees, but since in this case $m_2=m_3=m_4$ the result follows nevertheless.

For n=7 consider the tree as depicted in Figure 4.3.9.

Fig. 4.3.9.

Here there is an unique walk of maximal length (=4). Thus there are at least two non-isomorphic extended trees (in fact there are two, see Lemma 4.3.6 - ad.2). Note that the two corresponding non-equivalent phase-portraits are just the ones given as example by Peixoto (cf.[21]). Obviously an analogous situation as in the case n=7 can be realized for all n>7. □

Given a polynomial f of degree n. In view of Theorems 4.3.1,2 it is possible - by introducing an extra zero - to construct a polynomial g of degree (n+1) such that, on the level of the trees, $\mathcal{G}(g)$ is derived from $\mathcal{G}(f)$ by attaching an edge to a certain (arbitrary) vertex of $\mathcal{G}(f)$. The same vertex of $\mathcal{G}(f)$, however, possibly gives rise to different phase-portraits $\mathcal{N}(g)$. In view of Lemma 4.3.5 (a) it is most likely that this phenomenon takes place if a vertex of maximal degree is used for the "attaching-process".

This will be illustrated in the following example:

Example 4.3.2.

n=6 :

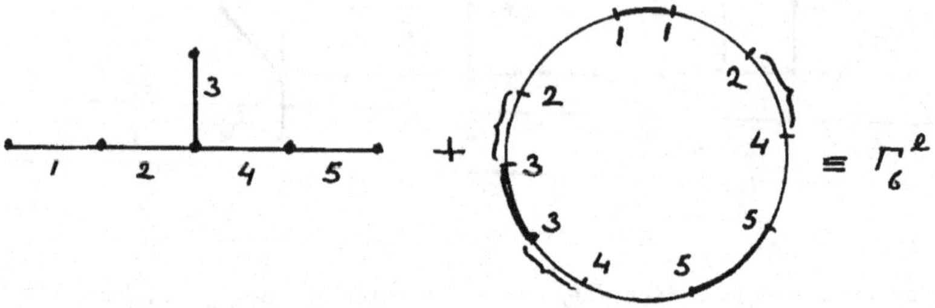

Attaching an "extra" edge to the vertex with maximal degree yields to the following extended trees of order 7:

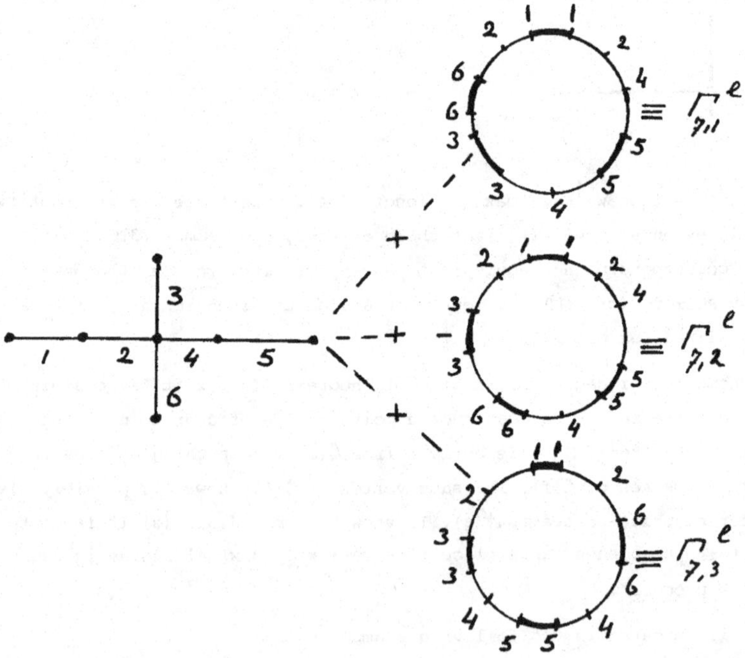

Obviously: $\rho(\Gamma^e_{7,1})=(3,2,3,4)$; $\rho(\Gamma^e_{7,2})=(3,2,3,4)$; $\rho(\Gamma^e_{7,3})=(3,3,3,3)$.

$\Gamma^e_{7,1} \cong \Gamma^e_{7,2}$ (interchange 3 and 6!) ; $\Gamma^e_{7,3} \ncong \Gamma^e_{7,1}$ (different ρ's!).

4.4. The relation with the theory on asymptotic values.

Definition 4.4.1: An asymptotic path for the meromorphic function f, with asymptotic value α, is a continuous curve γ tending to $z = \infty$, with the property that if $z \to \infty$ along γ then $f(z) \to \alpha$. (Note that $\alpha = \infty$ is possible).

An asymptotic path for f, which is also a trajectory for $\mathcal{N}(f)$ is called a Newton-asymptotic path.

Lemma 4.4.1. Let $f \in \mathcal{E}$; the trajectory $\gamma(\gamma^{\perp})$ of $\mathcal{N}(f)$ $(\mathcal{N}^{\perp}(f))$ is given by the solution $z(t)$, $t \in]a,b[$ $(z^{\perp}(t),\ t \in]c,d[)$. Then:

1. If γ emanates from (tends to) $z = \infty$ we have: $a = -\infty$ $(b = +\infty)$.
2. If $\gamma^{\perp}$ emanates from (tends to) $z = \infty$ we have: $c = -\infty$ $(d = +\infty)$.

Proof: The proofs of 1 and 2 being similar, we only prove 1.

Since f is of the form $f(z) = p(z)e^{q(z)}$, p and q polynomials it follows from (2.1) that:

$$\log p(z(t)) + q(z(t)) = -t + c \qquad , t \in]a,b[.$$

From the proof of Lemma 2.5 we know that $\lim_{t\downarrow a} \arg z(t)$ $(\lim_{t\uparrow b} \arg z(t))$ exists.

Consequently $\lim_{t\downarrow a} z(t) = \infty$ $(\lim_{t\uparrow b} z(t) = \infty)$ is only possible if $a = -\infty$ $(b = +\infty)$. □

Corollary 4.4.1. In view of (2.1) resp. Lemma 2.9 (1) it follows:

1. If γ emanates from (tends to) $z = \infty$ it is a Newton-asymptotic path for f with asymptotic value $\infty(0)$.
2. There are no trajectories of $\mathcal{N}^{\perp}(f)$ which are also asymptotic paths for f.

Definition 4.4.2: Let f be an entire function. A ray ρ, emanating from the origin $z = 0$ is called a line of Julia for f if: Given any $z_0 \in \mathbb{C}$ (with one possible exception) and any angle with ρ as its bisector, f takes the value z_0 at every point of an infinite sequence, converging to $z = \infty$ and lying within this angle.

As an application of the theory developed in Section 2 we prove a property, which is for a simple class of entire functions, a refinement of the theorem of Denjoy-Ahlfors and the theorem of Julia (cf. [18], [19]).

Theorem 4.4.1. Let f be an entire function of order $m < \infty$ with α as a Picard-exceptional value, then:

1. There are exactly 2m lines of Julia for f, two adjacent lines intersecting under equal angles $\frac{\pi}{m}$.

2. The only asymptotic values for f are α and ∞.

Proof: Suppose $f \in \mathcal{E}$ (i.e. $\alpha = 0$).

We adopt the notations of Section 2 (especially Lemmas 2.4,5).

1. By the L-rays (R-rays) we denote the 2m rays emanating from $w = 0$ with the property that any trajectory in an elliptic sector at $w = 0$ w.r.t. $\bar{\bar{\mathcal{N}}}(f)$ ($\bar{\bar{\mathcal{N}}}^{\perp}(f)$) is tangent in $w = 0$ to two of them which are adjacent. (Compare also Remark 2.5). From the calculations of Lemma 2.5 (Proof) it follows that the L-rays are given by:

$$\{w = \tau.\ \exp\ [\frac{i}{m}(\pi k + \arg b_m)]\ |\ \tau > 0\}, \qquad k = 1, \ldots, 2m.$$

Using similar calculations one shows that the R-rays are given by:

$$\{w = \tau.\ \exp\ [\frac{i}{m}(\ \frac{\pi}{2} + \pi k + \arg b_m)]\ |\ \tau > 0\}, \qquad k = 1, \ldots, 2m.$$

Consequently, each R-ray is the bisector of two (adjacent) L-rays. Let R_k be a arbitrary R-ray; R_k is the bisector of the L-rays $L_{k,1}$ $L_{k,2}$. Let γ_k be a trajectory of $\bar{\bar{\mathcal{N}}}(f)$ in the elliptic sector ε_k determined by $L_{k,1}$ and $L_{k,2}$. The connected component of $S^2 \setminus \bar{\gamma}_k$ which is contained in ε_k will be denoted by Λ_k. Let v be an arbitrary complex-number $\neq 0$. In view of Lemma 4.4.1 and (2.1) $|f(z)|$ takes the value $|v|$ in exactly one point of γ_k, say z_0. Since $\bar{\bar{\mathcal{N}}}(f)$ and $\bar{\bar{\mathcal{N}}}^{\perp}(f)$ are orthogonal the trajectory $\gamma^{\perp}(z_0)$- which "enters" Λ_k for (increasing or decreasing t)-cannot "leave" Λ_k and thus approaches $w = 0$ (cf. Remark 2.6) in a definite direction (the latter assertion follows from an argument, analogous to the argument used in Lemma 2.5 (Proof)). See Figure 4.4.1.

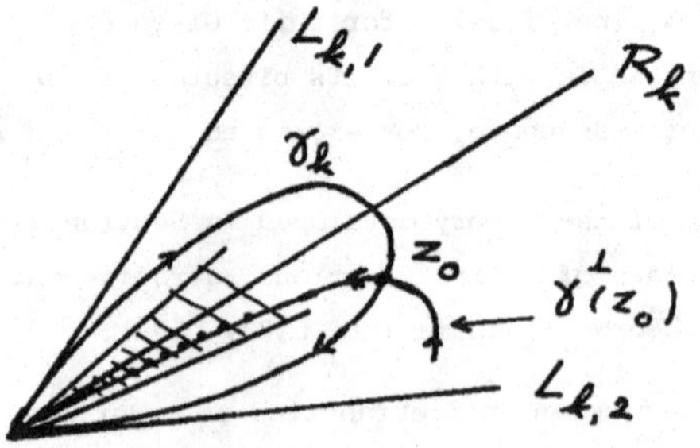

Fig. 4.4.1.

Consequently (cf. Remark 2.5) $\gamma^{\perp}(z_0)$ approaches $w = 0$ tangent to R_k. From Lemmas 4.4.1; 2.9(1) it follows that R_k is a line of Julia for f. Since any ray emanating from $w = 0$ which is not an R-ray must approach $w = 0$ within one elliptic sector μ_k of $\bar{\mathcal{N}}^{\perp}(f)$, the R-rays are the only lines of Julia (see also proof of part (2)).

2. Suppose γ is a asymptotic path for f with asymptotic value β, $\beta \neq 0$, $\neq \infty$. We denote the elliptic sectors of $\bar{\mathcal{N}}^{\perp}(f)$ at $w = 0$ by μ_k, $k = 1, \ldots, 2m$. We choose in each μ_k a trajectory $\gamma_k^{\perp}$ with the property that on $\gamma_k^{\perp}$: $|f(z)| < |\beta|$ or $|f(z)| > |\beta|$. This is possible in view of Lemmas 4.4.1;2.9(1). The connected component of $S^2 \setminus \bar{\gamma}_k^{\perp}$ which is contained in μ_k will be denoted by $\Lambda_k^{\perp}$. Let C_r be a circle around $w = 0$ with radius r. For r sufficiently small $C_r \setminus \bigcup_{k=1}^{2m} \Lambda_k^{\perp}$ consists of exactly 2m connected components. Consequently γ approaches $w = 0$ within one of these components, say the component determined by $\gamma_k^{\perp}$ and $\gamma_{k+1}^{\perp}$ (if $k = 2m$, then $k+1 = 1$). See Figure 4.4.2. Since $\gamma_k^{\perp}$ and $\gamma_{k+1}^{\perp}$ are tangent to an R-ray at $w = 0$, γ approaches $w = 0$ in a definite direction namely:

$$\Theta = \frac{\pi}{2m} + \frac{\pi}{m} k_0 + \frac{1}{m} \arg b_m, \qquad \text{some } k_0 \in \mathbb{Z}. \qquad (*)$$

On the other hand, if γ is described by $w(t)$, $t \in]0,\infty[$, we have $\lim_{t\to\infty} \arg f(w(t)) = \arg \beta$. Essentially the same calculations as used in the proof of Lemma 2.5 yield to: $\sin(\arg b_m - m\Theta) = 0$.
This contradicts (*).

If $f \notin \mathcal{E}$ (i.e. $\alpha \neq 0$) the assertion follows by considering the function $f(z) - \alpha$.

□

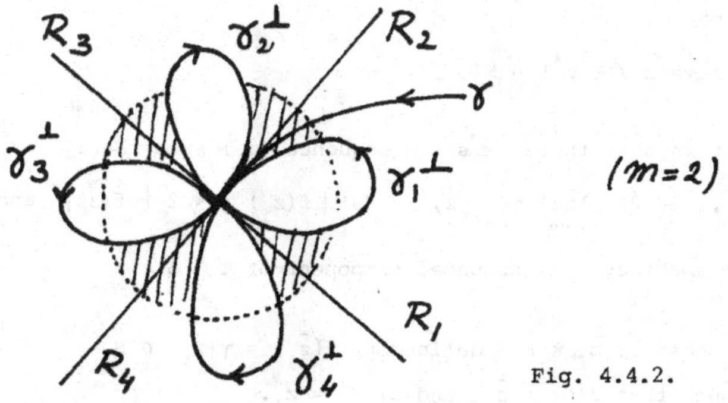

Fig. 4.4.2.

We proceed by giving some results on the Newton-asymptotic paths for meromorphic functions f.

<u>Theorem 4.4.2.</u> Suppose that $z^* \in N(f)$ and $\partial B(z^*) \neq \emptyset$. Then

(i) $\partial B(z^*) \cap C(f) \neq \emptyset$ or

(ii) there exists a Newton-asymptotic-path $\gamma(z_0)$ for some $z_0 \in B(z^*)$ having a finite asymptotic value $\alpha \neq 0$. (Note f is a <u>meromorphic</u> function here).

<u>Proof:</u> If $\mathcal{N}(f)$ is not a complete vectorfield on $B(z^*)$ then the assertion in the theorem is true. In fact assume

$$\gamma(z_0) = \{z(t) \mid t \in]a, +\infty[\}, \; z_0 \in B(z^*) \setminus \{z^*\}, \quad a > -\infty.$$

Then Lemma 2.2 implies (1) $\lim_{t \downarrow a} z(t) = \infty$ or (2) $\lim_{t \downarrow a} z(t) \in C(f)$.

In view of (2.1) we have in case (1) a Newton-asymptotic-path with finite asymptotic value $\alpha \neq 0$ and in case (2) $\lim_{t \downarrow a} z(t) \in \partial B(z^*) \cap C(f)$.

The theorem is a consequence of the following assertion:

$\mathcal{N}(f)$ is a complete vectorfield on $B(z^*)$ iff $\partial B(z^*) = \emptyset$. (*)

One implication of (*) follows directly from Lemma 2.3 (1).

So now suppose that $\mathcal{N}(f)$ is complete on $B(z^*)$. Furthermore, let S be a circle around z^* within $B(z^*)$ transversal to $\mathcal{N}(f)$;

$$M := \min_{z \in S} |f(z)| ; \; M > 0.$$

Firstly we prove that assumption of the existence of $\tilde{z} \in \partial B(z^*) \setminus P(f)$ leads to a contradiction.

Obviously we have $\tilde{z} \notin B(z^*) \cup N(f)$.

It is easily seen that there exists a sequence (z_i) such that:

1) $z_i \in B(z^*)$, 2) $\lim_{i \to \infty} z_i = \tilde{z}$, 3) $|f(z_i)| < 2 |f(\tilde{z})|$ and

4) z_i is an element of the unbounded component of $\mathbb{C} \setminus S$.

For all i we consider $\hat{z}_i \in S$ - defined by $\{\hat{z}_i\} = \gamma(z_i) \cap S$ - and $\hat{t}_i < 0$ such that $z(0) = \hat{z}_i$ and $z(\hat{t}_i) = z_i$.

In view of (2.1) it is easily seen that $\hat{t}_i > A$ where $A := \log M - \log(2|f(\tilde{z})|)$.

Consequently $(\hat{z}_i, \hat{t}_i) \in S \times [A,0]$ and so without loss of generality we may assume that the $(\hat{z}_i, \hat{t}_i)$ converge to a point $(\hat{z}, \hat{t}) \in S \times [A, 0]$.

With $z(0) = z'$ we put $\gamma(z';t) = z(t)$ and consider the map $\gamma: B(z^*) \times \mathbb{R} \to B(z^*)$ defined by $(z',t) \mapsto \gamma(z';t)$. As is well-known from the general theory of ordinary differential equations γ is a bicontinuous map. Consequently:

$$\tilde{z} = \lim_{i \to \infty} z_i = \lim_{i \to \infty} \gamma(\hat{z}_i; \hat{t}_i) = \gamma(\hat{z}; \hat{t})$$

and thus $\tilde{z} \in B(z^*)$. Contradiction.

Now let $\overset{\sim}{\mathbb{C}} := \mathbb{C} \setminus P(f)$, then $\overset{\sim}{\mathbb{C}}$ is open and (path)-connected with $B(z^*)$ as an open and closed subset, so we have $B(z^*) = \overset{\sim}{\mathbb{C}}$. Consequently $N(f) = \{z^*\}$; $C(f) = \emptyset$. The unbounded component of $\mathbb{C} \setminus S$ will be denoted by U. Now let $z \in U \setminus P(f)$, then we may write $z = \gamma(z';t)$ with $z' \in S$ and $t < 0$. In view of (2.1) we have $|f(z)| > |f(z')| \geq M$, consequently the function $\frac{1}{f(z)}$ is analytic and bounded on U; it follows by the Weierstrass theorem that $f(z)$ is a rational function (compare the proof of Lemma 2.3 (1)).

According to Lemma 2.4 (i) $z = \infty$ is either a non-degenerate node or a k-fold saddle-point for $\overset{=}{\mathcal{N}}(f)$ on S^2.

In the latter case there exists a trajectory tending to $z = \infty$ for increasing t, contradicting $B(z^*) = \overset{\sim}{\mathbb{C}}$.

So $z = \infty$ must be a node (as well as $z = z^*$) for the vectorfield $\overset{=}{\mathcal{N}}(f)$; since $C(f) = \emptyset$ the Poincaré-Hopf index-theorem implies $P(f) = \emptyset$, thus $f(z) = \alpha(z-z^*)^n$. $\quad\square$

Remark 4.4.1.

In the proof of Theorem 4.4.2 we did not use Lemma 2.3(2); in particular consider in the proof of this Lemma case 2 (iv), $z \in P(f)$ is an isolated point of $\partial B(z^*)$. As in the proof of Lemma 2.3 we then have $\Omega \cap B(z^*) = \Omega$, consequently $\mathcal{N}(f)$ is a complete vectorfield on $B(z)$ and $z^* \in \partial B(z)$, in contradiction with (*).

Remark 4.4.2.

If $z^* \in N(f)$ is such that Theorem 4.4.2(ii) is false, then there exists $z^{**} \in \partial B(z^*) \cap C(f)$ "connected" to z^* by an unstable separatrix of z^{**} (w.r.t. the vectorfield $\mathcal{N}(f)$). This follows directly from the proof of the preceding theorem.

Lemma 4.4.2. Let $f \in \mathcal{E}$, $z^* \in N(f)$ and Λ a connected component of $\partial B(z^*)$. Then Λ is union of the closures of regular trajectories w.r.t. to $\mathcal{N}(f)$ and there exists $z^{**} \in C(f) \cap \Lambda$ such that $|f(z^{**})| = \inf_{z \in \Lambda} |f(z)|$.

Proof: Since the closure of a regular trajectory is a connected set, the first assertion is a direct consequence of Lemma 2.3 (2).

Define S, M and U as in the proof of Theorem 4.4.2. Let $\tilde{z} \in \Lambda$. There exists a sequence (z_i) with $z_i \in U \cap B(z^*)$ and $\lim_{i\to\infty} z_i = \tilde{z}$. We can write $z_i = \gamma(z_i';t_i')$ with $z_i' \in S$ and $t_i' < 0$.

Again in view of (2.1) we have $|f(z_i)| > |f(z_i')| \geq M$, consequently

$$|f(\tilde{z})| = \lim_{i\to\infty} |f(z_i)| \geq M \qquad \text{and} \quad \inf_{z\in\Lambda} |f(z)| > 0.$$

Now let $\tilde{z} \in \Lambda \setminus C(f)$. It follows that $\gamma(\tilde{z};t)$ cannot be defined for all $t > 0$, hence we conclude from Lemma 2.2 that either 1) $\lim_{t\uparrow b} \gamma(\tilde{z};t) = \infty$

or 2) $\lim_{t\uparrow b} \gamma(\tilde{z};t) = \hat{z} \in C(f)$.

The case 1) would yield a finite asymptotic value $\alpha \neq 0$ for f which is impossible in view of Theorem 4.4.1(2).

Consequently case 2) is valid and then in fact $\hat{z} \in C(f) \cap \Lambda$. From (2.1) it follows that $|f(\tilde{z})| > |f(\hat{z})|$ and this essentially completes the proof since $\# C(f) < \infty$.

□

Remark 4,4.3. Note that $\mathcal{E}$ provides an example of a class of functions f such that for all $z^* \in N(f)$ the assertion (ii) in Theorem 4.4.2 is false. In this case Remark 4.4.2 can be generalized as follows:

Corollary 4.4.2. Notations as in Lemma 4.4.2.

Then z^{**} is the only point of $C(f) \cap \Lambda$ which is "connected" to z^* by an unstable separatrix of z^{**} (w.r.t. $\overline{\mathcal{N}}(f)$).

Proof: In at least one hyperbolic sector H at z^{**} ($\in \partial B(z^*)$) there exists a sequence (z_n) such that $z_n \in B(z^*) \cap H$ and $\lim_{n\to\infty} z_n = z^{**}$ (compare (iii) in the proof of Lemma 2.3). It follows that both separatrices of H belong to $\overline{B(z^*)}$. The unstable separatrix either is in $\partial B(z^*)$ and consequently in Λ or in $B(z^*)$. The first possibility violates $|f(z^{**})| = \inf_{z\in\Lambda} |f(z)|$, thus z^{**} is "connected" by an unstable separatrix to z^*.

Let $\tilde{z} \in C(f) \cap \Lambda$ be another point "connected" to z^*, then a closed Jordan curve can be constructed consisting of separatrices, z^*, z^{**}, $\tilde{z}$ and possibly other points of $C(f) \cap \Lambda$. This however is in contradiction with $\overline{R}(\infty) = \mathbb{C}$ (cf. Lemma 2.6, proof of Part 2).

□

Theorem 4.4.3. Let f be entire of finite order with a Picard exceptional value $\alpha \neq 0$. Then - with the possible exception of a finite number - for $z^* \in N(f)$ with multiplicity k we have:

$$\#\{\gamma \mid \gamma \text{ is a Newton-asymptotic path, } z^* \in \overline{\gamma}\} \in \{1, \ldots, k\}.$$

Proof: Since $\alpha \neq 0$ is a Picard exceptional value there are infinitely many zeros for f. Furthermore since $(f(z) - \alpha) \in \mathcal{E}$, we have $\# C(f) < \infty$.

Let $z^* \in N(f)$ be such that Theorem 4.4.2(ii) is false (cf. Remark 4.4.2), then z^* is connected to $z^{**} \in \partial B(z^*) \cap C(f)$ by a separatrix. Since $\# C(f) < \infty$, there are finitely many separatrices and consequently - with the possible exception of a finite number - for $z^* \in N(f)$ the assertion (ii) must be true (note $\partial B(z^*) \neq \emptyset$ by Lemma 2.3). The corresponding finite asymptotic value must be α (cf. Theorem 4.4.1(2)). On such a path we have $\arg f(z) = \arg \alpha$ in view of (2.1). Since f(z) is "k-fold" conformal at z^*, there can be at most k of these paths "starting" at z^*. This completes the proof. □

Example 4.4.1. Let $f(z) = z \exp\int_0^z \frac{e^\xi - 1}{\xi}\, d\xi$.

We have $N(f) = \{0\}$ and by Lemma 2.3 the basin $B(0) \neq \mathbb{C}$ (or $\partial B(0) \neq \emptyset$). In fact an easy calculation shows that $\mathcal{N}(f)$ equals $-z\exp(-z)$ and consequently - w.r.t. the region $\{z \mid -\pi < \text{Im } z < \pi\}$ - the vectorfield $\mathcal{N}(f)$ points outward on the lines $l_j = \{z \mid \text{Im } z = (-1)^j \pi\}$, $j = 1,2$ from which it follows $B(0) \subset \{z \mid -\pi < \text{Im } z < \pi\}$.

Furthermore $C(f) = \emptyset$, thus f must be of infinite order in view of Lemma 4.4.2 and by Theorem 4.4.2 there must exist a Newton-asymptotic-path "starting" at 0 with finite asymptotic value $\alpha \neq 0$.

In fact $\{z = x + iy \mid x < 0, y = 0\}$ is such a path and the corresponding asymptotic value turns out to be $-\exp(-C)$, where C is the well-known Euler constant.

4.5. Some numerical illustrations.

We conclude this section by giving some computer-drawn pictures, illustrating the preceding theory. These pictures are obtained by plotting out - by a P.D.P.11-computer - the points of trajectories of $\mathcal{N}(f)$ which are approximated using the discrete Newton-method.

Figure 4.5.1. illustrates conjecture C_1 (see Section 4.1). Note that in this figure C_1 seems to be violated, but this is due to the different scales on the coordinate-axes.

Figure 4.5.2. illustrates the counterexample for conjecture C_2 (see Section 4.2).

In Figure 4.5.3 there is only one pole for f. Since in this case $R(\infty) = \emptyset$, it follows that $\overline{B}(0)=\mathbb{C}$ and consequently all zeros for f are contained in $\partial B(0)$, Thus every zero for f can be "reached" along a trajectory of $\mathcal{N}(f)$, which emanates from $z = 0$. Note that, in accordance to Lemma 2.4(b), there are two critical points. Moreover, exactly four trajectories emanating from $z=0$ tend to these critical points, only one tends to $z=\infty$ (along the positive real axis).

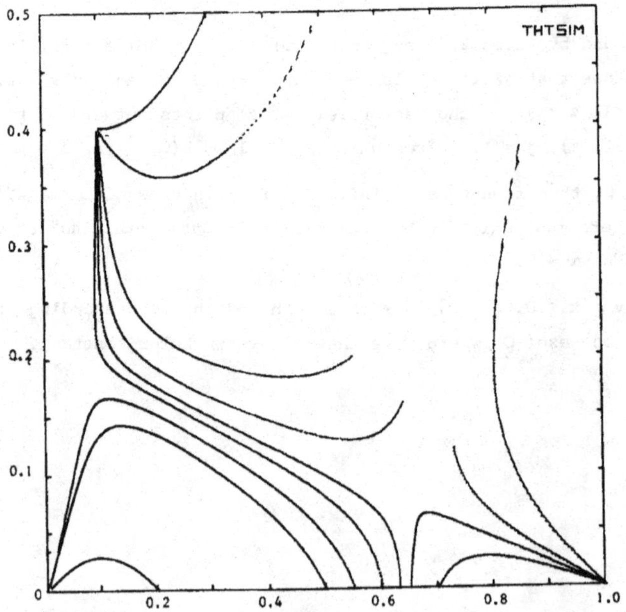

The phase-portrait of $\mathcal{N}(f)$; $f(z) = z(z-1)(z-a)$, $a = 0.1+0.4i$

Fig. 4.5.1.

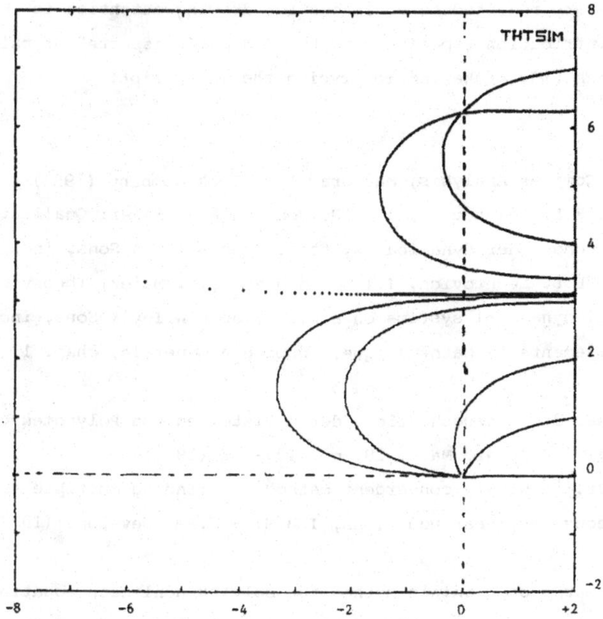

Phase-portrait of $\mathcal{N}(f)$; $f(z) = e^{z}-1$

Fig. 4.5.2.

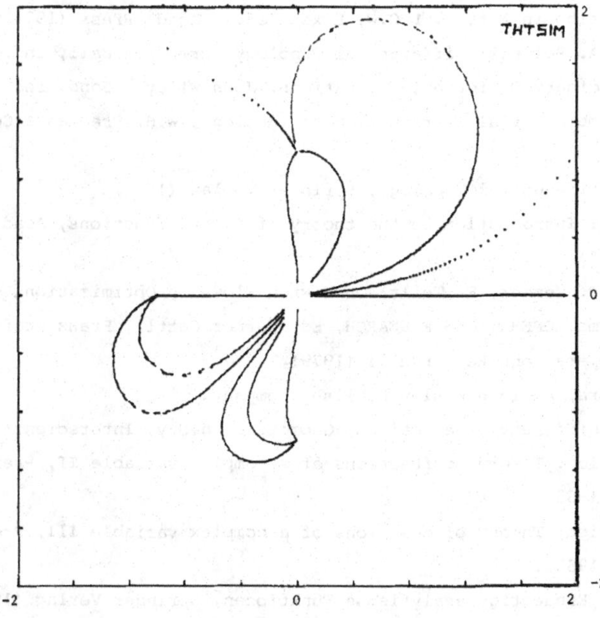

Phase-portrait of $\mathcal{N}(f)$; $f(z) = (z^2+1)\cdot(z+1)/z^3$

Fig. 4.5.3.

Acknowledgement: We are indepted to D. Braess for drawing attention to [5] and H. Koers for his contribution especially to the "computer-aspects" of this paper. We also want to thank Lidy Krukerink for typing the manuscript.

References:

[1] L.V. Alfohrs: Complex Analysis, Mac Graw Hill Book Company (1953).

[2] A.A. Andronov, E.L. Leontovich, I.I. Gordon and A.G. Maier: Qualitative theory of second-order dynamical systems, John Wiley & Sons, inc. (1973)

[3] A.A. Andronov, E.L. Leontovich, I.I. Gordon and A.G. Maier: Theory on bifurcations of dynamical systems on a plane, John Wiley & Sons, inc. (1973).

[4] N. Bourbaki: Eléments de Mathématique, Topologie Générale, chap. 1, Hermann (1961).

[5] D. Braess: Ueber die Einzugsbereiche der Nullstellen von Polynomen beim Newton Verfahren, Numer. Math. 29, pp. 123-132 (1977).

[6] F.H. Branin, Jr.: A widely convergent method for finding multiple solutions of simultaneous nonlinear equations, I.B.M. - J.Res.Develop. (1972), pp. 504-522.

[7] J. Gomulka: Remarks on Branin's method for solving nonlinear equations, In: Towards Global Optimization, Ed.: L.C.W. Dixon and G.P. Szegö, North-Holland Publishing Company (1975).

[8] L.C.W. Dixon, J. Gomulka and S.E. Hersom: Reflections on the Global Problem, In: Optimization in Action, L.C.W. Dixon (ed.), Acad. Press (1976).

[9] V. Guillemin, A. Pollack: Differential Topology, Prentice Hall, inc. (1974).

[10] P. Hartman: Ordinary Differential Equations, John Wiley & Sons, inc. (1964).

[11] M. Henle: A Combinatorial Introduction to Topology, W.H. Freeman & Company (1979).

[12] M.W. Hirsch: Differential Topology, Springer Verlag (1976).

[13] A.S.B. Holland: Introduction to the theory of entire functions, Acad. Press (1973).

[14] H.Th. Jongen, P. Jonker, F. Twilt: On Newton Flows in Optimization. In: Proc. III Symp. OPERATIONS RESEARCH, Ed. Werner Oettli, Franz Steffens, Methods of Operations Research 31 (1979).

[15] S. Lang, Algebra, Addison-Wesley Publish. Company.

[16] S. Lefschetz: Differential equations, Geometric theory, Interscience Publ.

[17] A.I. Markushevich: Theory of functions of a complex variable II, Prentice Hall, inc. (1965).

[18] A.I. Markushevich: Theory of functions of a complex variable III, Prentice Hall, inc. (1965).

[19] R. Nevanlinna: Eindeutige analytische Funktionen, Springer Verlag (1953).

[20] M.M. Peixoto: Structural stability on two-dimensional manifolds, Topology 1, pp. 101-120 (1962).

[21] M.M. Peixoto; On the Classification of Flows on 2-Manifolds, In: Dynamical Systems, Ed. M.M. Peixoto, Acad. Press (1973).

[22] S. Smale: A convergent process of price adjustment and global Newton methods, J.Math.Economics 3 (1976), pp. 107-120 .

[23] E.H. Spanier: Algebraic Topology, Mac Graw Hill Book Company (1966).

A PRECISE DEFINITION
OF SEPARATION OF VARIABLES

Tom H. Koornwinder

Mathematisch Centrum
Amsterdam, The Netherlands

1. INTRODUCTION

The work presented here originated from a review [8] I wrote on MILLER's [9] book "Symmetry and separation of variables". This book discusses the relationship between group theory and the coordinate systems for which a given partial differential equation is solvable by separation of variables. A remarkable omission in this book is that it does not contain a precise definition for one of the key words: separation of variables. However, when looking in older papers which discuss criteria for separation of variables (cf. ROBERTSON [14], EISENHART [3], MOON & SPENCER [10]), I could not find a precise definition either. Probably the applied mathematician will not be bothered very much by this omission, since he will have a fairly good informal notion of the method of separation of variables, which he can use in ad hoc cases. But if one wants to prove general theorems giving necessary and sufficient conditions for separation of variables or if one wants to classify all separable coordinates for a given partial differential equation then it becomes crucial to have a precise definition.

In section 2, after discussing some definitions from literature, I will propose a new definition of separation of variables which meets the three requirements of (i) being precise, (ii) being conceptual rather than formal, (iii) admitting a rigorous proof that the Stäckel-Robertson conditions (cf. ROBERTSON [14]) are necessary and sufficient for separability of the Helmholtz equation on a Riemannian manifold in given coordinates. In section 3 I derive Stäckel-Robertson type conditions which are necessary and sufficient for a linear homogeneous second order partial differential equation to separate into second order ordinary differential equations. Without proof I state some generalizations of this result for higher order equations and for non-linear equations. As a side result I show that a certain transformed version of the two-variable sine-Gordon equation, which is known to separate into first order o.d.e.'s, cannot separate into second order o.d.e.'s.

Section 4 deals with certain conditions equivalent to Stäckel's condition. The main result relates separability of $Lu = 0$ (L linear second order operator) with a family of $n-1$ linearly independent, mutually commuting second order operators commuting with ΦL for some function Φ. In section 5 we will see that it is no accident that classical and recent work on separation of variables is focused on equations living on (pseudo-) Riemannian manifolds: A general separable linear second order equation has an underlying (formal) Riemannian manifold. The paper concludes with a result stating that on an Einstein manifold the separability of $\Delta u + C(x)u = 0$ for some function C will imply the separability of $\Delta u + u = 0$.

2. THE DEFINITION OF SEPARATION OF VARIABLES

Let us start with a simple example. The Helmholtz equation in two variables

$$u_{xx} + u_{yy} + \omega^2 u = 0$$

is certainly separable as it stands. A slightly less trivial case occurs when we introduce polar coordinates $x = r\cos\theta$, $y = r\sin\theta$:

$$u_{rr} + r^{-1}u_r + r^{-2}u_{\theta\theta} + \omega^2 u = 0. \tag{2.1}$$

Now suppose u is a function of the form $u(r,\theta) = f(r)g(\theta)$, not identically zero. Then u is a solution of (2.1) if and only if there is some constant k such that f and g satisfy the ordinary differential equations

$$\begin{cases} r^2 f''(r) + rf'(r) + (\omega^2 r^2 - k^2)f(r) = 0 \\ g''(\theta) + k^2 g(\theta) = 0. \end{cases} \tag{2.2}$$

The general solutions of the o.d.e.'s (2.2) are

$$\begin{cases} f(r) = \alpha_1 J_k(\omega r) + \alpha_2 J_{-k}(\omega r), \\ g(\theta) = \beta_1 e^{ik\theta} + \beta_2 e^{-ik\theta}, \end{cases}$$

where J_k and J_{-k} are Bessel functions.

This example illustrates the way special functions arose in history: as factorized solutions of the p.d.e.'s of mathematical physics when written in separable coordinates. Suitable boundary conditions for (2.1) will restrict the generality of the solutions to be considered for (2.2). More general solutions of (2.1) can be obtained as linear combinations of factorized solutions. However, all these aspects will not bother us in this paper. We will concentrate on making precise the relationship between a p.d.e. like (2.1) and o.d.e.'s like (2.2).

Historically, a more systematic research on separability for p.d.e.'s was done in the context of a Riemannian manifold. For local coordinates $x_1, x_2, \ldots, x_n$ on the manifold let $g_{ij}(x)$ be the fundamental tensor and write $g_i := g_{ii}$. Assume the coordinates are orthogonal, i.e., the tensor g_{ij} is diagonal. STACKEL [16] proved in 1891 that the Hamilton-Jacobi equation

$$(2.3)\qquad \sum_{i=1}^{n} g_i^{-1}\left(\frac{\partial u}{\partial x_i}\right)^2 = \omega^2 \qquad (\omega>0)$$

is separable by solutions of the form $u(x) = X_1(x_1) + \ldots + X_n(x_n)$ if and only if the following condition (the so-called Stäckel condition) holds:

(i) There is a nonsingular $n \times n$ matrix $(c_{ij}(x_i))$ with inverse $(\gamma_{ij}(x))$ such that $(g_i(x))^{-1} = \gamma_{1i}(x)$.

For orthogonal local coordinates $(x_1,\ldots,x_n)$ the Laplace-Beltrami operator Δ on the Riemannian manifold takes the form

$$(2.4)\qquad \Delta = \sum_{i=1}^{n} g^{-\frac{1}{2}} \frac{\partial}{\partial x_i} \circ g^{\frac{1}{2}} g_i^{-\frac{1}{2}} \frac{\partial}{\partial x_i},$$

where $g := \det(g_{ij}) = \prod_{i=1}^{n} g_i$. ROBERTSON [14] proved in 1928 that the Helmholtz type equation

$$(2.5)\qquad \Delta u + \omega^2 u = 0 \qquad (\omega>0)$$

is separable by solutions of the form $u(x) = \prod_{i=1}^{n} X_i(x_i)$ if and only if the Stäckel condition (i) holds and, furthermore:

(ii) There are functions f_k $(k = 1,\ldots,n)$ in one variable such that

$$\frac{(g(x))^{\frac{1}{2}}}{\det(c_{ij}(x_i))} = \prod_{k=1}^{n} f_k(x_k).$$

EISENHART [3] (see also [4, Appendix 13]) observed in 1934 that Robertson's condition (ii) can be replaced by the condition (cf. section 5):

(ii)' The Ricci tensor $R_{ij}(x)$ is diagonal.

For certain classes of Riemannian manifolds, for instance for flat spaces, this condition is always satisfied if g_{ij} is diagonal. By way of application, EISENHART [3] classified the eleven separable coordinate systems for the three-variable Helmholtz equation

$$(2.6)\qquad u_{xx} + u_{yy} + u_{zz} + \omega^2 u = 0.$$

Although the above-mentioned authors state and prove necessary and sufficient conditions for the separability of (2.3) or (2.5), they do not give a precise definition of separation of variables. Therefore, in order to learn about the definition they have in mind, we have to look for the implicit assumptions they make in their proofs. Let us consider ROBERTSON [14] (with his potential energy V being zero), see also the same proof in MOON & SPENCER [10, Theorem 1]. In the necessity proof of conditions (i) and (ii) Robertson states that separability of (2.5) implies that the coefficients of $\partial^2u/\partial x_i^2$ and $\partial u/\partial x_i$ in (2.5) are proportional by a factor only depending on x_i. Furthermore, he concludes from the separability assumption that there must be a family of factorized solutions $u(x) = \prod_{i=1}^n X_i(x_i)$ of (2.5) depending on n parameters $\alpha_1 := \omega^2, \alpha_2, \ldots, \alpha_n$ such that

$$\det\left(\frac{\partial}{\partial \alpha_j}[X_i^{-1}(f_i(x_i)X_i'(x_i))']\right) \neq 0$$

(f_i as in condition (ii)) for some $(\alpha_1, \ldots, \alpha_n)$. Apparently, Robertson assumes that (2.5) is simultaneously separable for all values of ω. On the other hand, Robertson proves the sufficiency of conditions (i), (ii) by showing that there are n o.d.e.'s jointly depending on parameters $\alpha_1 := \omega^2, \alpha_2, \ldots, \alpha_n$ such that, if $u(x) = \prod_{i=1}^n X_i(x_i)$ satisfies (2.5), then $X_1, \ldots, X_n$ satisfy the o.d.e.'s for some choice of $(\alpha_1, \ldots, \alpha_n)$. I think it is difficult to extract from the above elements a clear and precise picture of the definition of separation of variables Robertson had in mind.

Let me next discuss three different definitions of separation of variables I met in literature.

DEFINITION A. Cf. MORSE & FESHBACH [11, §5.1, p.497].
Write the three-variable Helmholtz equation (2.6) in new coordinates ξ_1, ξ_2, ξ_3. These coordinates are called *separable* for equation (2.6) if each solution of (2.6) is a linear combination of solutions of the form $F_1(\xi_1)F_2(\xi_2)F_3(\xi_3)$.

Discussion. This is typically a definition from the user's point of view: one can use the method of separation of variables if one can build the general solution as a linear combination of factorized solutions. However, the definition is not precise, since it is not specified in which topology these linear combinations have to converge. It will also be hard to derive a separability criterium or a classification result starting from this definition. Finally it is remarkable that the definition does not require that the functions F_i satisfy certain o.d.e.'s.

DEFINITION B. Cf. SNEDDON [15, Ch.3, §9, p.123].
A second order homogeneous linear p.d.e. in two variables

$$A_{11}u_{xx} + 2A_{12}u_{xy} + A_{22}u_{yy} + B_1u_x + B_2u_y + Cu = 0 \tag{2.7}$$

is called *separable* if, for each solution u of the form $u(x,y) = X(x)Y(y)$, equation (2.7) can be written as

$$X^{-1}D_1X = Y^{-1}D_2Y, \tag{2.8}$$

where D_1 and D_2 are second order ordinary differential operators in x and y, respectively.

Discussion. This is a formal criterium for separability. By manipulation of (2.7) one has to try to achieve an equation of the form (2.8). It is apparent that (2.8) separates into two o.d.e.'s, but it is left unspecified what is meant by this last and most important step.

DEFINITION C. Cf. NIESSEN [12, p.329].
A linear partial differential operator L in $x_1,\ldots,x_n$ is called *separable* if there is an $n \times n$ matrix (L_{ij}), with the matrix element L_{ij} being an ordinary (possibly zero order) linear differential operator in x_j, such that, for all sufficiently often differentiable functions $x_i \to X_i(x_i)$ $(i = 1,\ldots,n)$ we have

$$(2.9) \qquad L\Big(\prod_{i=1}^{n} X_i\Big) = \det(L_{ij}X_j).$$

Discussion. Call the equation $Lu = 0$ separable if the operator L is separable according to Definition C. Definition B can be viewed as a special case of Definition C if we put

$$(L_{ij}X_j) := \begin{pmatrix} D_1X_1 & D_2X_2 \\ X_1 & X_2 \end{pmatrix}.$$

The criticism to Definition B also applies here. Further objections are that the definition does not cover nonlinear equations or the case of a linear second order p.d.e. separating into first order o.d.e.'s. However, positive aspects of Definitions B and C are their preciseness and the fact that they can easily be used for the derivation of separability criteria.

I will conclude this section by proposing yet another definition of separation of variables, which will meet the three requirements mentioned in the introduction.

Let $m \in \mathbb{N}$. Consider for real $x_1,x_2,\ldots,x_n$ the p.d.e.

$$(2.10) \qquad F\left(\frac{\partial^{i_1+\ldots+i_n}}{\partial x_1^{i_1}\ldots x_n^{i_n}}\, u(x_1,\ldots,x_n), x_1,\ldots,x_n\right) = 0,$$

where the derivatives of u are running over all orders such that $i_1+\ldots+i_n \le m$, where u is allowed to be a complex-valued function, and where F is assumed to be analytic in all its arguments.

DEFINITION 2.1. The p.d.e. (2.10) is called *separable* for $(x_1,\ldots,x_n)$ lying in some open connected region $\Omega \subset \mathbb{R}^n$ if there are n analytic o.d.e.'s

$$(2.11) \qquad X_i^{(k_i)}(x_i) + f_i(X_i^{(k_i-1)}(x_i),\ldots,X_i(x_i),x_i,\alpha_2,\ldots,\alpha_n) = 0, \quad i = 1,\ldots,n,$$

jointly depending in an analytic way on n-1 independent complex parameters $\alpha_2,\ldots,\alpha_n$, such that, for each $(\alpha_2,\ldots,\alpha_n)$ and for each set of solutions $(X_1,\ldots,X_n)$ of (2.11) with arguments $(x_1,\dots,x_n)$ in Ω, the function

$$u(x_1,\dots,x_n) := \prod_{i=1}^{n} X_i(x_i) \tag{2.12}$$

is a solution of (2.10).

DEFINITION 2.2. The n-1 complex parameters $\alpha_2,\dots,\alpha_n$ in (2.11) are called *independent* if the $n\times(n-1)$ matrix

$$\left(\frac{\partial f_i}{\partial \alpha_j}(y_{k_i-1,i},y_{k_i-2,i},\dots,y_{0,i},x_i,\alpha_2,\dots,\alpha_n)\right)$$

has rank n-1 whenever $\prod_{i=1}^{n} y_{0,i} \neq 0$.

REMARK 2.3. If the function F in (2.10) is not defined globally as a function of u and its derivatives then suitable modifications have to be made in Definition 2.1, such that it can be understood locally.

REMARK 2.4. The requirement of analyticity for (2.10) and (2.11) is not very stringent, but just for convenience. Because an analytic function in one variable, not identically zero, has the properties that it is completely determined by its restriction to some real interval and that its zeros are isolated, we will be able in later proofs to divide by such a function, neglecting possible zeros.

REMARK 2.5. In certain circumstances, the condition $\prod_{i=1}^{n} y_{0,i} \neq 0$ in Definition 2.2 may not be the right choice. Anyhow, for each value of $x_1,\dots,x_n$, $\alpha_2,\dots,\alpha_n$, the rank of $(\partial f_i/\partial\alpha_j)$ should be n-1 for generic values of $y_{k_i-1,i},\dots,y_{0,i}$ $(i = 1,\dots,n)$.

REMARK 2.6. Under the terms of Definition 2.1 a converse implication often holds: If u is a function of the form (2.12), analytic and not identically zero, and if u satisfies (2.10) then the functions X_i, $i = 1,\dots,n$, satisfy (2.11) for some choice of the parameters $\alpha_2,\dots,\alpha_n$. In section 3 we will prove this converse implication in the case of a linear second order p.d.e. which separates into second order o.d.e.'s.

REMARK 2.7. It is easy to make a connection between our Definition 2.1 and NIESSEN's [12] Definition C. Let the linear partial differential operator L have the property of formula (2.9). Now, for each value of $(\alpha_1,\alpha_2,\dots,\alpha_n)$, if the functions $X_1,\dots,X_n$ satisfy the o.d.e.'s

$$\sum_{i=1}^{n} \alpha_i L_{ij} X_j(x_j) = 0$$

then $u := \prod_{i=1}^{n} X_i$ satisfies $Lu = 0$. However, without further assumptions on the L_{ij}'s it is not clear whether n-1 of the parameters $\alpha_1,\dots,\alpha_n$ form a set of independent parameters for this set of o.d.e.'s.

EXAMPLE 2.8. Clearly the p.d.e. (2.1) separates into the o.d.e.'s (2.2) according to Definition 2.1 and also the converse implication of Remark 2.6 holds. Similarly, the p.d.e.

$$u_{xx} + u_{yy} + \omega^2 u = 0 \tag{2.13}$$

separates, under the Ansatz $u(x,y) = X(x)Y(y)$, into the o.d.e.'s

$$\begin{cases} X''(x) + (\omega^2-k^2)X(x) = 0, \\ Y''(y) + k^2Y(y) = 0, \end{cases}$$

and again the converse implication holds. However note that (2.13) also separates into the first order o.d.e.'s

$$\begin{cases} X'(x) + i\sqrt{\omega^2-k^2}\, X(x) = 0, \\ Y'(y) + ikY(y) = 0. \end{cases} \tag{2.14}$$

The converse implication of Remark 2.6 no longer holds in this case, but it does hold for a three-parameter family of pairs of o.d.e.'s extending (2.14): Let u be a function of the form $u(x,y) = X(x)Y(y)$, not identically zero. Then u satisfies (2.13) if and only if X and Y satisfy the o.d.e's

$$\begin{cases} (X'(x))^2 + (\omega^2-k^2)(X(x))^2 = A^2, \\ (Y'(y))^2 + k^2(Y(y))^2 = B^2 \end{cases}$$

for some value of (k,A,B).

EXAMPLE 2.9. Consider the sine-Gordon equation

$$\phi_{xx} - \phi_{tt} = \sin \phi.$$

Put $u(x,t) := tg(\frac{1}{4}\phi(x,t))$. The transformed equation reads

$$(1+u^2)(u_{xx}-u_{tt}) - 2u(u_x^2-u_t^2) = u(1-u^2). \tag{2.15}$$

Under the Ansatz $u(x,t) = X(x)T(t)$ this nonlinear p.d.e. separates into the first order o.d.e.'s

$$\begin{cases} X'(x) = \alpha^{\frac{1}{2}}X(x) \\ T'(t) = \sqrt{\alpha-1}\; T(t), \end{cases}$$

according to Definition 2.1. On the other hand, let $u(x,t) = X(x)T(t)$ be an analytic solution of (2.15) such that u_x and u_t are not identically zero. Then it can be shown that, for some choice of α,β,γ, the functions X and T satisfy the o.d.e.'s

$$\begin{cases} (X')^2 = \beta X^4 + \alpha X^2 + \gamma, \\ (T')^2 = -\gamma T^4 + (\alpha-1)T^2 - \beta, \end{cases}$$

see also OSBORNE & STUART [13]. It will follow from Lemma 3.6 that equation (2.15) does not separate into second order o.d.e.'s.

3. SEPARATION OF VARIABLES FOR LINEAR SECOND ORDER EQUATIONS

In this section we will derive criteria for separability of a general linear homogeneous second order p.d.e.

$$\sum_{i,j=1}^{n} A_{ij}(x) \frac{\partial^2 u}{\partial x_i \partial x_j} + \sum_{i=1}^{n} B_i(x) \frac{\partial u}{\partial x_i} + C(x)u = 0. \tag{3.1}$$

Here A_{ij}, B and C are complex-valued analytic functions of $x = (x_1,\ldots,x_n)$ on some open connected region Ω in $\mathbb{R}^n$. We may assume $A_{ij} = A_{ji}$ and we will write $A_i := A_{ii}$. Furthermore, we require that, for each i, (3.1) contains some nonvanishing term involving a derivative with respect to x_i, i.e., for each $x \in \Omega$ and for each $i \in \{1,\ldots,n\}$ not all of the numbers $A_{i1}(x), A_{i2}(x),\ldots,A_{in}(x), B_i(x)$ are zero. Let us formulate the main theorem.

<u>THEOREM 3.1</u>. *The p.d.e.* (3.1) *separates on* Ω *into* n *second order o.d.e.'s if and only if the following three conditions hold*:

(i) $A_{ij} = 0$ *if* $i \neq j$ *and* $A_i(x) \neq 0$ *for all* x *and* i.

(ii) *There are analytic functions* b_i $(i = 1,\ldots,n)$ *in one real variable such that*

$$B_i(x) = b_i(x_i)A_i(x). \tag{3.2}$$

(iii) *There are analytic functions* c_{ij} $(i,j = 1,\ldots,n)$ *in one real variable such that the* $n \times (n-1)$ *matrix*

$$\begin{pmatrix} c_{12}(x_1)\ldots c_{1n}(x_1) \\ c_{n2}(x_n)\ldots c_{nn}(x_n) \end{pmatrix} \tag{3.3}$$

has rank n-1 *for all* $x = (x_1,\ldots,x_n) \in \Omega$ *and*

$$(3.4)\qquad \sum_{i=1}^{n} c_{ij}(x_i)A_i(x) = C(x)\delta_{j,1},\ j = 1,\dots,n.$$

Under the assumption of conditions (i), (ii), (iii), *the p.d.e.* (3.11) *takes the form*

$$(3.5)\qquad \sum_{i=1}^{n} A_i(x)\left(\frac{\partial^2 u}{\partial x_i^2} + b_i(x_i)\frac{\partial u}{\partial x_i} + c_{i1}(x_i)u\right) = 0.$$

Furthermore, if a function u *has the form* $u(x) = \Pi_{i=1}^{n} X_i(x_i)$ *and if* u *is not identically zero on* Ω, *then* u *is a solution of* (3.5) *if and only if, for some* $(\alpha_2,\dots,\alpha_n) \in \mathbb{C}^{n-1}$, *the functions* X_i *are solutions of the o.d.e.'s*

$$(3.6)\qquad X_i''(x_i)+b_i(x_i)X_i'(x_i)+\left(c_{i1}(x_i)+\sum_{j=2}^{n} c_{ij}(x_i)\right)X_i(x_i) = 0,\qquad i = 1,\dots,n.$$

For the proof we will need a lemma, see Lemma 3.2. First we introduce some notation and we formulate alternatives to condition (iii). Consider an (n×n) matrix-valued function

$$(3.7)\qquad c(x) := (c_{ij}(x_i)).$$

Let

$$(3.8)\qquad M_{ij}(x) := \text{cofactor of } c(x) \text{ for entry}(i,j),$$

$$(3.9)\qquad S(x) := \det c(x).$$

$M_{ij}(x)$ only depends on $x_1,\dots,x_{i-1},x_{i+1},\dots,x_n$. If conditions (i) and (iii) of the theorem hold then, by (3.4),

$$(3.10)\qquad \frac{A_i(x)}{A_j(x)} = \frac{M_{i1}(x)}{M_{j1}(x)},\qquad i,j = 1,\dots,n,$$

and, hence, $M_{i1}(x) \neq 0$ for all x and i.

Let condition (i) hold. If $C(x) = 0$ for all x then (iii) is equivalent to:

(iii)' *There is an analytic matrix-valued function* c *of the form* (3.7) *such that, for all* $x \in \Omega$, *the matrix* (3.3) *has rank* n−1, $c_{i1}(x_i) = 0$ $(i = 1,\dots,n)$ *and* (3.10) *holds.*

If $C(x) \neq 0$ for all $x \in \Omega$ then (iii) is equivalent to:

(iii)" *There is an analytic matrix-valued function* c *of the form* (3.7) *such that, for all* $x \in \Omega$, $S(x) \neq 0$ *and*

$$(3.11)\qquad A_i(x) = \frac{C(x)M_{i1}(x)}{S(x)}.$$

Note that (3.11) can also be written as $A_i(x) = C(x)(c(x)^{-1})_{1i}$, where $c(x)^{-1}$ is the matrix inverse of $c(x)$.

<u>LEMMA 3.2</u>. *Suppose there are* n *second order o.d.e.'s*

$$X_i''(x_i) + f_i(X_i'(x_i),\ X_i(x_i),x_i) = 0, \qquad i = 1,\dots,n, \tag{3.12}$$

with f_i *analytic, such that for each set of solutions* $(X_1,\dots,X_n)$ *of* (3.12) *the function* u *given by* $u(x) = \prod_{i=1}^n X_i(x_i)$ *is a solution of* (3.1). *Then condition* (i) *of Theorem* 3.1 *holds and there are analytic functions* b_i *and* c_i $(i = 1,\dots,n)$ *in one real variable such that* (3.2) *is valid and also*

$$C(x) = \sum_{i=1}^n c_i(x_i)A_i(x), \tag{3.13}$$

$$f_i(X_i'(x_i),X_i(x_i),x_i) = b_i(x_i)X_i'(x_i)+c_i(x_i)X_i(x_i), \qquad i = 1,\dots,n. \tag{3.14}$$

<u>PROOF</u>. Substitute (3.12) into (3.1) with $u = \prod_{i=1}^n X_i$. Then

$$- \sum_{i=1}^n A_i(x) \frac{f_i(X_i'(x_i),X_i(x_i),x_i)}{X_i(x_i)} + \tag{3.15}$$

$$+ \sum_{i\neq j} A_{ij}(x) \frac{X_i'(x_i)X_j'(x_j)}{X_i(x_i)X_j(x_j)} + \sum_{i=1}^n B_i(x) \frac{X_i'(x_i)}{X_i(x_i)} + C(x) = 0.$$

It follows from the assumptions in the lemma and from the theory of second order ordinary differential equations that, for each $x \in \Omega$, equation (3.15) will be satisfied for all complex values of $X_i'(x_i)$ $(i = 1,\dots,n)$ and for all nonzero complex values of $X_i(x_i)$ $(i = 1,\dots,n)$. Fix $x \in \Omega$. Successive differentiation of (3.15) with respect to $X_i'(x_i)$ and $X_j'(x_j)$ $(i \neq j)$ yields that $A_{ij}(x) = 0$ for $i \neq j$. Next, by differentiation of (3.15) with respect to $X_i'(x_i)$ we obtain

$$B_i(x) = A_i(x) \frac{\partial f_i(X_i'(x_i),X_i(x_i),x_i)}{\partial X_i'(x_i)} .$$

$A_i(x) = 0$ would imply $B_i(x) = 0$, contradicting the original assumptions about (3.1). So $A_i(x) \neq 0$ and condition (i) of Theorem 3.1 is proved. It also follows that

$$f_i(X_i'(x_i),X_i(x_i),x_i) = b_i(x_i)X_i'(x_i)+c_i(X_i(x_i),x_i)X_i(x_i) \tag{3.16}$$

for certain analytic functions b_i and c_i, with b_i satisfying (3.2). Substitution of (3.16) and (3.2) into (3.15) yields

$$C(x) = \sum_{i=1}^n c_i(X_i(x_i),x_i)A_i(x).$$

By differentiating this formula with respect to $X_i(x_i)$ we obtain

$\partial c_i(X_i(x_i),x_i)/\partial X_i(x_i) = 0$. Thus c_i only depends on x_i. Now (3.13) and (3.14) are proved. □

Lemma 3.2 states that, if a linear homogeneous second order p.d.e. (3.1) "separates" into one set of n o.d.e.'s (3.12), not a priori linear and not depending on additional parameters, then this assumption already forces the o.d.e.'s to be linear and also imposes strong restrictions on the coefficients A_i, B_i, C in (3.1). However, for the proof of this lemma it seems to be crucial to assume that $u = \prod_{i=1}^{n} X_i$ satisfies (3.1) for all possible solutions $(X_1,\dots,X_n)$ of the o.d.e.'s, not just for one set of solutions.

PROOF of Theorem 3.1.

(a) *Necessity of the conditions* (i), (ii), (iii).
Suppose (3.1) separates into n second order o.d.e.'s. Then we know from Lemma 3.2 that conditions (i) and (ii) are satisfied and that (3.1) separates into o.d.e.'s of the form

$$X_i''(x_i)+b_i(x_i)X_i'(x_i)+c_i(x_i,\alpha_2,\dots,\alpha_n)X_i(x_i) = 0, \qquad i = 1,\dots,n,$$

where b_i satisfies (3.2) and

$$\sum_{i=1}^{n} c_i(x_i,\alpha_2,\dots,\alpha_n)A_i(x) = C(x).$$

Differentiate the last identity with respect to α_j:

$$\sum_{i=1}^{n} \frac{\partial c_i}{\partial \alpha_j}(x_i,\alpha_2,\dots,\alpha_n)A_i(x) = 0, \qquad j = 2,\dots,n.$$

By Definitions 2.1 and 2.2 the $n \times (n-1)$ matrix $\left(\frac{\partial c_i}{\partial \alpha_j}(x_i,\alpha_2,\dots,\alpha_n)\right)$ has rank n-1. Choose a fixed $(\alpha_2,\dots,\alpha_n)$ and define

$$c_{i1}(x_i) := c_i(x_i,\alpha_2,\dots,\alpha_n), \qquad i = 1,\dots,n,$$

$$c_{ij}(x_i) := \frac{\partial c_i}{\partial \alpha_j}(x_i,\alpha_2,\dots,\alpha_n), \qquad i = 1,\dots,n,\ j = 2,\dots,n.$$

With this choice of the functions c_{ij} condition (iii) holds.
(b) *Sufficiency of the conditions* (i), (ii), (iii).
Assume conditions (i), (ii), (iii) hold. Then (3.1) takes the form (3.5). Let, for some $(\alpha_2,\dots,\alpha_n) \in \mathbb{C}^{n-1}$, $(X_1,\dots,X_n)$ be a set of solutions of (3.6), with $X_i(x_i) \neq 0$ for all x and i. Multiply (3.6) by $A_i(x)/X_i(x_i)$ and sum over i. Using (3.4) we obtain

$$(3.17) \qquad \sum_{i=1}^{n} A_i(x)\left(\frac{X_i''(x_i)}{X_i(x_i)} + b_i(x_i)\frac{X_i'(x_i)}{X_i(x_i)} + c_{i1}(x_i)\right) = 0.$$

If $u = \prod_{i=1}^n X_i$ then (3.17) implies (3.5).

(c) *Proof that the factors in the factorized solutions of* (3.5) *satisfy* (3.6). Without loss of generality we may assume $X_i(x_i) \neq 0$ (cf. Remark 2.4). Compare (3.17) with the n-1 equations (3.4) for $j = 2,\ldots,n$. Since the matrix (3.3) has rank n-1 we conclude that

$$\frac{X_i''(x_i)}{X_i(x_i)} + b_i(x_i)\frac{X_i'(x_i)}{X_i(x_i)} + c_{i1}(x_i) + \sum_{j=2}^n \alpha_j(x)c_{ij}(x_i) = 0$$

for certain coefficients α_j $(j = 1,\ldots,n)$ depending on x. Hence, for fixed ℓ:

$$\sum_{j=2}^n (\alpha_j(x_1,\ldots,\tilde{x}_\ell,\ldots,x_n)-\alpha_j(x_1,\ldots,x_\ell,\ldots,x_n))\, c_{ij}(x_i) = 0,$$

$$i = 1,\ldots,\ell-1,\ell+1,\ldots,n.$$

The determinant of the $(n-1)\times(n-1)$ matrix which is obtained by deleting the ℓth row and the first column in $(c_{ij}(x_i))$, equals $(-1)^{\ell+1}\, M_{\ell 1}(x)$ (cf. (3.8)). Because of (3.10) this determinant is nonzero. Hence $\alpha_j(x_1,\ldots,\tilde{x}_\ell,\ldots,x_n) = \alpha_j(x_1,\ldots,x_\ell,\ldots,x_n)$, so α_j does not depend on x_ℓ $(\ell = 1,\ldots,n)$. Thus (3.6) holds for these α_j. □

REMARK 3.3. The relationship between our Theorem 3.1 and NIESSEN's [12] definition C is easily established (cf. also Remark 2.7). Indeed, the p.d.e. (3.1) satisfies the conditions (i), (ii), (iii) of Theorem 3.1 if and only if the left hand side of (3.1), after multiplication by some function of x and for functions u of the form $u(x) = \prod_{i=1}^n X_i(x_i)$, can be written as $\det(L_{ij}\, X_j(x_j))$, where

$$L_{1j} := \frac{d^2}{dx_j^2} + b_j(x_j)\frac{d}{dx_j} + c_{j1}(x_j),$$

$$L_{ij} := c_{ji}(x_j), \qquad i \neq 1,$$

and the matrix (3.3) has rank n-1.

REMARK 3.4. Consider the Helmholtz equation (2.5) on a Riemannian manifold. On comparing (2.5) with (3.1) we have:

$$A_i = \frac{1}{g_i}, \; A_{ij} = 0 \text{ if } i \neq j, \qquad C = \omega^2,$$

$$B_i = \frac{1}{g_i}\frac{\partial}{\partial x_i}(\log \frac{g^{\frac{1}{2}}}{g_i}).$$

If A_i takes the form (3.11) then

$$\frac{B_i}{A_i} = \frac{\partial}{\partial x_i}\log(\frac{g^{\frac{1}{2}}}{S}).$$

Hence our separability conditions (i), (ii), (iii) of Theorem 3.1, if applied to the p.d.e. (2.5), are equivalent to the Stäckel-Robertson conditions (i) and (ii), mentioned in §2.

Next we mention some generalizations of Theorem 3.1 and Lemma 3.2 to higher order and nonlinear p.d.e.'s. The proofs, which are omitted, are similar to the proofs earlier in this section.

LEMMA 3.5. *Consider a linear homogeneous analytic p.d.e. of* m^{th} *order*

$$\sum_{|\alpha|\leq m} A_\alpha(x) \frac{\partial^{\alpha_1+\ldots+\alpha_n} u}{\partial x_1^{\alpha_1}\ldots\partial x_n^{\alpha_n}} = 0 \tag{3.18}$$

such that, for each x *and* i, *not all terms involving derivatives with respect to* x_i *will vanish. Suppose there are* n *analytic o.d.e.'s of* m^{th} *order*

$$X_i^{(m)}(x_i)+f_i(X_i^{(m-1)}(x_i),\ldots,X_i(x_i),x_i) = 0, \qquad i = 1,\ldots,n, \tag{3.19}$$

such that, for each set of solutions $(X_1,\ldots,X_n)$ *of* (3.19), *the function* $u(x) := \prod_{i=1}^n X_i(x_i)$ *is a solution of* (3.18). *Then* (3.18) *and* (3.19) *must have the form*

$$\sum_{i=1}^n A_i(x)\left(\frac{\partial^m u}{\partial x_i^m} + \sum_{j=1}^{m-1} b_i^j(x_i)\frac{\partial^j u}{\partial x_i^j}\right) + C(x)u = 0,$$

$$X_i^{(m)}(x_i) + \sum_{j=1}^{m-1} b_i^j(x_i)X_i^{(j)}(x_i)+c_i(x_i)X_i(x_i) = 0, \qquad i = 1,\ldots,n,$$

respectively, for certain analytic functions b_i^j, c_i, *where*

$$C(x) = \sum_{i=1}^n A_i(x)c_i(x_i)$$

and $A_i(x) \neq 0$ *for all* x *and* i.

LEMMA 3.6. *Consider a (generally nonlinear) second order p.d.e. of the form*

$$\sum_{i=1}^n A_i(x)\frac{\partial^2 u}{\partial x_i^2} + F\left(\frac{\partial u}{\partial x_1},\ldots,\frac{\partial u}{\partial x_n},u,x\right) = 0, \tag{3.20}$$

with A_i *and* F *analytic and* $A_i(x) \neq 0$ *for all* x *and* i. *Suppose there are* n *analytic second order o.d.e.'s*

$$X_i''(x_i)+f_i(X_i'(x_i),X_i(x_i),x_i) = 0, \qquad i = 1,\ldots,n, \tag{3.21}$$

such that, for each set of solutions $(X_1,\ldots,X_n)$ *of* (3.21), *the function* $u(x) := \prod_{i=1}^n X_i(x_i)$ *is a solution of* (3.22). *Then* (3.20) *and* (3.21) *must have the form*

$$\sum_{i=1}^{n} A_i(x)\left(\frac{\partial^2 u}{\partial x_i^2} + \beta_i\left(\frac{\partial u}{\partial x_i}, u, x_i\right)\right) + D(x)\ u \log u = 0,$$

$$X_i''(x_i) + \beta_i(X_i'(x_i), X_i(x_i), x_i) + d_i(x_i)X_i(x_i)\log X_i(x_i) = 0, \quad i = 1,\ldots,n,$$

respectively, for certain analytic β_i, d_i *and* D, *where* β_i *is homogeneous of degree* 1 *in its first two arguments and*

$$(3.22) \qquad A_i(x)d_i(x_i) = D(x), \qquad i = 1,\ldots,n.$$

Note that, in case D is not identically zero, equation (3.22) highly restricts the possible choices for the A_i's.

REMARK 3.7. It follows from Lemma 3.6 that the nonlinear second order p.d.e. (2.15) does not separate into second order o.d.e.'s.

THEOREM 3.8. *Consider a (generally nonlinear)* m^{th} *order p.d.e.*

$$(3.23) \qquad \sum_{i=1}^{n} A_i(x)\left(\frac{\partial^m u}{\partial x_i^m} + \beta_i\left(\frac{\partial^{m-1} u}{\partial x_i^{m-1}}, \ldots, \frac{\partial u}{\partial x_i}, u, x_i\right)\right) = 0,$$

where A_i *and* β_i *are analytic,* $A_i(x) \neq 0$ *for all* x *and* i, *and* β_i *is homogeneous of degree* 1 *in* $\partial^{m-1}u/\partial x_i^{m-1}, \ldots, \partial u/\partial x_i, u$. *Then* (3.23) *separates into* n m^{th} *order o.d.è.'s if and only if there are analytic functions* c_{ij} $(i = 1,\ldots,n,\ j = 2,\ldots,n)$ *in one real variable such that the* $n \times (n-1)$ *matrix* $(c_{ij}(x_i))$ *has rank* n-1 *for all* x *and*

$$\sum_{i=1}^{n} c_{ij}(x_i)A_i(x) = 0, \qquad j = 2,\ldots,n.$$

In case of separability a function $u(x) = \prod_{i=1}^{n} X_i(x_i)$, *not identically zero, is a solution of* (3.23) *if and only if, for some* $(\alpha_2,\ldots,\alpha_n) \in \mathbb{C}^{n-1}$, *the functions* X_i *are solutions of the o.d.e.'s*

$$X_i^{(m)}(x_i) + \beta_i(X_i^{(m-1)}(x_i), \ldots, X_i'(x_i), X_i(x_i), x_i) + \\ + \left(\sum_{j=2}^{n} \alpha_j c_{ij}(x_i)\right)X_i(x_i) = 0, \qquad i = 1,\ldots,n.$$

Finally we turn to the case of the Hamilton-Jacobi equation (2.3) considered by STÄCKEL [16]. We will now use Definition 2.1 with the Ansatz (2.12) replaced by

$$(3.24) \qquad u(x) = \sum_{i=1}^{n} X_i(x_i).$$

In Definition 2.2 the condition $\prod_{i=1}^{n} y_{0,i} \neq 0$ can then be omitted. It is easy to prove the following analogue of Theorem 3.1.

THEOREM 3.9. *Consider the first order nonlinear p.d.e.*

$$\sum_{i=1}^{n} A_i(x)\left(\frac{\partial u}{\partial x_i}\right)^2 = C(x), \tag{3.25}$$

where A_i *and* C *are analytic and* $A_i(x) \neq 0$ *for all* x *and* i. *Then the p.d.e.* (3.25) *separates into first order o.d.e.'s under the Ansatz* (3.24) *if and only if condition* (iii) *of Theorem* 3.1 *holds.*

4. ON CONDITIONS EQUIVALENT TO STACKEL's CRITERIUM

The main result in this section is Theorem 4.5, which associates n linearly independent, mutually commuting partial differential operators with a separable second order p.d.e.. Theorem 4.6 and Corollary 4.7 will be needed in section 5.

Lemma 4.2 and Theorems 4.3 and 4.4 will involve:

ASSUMPTION 4.1. The functions c_{ij} $(i,j = 1,\dots,n)$ are analytic on an open connected region $\Omega \subset \mathbb{R}^n$ such that $\det(c_{ij}(x)) \neq 0$ on Ω. The matrix inverse of $(c_{ij}(x))$ is denoted by $(\gamma_{ij}(x))$. Assume that $\gamma_{1i}(x) \neq 0$ $(i = 1,\dots,n)$ on Ω.

LEMMA 4.2. *Let* c_{ij} *and* γ_{ij} *be as in Assumption* 4.1. *Then the following three statements are equivalent*:

(a) c_{ij} *only depends on* x_i $(i,j = 1,\dots,n)$.

(b) $\gamma_{kp}\dfrac{\partial\gamma_{1j}}{\partial x_p} = \gamma_{1p}\dfrac{\partial\gamma_{kj}}{\partial x_p}$ $(k = 2,\dots,n;\ j,p = 1,\dots,n)$.

(c) $\gamma_{kp}\dfrac{\partial^m\gamma_{\ell j}}{\partial x_p^m} = \gamma_{\ell p}\dfrac{\partial^m\gamma_{kj}}{\partial x_p^m}$ $(j,k,\ell,p = 1,\dots,n;\ m = 1,2,3,\dots)$.

PROOF. We prove (a) ⇒ (c) ⇒ (b) ⇒ (a).

(a) ⇒ (c): $\sum_k c_{ik}(x_i)\gamma_{kj}(x) = \delta_{ij}$.

Hence

$$\sum_k c_{ik}(x_i)\,\frac{\partial^m\gamma_{kj}(x)}{\partial x_p^m} = 0, \qquad i \neq p.$$

So

$$\sum_k c_{ik}(x_i)\left(\gamma_{kp}\frac{\partial^m\gamma_{\ell j}}{\partial x_p^m} - \gamma_{\ell p}\frac{\partial^m\gamma_{kj}}{\partial x_p^m}\right) = 0, \quad i \neq p.$$

Since also

$$\sum_k c_{ik}(x_i)\gamma_{kp}(x) = 0, \qquad i \neq p,$$

it follows that

$$\frac{1}{\gamma_{kp}}\left(\gamma_{kp}\frac{\partial^m\gamma_{\ell j}}{\partial x_p^m}-\gamma_{\ell p}\frac{\partial^m\gamma_{kj}}{\partial x_p^m}\right)$$

is independent of k. So (c) will follow if $\gamma_{\ell p}\neq 0$. Suppose $\gamma_{\ell p}(y)=0$ for some $y\in\Omega$. Then $\ell\neq 1$. For $\alpha\in\mathbb{C}$ let

$$\tilde{\gamma}_{ij}:=\gamma_{ij}+\alpha\delta_{i\ell}\gamma_{1j},$$

$$\tilde{c}_{ij}:=c_{ij}-\alpha\delta_{j1}c_{i\ell}.$$

Then $(\tilde{\gamma}_{ij}(x))$ and $(\tilde{c}_{ij}(x_i))$ are matrix inverses of each other and $\tilde{\gamma}_{\ell p}(y)\neq 0$ if $\alpha\neq 0$. Hence (c) is valid for $(\tilde{\gamma}_{ij}(y))$ if $\alpha\neq 0$ and the case $\alpha=0$ follows by continuity.

(b) ⇒ (a): Let $p\neq i$. Differentiation of

$$\sum_k c_{ik}\gamma_{kj}=\delta_{ij}$$

yields

$$\sum_k\frac{\partial c_{ik}}{\partial x_p}\gamma_{kj}=-\sum_k c_{ik}\frac{\partial\gamma_{kj}}{\partial x_p}=-\frac{1}{\gamma_{1p}}\frac{\partial\gamma_{1j}}{\partial x_p}\sum_k c_{ik}\gamma_{kp}=0. \quad \square$$

THEOREM 4.3. *Let* c_{ij} *and* γ_{ij} *be as in Assumption 4.1. Let* b_i, c_i $(i=1,\ldots,n)$ *be analytic functions in one real variable. Consider the* n *linearly independent partial differential operators*

$$L_k:=\sum_{i=1}^n\gamma_{ki}(x)\left(\frac{\partial^2}{\partial x_i^2}+b_i(x_i)\frac{\partial}{\partial x_i}+c_i(x_i)\right),\qquad k=1,\ldots,n.$$

Then the following three statements are equivalent:

(a) c_{ij} *only depends on* x_i $(i,j=1,\ldots,n)$.

(b) L_k *commutes with* L_1 *for* $k=2,\ldots,n$.

(c) *The operators* $L_1,\ldots,L_n$ *mutually commute.*

PROOF. A calculation shows that

$$[L_k,L_\ell]:=L_kL_\ell-L_\ell L_k=\sum_{i,j=1}^n\left[\left(\gamma_{ki}\frac{\partial\gamma_{\ell j}}{\partial x_i}-\gamma_{\ell i}\frac{\partial\gamma_{kj}}{\partial x_i}\right)\left(2\frac{\partial}{\partial x_i}+b_i(x_i)\right)+\right.$$

$$\left.+\left(\gamma_{ki}\frac{\partial^2\gamma_{\ell j}}{\partial x_i^2}-\gamma_{\ell i}\frac{\partial^2\gamma_{kj}}{\partial x_i^2}\right)\right]\left(\frac{\partial^2}{\partial x_j^2}+b_j(x_j)\frac{\partial}{\partial x_j}+c_j(x_j)\right).$$

Now apply Lemma 4.2. $\square$

The implication (a) ⇒ (c) in the above theorem formally coincides with a result in KÄLLSTRÖM & SLEEMAN [6, Theorem 1]. (See also ATKINSON [1, Theorem 6.7.2].) These authors are working in a Hilbert space context. Our result is obtained from formula (0.1) in [6] by putting $\alpha_0 := 1$, $\alpha_i := 0$ $(i = 1,\dots,n)$, $A_i := -(\partial^2/\partial x_i^2 + b_i(x_i)\partial/\partial x_i + c_i(x_i))$, $S_{ij} := c_{ij}(x_i)$.

Let F, G be C^∞-functions in the 2n real variables $x = (x_1,\dots,x_n)$, $p = (p_1,\dots,p_n)$. The *Poisson bracket* of F and G is defined by

$$\{F,G\} := \sum_{k=1}^{n} \left(\frac{\partial F}{\partial x_k}\frac{\partial G}{\partial p_k} - \frac{\partial F}{\partial p_k}\frac{\partial G}{\partial x_k}\right).$$

THEOREM 4.4. *Let* c_{ij} *and* γ_{ij} *be as in Assumption* 4.1. *Consider the* n *functions*

$$F_k(x,p) := \sum_{i=1}^{n} \gamma_{ki}(x)p_i^2.$$

Then the following three statements are equivalent:

(a) c_{ij} *only depends on* x_i $(i,j = 1,\dots,n)$.

(b) $\{F_k,F_1\} = 0$ $(k = 2,\dots,n)$.

(c) $\{F_k,F_\ell\} = 0$ $(k,\ell = 1,\dots,n)$.

PROOF. We have

$$\{F_k,F_\ell\} = 2\sum_{i,j=1}^{n} \left(\gamma_{\ell i}\frac{\partial \gamma_{kj}}{\partial x_i} - \gamma_{ki}\frac{\partial \gamma_{\ell j}}{\partial x_i}\right) p_i p_j^2.$$

Now apply Lemma 4.2. □

Theorem 4.4 is contained in a result by EISENHART [3, p.289], who gives necessary and sufficient conditions for the existence of orthogonal separable coordinate systems for the Hamilton-Jacobi equation on a Riemannian manifold. These conditions involve the existence of n-1 independent quadratic first integrals for the equations of geodesics. Eisenhart's conditions were considerably improved by KALNINS & MILLER [7, Theorem 6].

THEOREM 4.5. *Consider the analytic p.d.e.*

$$(4.1)\qquad \sum_{i=1}^{n} A_i(x)\left(\frac{\partial^2 u}{\partial x_i^2} + b_i(x_i)\frac{\partial u}{\partial x_i}\right) + C(x)u = 0,$$

with $A_i(x) \neq 0$ *for all* x *and* i. *Suppose* (4.1) *separates into* n *second order o.d.e.'s. Then there are* n *linearly independent, mutually commuting linear partial differential operators* $L_1,\dots,L_n$ *of second order such that* $L_1 = \Phi L$ *for some analytic function* Φ $(\Phi(x)\neq 0)$ *and such that all solutions* u *of* (4.1) *of the form* $u(x) = \prod_{i=1}^{n} X_i(x_i)$ *are joint eigenfunctions of* $L_2,\dots,L_n$.

PROOF. By the separability assumption, condition (iii) of Theorem 3.1 holds. Let $(c_{ij}(x_i))$ be the matrix considered there and put

$$\tilde{c}_{i1}(x_i) := \delta_{i1}, \quad \tilde{c}_{ij}(x_i) := c_{ij}(x_i), \quad i = 1,\dots,n, \quad j = 2,\dots,n.$$

Let $(\gamma_{ij}(x))$ be the inverse of the matrix $(\tilde{c}_{ij}(x_i))$. Then, by (3.4), $\gamma_{1i}(x) = A_i(x)/A_1(x)$. Define the operators L_k by

$$L_k := \sum_{i=1}^{n} \gamma_{ki}(x)\left(\frac{\partial^2}{\partial x_i^2} + b_i(x_i)\frac{\partial}{\partial x_i} + c_{i1}(x_i)\right), \qquad i = 1,\dots,n.$$

Then $L_1 = A_1^{-1} L$, the operators L_k are linearly independent and, by Theorem 4.3, they mutually commute. Let $u(x) = \prod_{i=1}^n X_i(x_i) \neq 0$ be a solution of (4.1). By Theorem 3.1, there is $(\alpha_2,\dots,\alpha_n) \in \mathbb{C}^{n-1}$ such that $(X_1,\dots,X_n)$ satisfy the o.d.e.'s (3.6). Hence

$$\frac{\partial^2 u}{\partial x_i^2} + b_i(x_i)\frac{\partial u}{\partial x_i} + (c_{i1}(x_i) + \sum_{j=2}^{n} \alpha_j c_{ij}(x_i))u = 0, \qquad i = 1,\dots,n.$$

Multiplication of this equation by $\gamma_{ki}(x)$ $(k = 2,\dots,n)$ and summation over i yields $L_k u + \alpha_k u = 0$. □

Theorem 4.5 is well-known for many special separable second order p.d.e.'s (cf. MILLER [9]), but it seems that the general statement of the theorem has not been proved before.

Obviously, there still exists a commuting family of operators $L_1,\dots,L_n$ with $L_1 = \Phi L$ if $Lu = 0$ is separable after a transformation of coordinates. Conversely assume that a second order operator L can be extended to a family of n linearly independent, mutually commuting second order operators $L_1 := L, L_2,\dots,L_n$. It would be interesting to formulate a criterium, under which conditions the existence of such a commuting set implies that L is separable into second order o.d.e.'s after some transformation of coordinates. The result by KALNINS & MILLER [7, Theorem 6], which we already mentioned, may be helpful in achieving such a criterium.

The next theorem was already proved by EISENHART [3, p.289]. We include the proof in order to make the paper more self-contained. The subsequent corollary may be new.

<u>THEOREM 4.6</u>. *Let $A_1,\dots,A_n$ be analytic functions on an open connected region Ω in $\mathbb{R}^n$, not taking the value zero. Then, locally, the following two statements are equivalent*:

(a) *There are analytic functions c_{ij} $(i,j = 1,\dots,n)$ in one real variable with $\det(c_{ij}(x_i)) \neq 0$ such that*

$$\sum_{i=1}^{n} c_{ij}(x_i)A_i(x) = \delta_{j,1}, \qquad j = 1,\dots,n.$$

(b) *The functions A_i satisfy the system of p.d.e.'s*

$$A_{j;p,q} = 0, \qquad p,q,j = 1,\dots,n, \qquad p \neq q, \tag{4.2}$$

where

$$(4.3) \qquad A_{j;p,q} := \frac{\partial^2 \log A_j}{\partial x_p \partial x_q} + \frac{\partial \log A_j}{\partial x_p}\frac{\partial \log A_j}{\partial x_q} - \frac{\partial \log A_p}{\partial x_q}\frac{\partial \log A_j}{\partial x_p} + - \frac{\partial \log A_q}{\partial x_p}\frac{\partial \log A_j}{\partial x_q}.$$

PROOF. First we prove (a) ⇒ (b). Suppose that (a) holds. Let $(\gamma_{ij}(x))$ be the matrix inverse of $(c_{ij}(x_i))$. Then (b) of Lemma 4.2 holds. Put

$$\rho_{kj}(x) := \gamma_{kj}(x)/\gamma_{1j}(x).$$

Then, for each $k = 1,\ldots,n$, the functions $\rho_{k1},\ldots,\rho_{kn}$ satisfy the system

$$(4.4) \qquad \frac{\partial \rho_{kj}}{\partial x_p} = (\rho_{kp}-\rho_{kj})\frac{\partial \log A_j}{\partial x_p}, \qquad j,p = 1,\ldots,n.$$

Hence

$$0 = \frac{\partial}{\partial x_p}\left(\frac{\partial \rho_{kj}}{\partial x_q}\right) - \frac{\partial}{\partial x_q}\left(\frac{\partial \rho_{kj}}{\partial x_p}\right) = (\rho_{kp}-\rho_{kq})A_{j;p,q}.$$

Fix p and q with $p \neq q$. Then $\rho_{kp} \neq \rho_{kq}$ for some k, because, otherwise, $\gamma_{kp} = (\gamma_{1p}/\gamma_{1q})\gamma_{kq}$ $(k = 1,\ldots,n)$, contradicting the nonsingularity of the matrix $(\gamma_{ij}(x))$. Thus $A_{j;p,q} = 0$ for $p \neq q$.

Next we prove (b) ⇒ (a). Suppose (4.2) holds. By a theorem of Frobenius (cf. DIEUDONNÉ [2,Ch.X, p.314]) this implies the complete integrability of the system

$$\frac{\partial \rho_j}{\partial x_p} = (\rho_p-\rho_j)\frac{\partial \log A_j}{\partial x_p}.$$

Hence, locally, there are n linearly independent analytic solutions $(\rho_{k1},\ldots,\rho_{kn})$, $k = 1,\ldots,n$, of this system, including the trivial solution $(\rho_{11},\ldots,\rho_{1n}) :=$ $:= (1,\ldots,1)$. Put $\gamma_{kj} := A_j\,\rho_{kj}$. Then (4.4) implies (b) of Lemma 4.2. Since $(\gamma_{ij}(x))$ is a nonsingular matrix, we conclude that (a) of Lemma 4.2 is valid for its matrix inverse $(c_{ij}(x_i))$. This proves (a) of the present theorem. □

COROLLARY 4.7. *Let* $A_1,\ldots,A_n$ *be analytic functions on an open connected region* Ω *in* $\mathbb{R}^n$, *not taking the value zero. Then, locally, the following two statements are equivalent:*

(a) *There are analytic functions* c_{ij} $(i = 1,\ldots,n;\ j = 2,\ldots,n)$ *in one real variable with rank* $(c_{ij}(x_i)) = n-1$ *such that*

$$\sum_{i=1}^{n} c_{ij}(x_i)A_i(x) = 0, \qquad j = 2,\ldots,n.$$

(b) *Let* $A_{j;p,q}$ *be defined by* (4.3). *Then*

$$A_{1;p,q} = A_{2;p,q} = \dots = A_{n;p,q}, \qquad p,q = 1,\dots,n, \quad p \neq q.$$

PROOF. First we prove (a) ⇒ (b). Suppose that (a) holds. Let $c_{i1}(x_i) := \delta_{i1}$ $(i = 1,\dots,n)$. Let, $\tilde{A}_i := A_i/A_1$ $(i = 1,\dots,n)$. Then the $\tilde{A}_i$'s satisfy (a) of Theorem 4.6. Hence $\tilde{A}_{j;p,q} = 0$ $(p\neq q)$. A calculation shows that $\tilde{A}_{j;p,q} = A_{j;p,q} - A_{1;p,q}$. Hence $A_{j;p,q} = A_{1;p,q}$ $(j = 2,\dots,n;\ p \neq q)$.

Next assume (b) and put again $\tilde{A}_j := A_j/A_1$. Then

$$0 = A_{j;p,q} - A_{1;p,q} = \tilde{A}_{j;p,q} \qquad (p\neq q),$$

so (a) of Theorem 4.6 is valid for the $\tilde{A}_i$'s, hence (a) of the corollary holds for the A_i's.

5. SEPARATION OF VARIABLES AND EQUATIONS ON RIEMANNIAN MANIFOLDS

In (2.4) we introduced the Laplace-Beltrami operator

$$(5.1) \qquad \Delta = \sum_{i=1}^{n} g^{-\frac{1}{2}} \frac{\partial}{\partial x_i} \circ g^{\frac{1}{2}} g_i^{-\frac{1}{2}} \frac{\partial}{\partial x_i} = \sum_{i=1}^{n} g_i^{-1}\left(\frac{\partial^2}{\partial x^2} + \left(\frac{\partial}{\partial x_i} \log(g^{\frac{1}{2}}/g_i)\right)\frac{\partial}{\partial x_i}\right)$$

for orthogonal local coordinates on a Riemannian manifold. More generally, we will consider operators Δ of the form (5.1) without this geometric interpretation, so it is no longer required that $g_i(x) > 0$, but g_i may be a complex-valued analytic function on an open connected region Ω in $\mathbb{R}^n$, with $g_i(x) \neq 0$ on Ω. We still put $g := \prod_{i=1}^{n} g_i$. Note that the second expression for Δ in (5.1) is independent of the choice of the branches for the square root or the logarithm.

Our first result, contained in Theorem 5.1 below, shows that a linear second order p.d.e. in $n \geq 3$ variables which separates into second order o.d.e.'s can always be written in the form $\Delta u + V(x)u = 0$, with Δ as in (5.1). Historically, a general theory of separation of variables was usually given in the context of a Riemannian manifold. Our theorem shows that this meant no loss of generality.

THEOREM 5.1. *Let*

$$(5.2) \qquad L := \sum_{i=1}^{n} A_i(x)\left(\frac{\partial^2}{\partial x_i^2} + b_i(x_i)\frac{\partial}{\partial x_i}\right)$$

be a partial differential operator on an open connected region Ω in $\mathbb{R}^n$, where A_i and b_i are analytic and $A_i(x) \neq 0$ on Ω. Suppose that the p.d.e. $Lu = 0$ separates into second order o.d.e.'s. Then for each analytic function Φ on Ω ($\Phi(x) \neq 0$ on Ω) there is an analytic function R on Ω ($R(x) \neq 0$ on Ω) such that

$$(5.3) \qquad \Phi L = R^{-1}\Delta \circ R - R^{-1}(\Delta R),$$

where Δ *is given by* (5.1) *and*

$$(5.4)\qquad g_i := \frac{1}{\Phi A_i}.$$

If $n \geq 3$ *then, in particular, we can choose* Φ *such that* $R \equiv 1$, *i.e.*

$$\Phi L = \Delta, \qquad g_i = 1/\Phi A_i.$$

PROOF. By the separability of $Lu = 0$, condition (iii)' of Theorem 3.1 holds. For the matrix $c_{ij}(x_i)$ introduced there, let $M_{i1}(x)$ be defined by (3.8) and let R and Φ be functions related by

$$(5.5)\qquad R^2\Phi^{1-n/2} = \exp\left(\sum_i \int_{x_{0i}}^{x_i} b(y_i)\,dy_i\right) \frac{(\prod_{i=1}^n M_{i1})^{1/n}}{(\prod_{i=1}^n A_i)^{1/n-1/2}}$$

for some $(x_{01},\ldots,x_{0n}) \in \Omega$. Let g_i be given by (5.4). Then

$$R^{-1}\Delta \circ R - R^{-1}(\Delta R) = \sum_{i=1}^n g_i^{-1}\left(\frac{\partial^2}{\partial x_i^2} + \left(\frac{\partial}{\partial x_i}\log\frac{R^2 g_i^{\frac12}}{g_i}\right)\frac{\partial}{\partial x_i}\right)$$

and

$$\frac{\partial}{\partial x_i}\log\frac{R^2 g^{\frac12}}{g_i} = \frac{\partial}{\partial x_i}\log\frac{R^2\Phi^{1-n/2}A_i}{(\prod_{i=1}^n A_i)^{\frac12}} = b_i(x_i) + \frac{\partial}{\partial x_i}\log\left(A_i\prod_{i=1}^n (M_{i1}/A_i)^{1/n}\right) =$$

$$= b_i(x_i) + \frac{\partial}{\partial x_i}\log M_{i1} = b_i(x_i),$$

where we used (5.4), (5.5), (3.10) and the fact that M_{i1} does not depend on x_i. Formula (5.3) now follows immediately. □

The second topic of this section deals with EISENHART's [3] condition on the vanishing of the Ricci tensor off the diagonal, which is necessary for separability of (2.5). Let Ω be an open connected region in $\mathbb{R}^n$ on which a fundamental tensor g_{ij} is defined, where the functions g_{ij} are complex-valued and analytic on Ω and $\det(g_{ij}(x)) \neq 0$. We may call Ω, together with the tensor g_{ij}, a formal Riemannian manifold. The Christoffel symbol of the second kind associated with this formal Riemannian structure, which we denote by Γ^i_{kj}, is defined by

$$2\sum_\ell \Gamma^\ell_{kj} g_{i\ell} = \frac{\partial}{\partial x_k} g_{ij} + \frac{\partial}{\partial x_j} g_{ik} - \frac{\partial}{\partial x_i} g_{jk}$$

(cf. EISENHART [4, (7.2)]). Next, theRiemannian curvature tensor is expressed in terms of these Christoffel symbols by

$$R^k_{\ell ij} := \sum_p (\Gamma^p_{j\ell}\Gamma^k_{ip} - \Gamma^p_{i\ell}\Gamma^k_{jp}) + \frac{\partial}{\partial x_i}\Gamma^k_{j\ell} - \frac{\partial}{\partial x_j}\Gamma^k_{i\ell}$$

(cf. EISENHART [4, (8.3)]). The Ricci tensor R_{ij} is obtained by contraction of the Riemannian curvature tensor:

$$R_{ij} := \sum_k R^k_{ijk}$$

(cf. EISENHART [4, p.21]); it is a symmetric tensor. Ω is called a (formal) *Einstein space* if

$$R_{ij}(x) = f(x) g_{ij}(x)$$

for some scalar function f. Clearly, on an Einstein space the Ricci tensor is diagonal if the fundamental tensor is diagonal. The class of Einstein spaces includes all Riemannian spaces of dimension 2, the flat spaces and the spaces of constant curvature.

If g_{ij} is diagonal and $g_i := g_{ii}$ then

$$(5.6) \qquad R_{pq} = \tfrac{1}{4} \sum_{j \neq p,q} \left[2 \frac{\partial^2 \log g_j}{\partial x_p \partial x_q} + \frac{\partial \log g_j}{\partial x_p} \frac{\partial \log g_j}{\partial x_q} - \frac{\partial \log g_p}{\partial x_q} \frac{\partial \log g_j}{\partial x_p} + \right.$$
$$\left. - \frac{\partial \log g_q}{\partial x_p} \frac{\partial \log g_j}{\partial x_q} \right], \quad p \neq q,$$

which follows from EISENHART [4, p.119].

THEOREM 5.2. *Let* $n \geq 3$. *Consider the p.d.e.*

$$(5.7) \qquad \Delta u + Cu := \sum_{i=1}^{n} \frac{1}{g_i} \left(\frac{\partial^2 u}{\partial x_i^2} + \left(\frac{\partial}{\partial x_i} \log \frac{g^{\frac{1}{2}}}{g_i} \right) \frac{\partial u}{\partial x_i} \right) + Cu = 0.$$

(a) *Suppose that condition* (iii)" *of Theorem* 3.1 *holds for* $A_i := g_i^{-1}$, $C(x) \equiv 1$. *Then* (5.7) *with* $C(x) \equiv 1$ *separates into second order o.d.e.'s if and only if the corresponding Ricci tensor is diagonal.*

(b) *If the Ricci tensor is diagonal and if* (5.7) *separates into second order o.d.e.'s for some specific function* $C = C_0$ *then* (5.7) *separates into second order o.d.e.'s for the function* $C(x) \equiv 1$.

PROOF. Let $A_i := g_i^{-1}$ and let $A_{j;p,q}$ be defined by (4.3). Under the assumptions of (a) we have $A_{j;p,q} = 0$ $(p \neq q)$, cf. Theorem 4.5, and the separation assumption of (b) implies $A_{j;p,q} = A_{1;p,q}$ $(p \neq q)$, cf. Theorem 3.1 and Corollary 4.7. So we have $A_{j;p,q} = A_{1;p,q}$ $(p \neq q)$ in both cases. It follows from (5.6) that, for $p \neq q$,

$$R_{pq} = \tfrac{1}{4}(n-2) A_{1;p,q} + \tfrac{1}{4} \sum_{k \neq p,q} \frac{\partial^2 \log g_k}{\partial x_p \partial x_q}.$$

Putting $b_i(x) := \dfrac{\partial \log(g^{\frac{1}{2}}/g_i)}{\partial x_i}$ we also have

$$\sum_{k\neq p,q} \frac{\partial^2 \log g_k}{\partial x_p \partial x_q} = \frac{\partial^2}{\partial x_p \partial x_q} \log\left(\frac{g}{g_p g_q}\right) = 2\,\frac{\partial^2 \log(g^{\frac{1}{2}}/g_p)}{\partial x_p \partial x_q} = 2\,\frac{\partial b_p}{\partial x_q},$$

since $M_{p1}\, g_p = M_{q1}\, g_q$ (cf. (3.10)) and M_{i1} does not depend on x_i. Hence

$$R_{pq} = \tfrac{1}{4}(n-2)A_{1;p,q} + \frac{3}{2}\frac{\partial b_p}{\partial x_q}. \tag{5.8}$$

Formula (5.8) implies that, if two of the three expressions $R_{p,q}$, $A_{1;p,q}$ and $\partial b_p/\partial x_q$ vanish then also the third one vanishes. This yields precisely the three required implications in (a) and (b) of the theorem (use Theorems 3.1 and 4.6). □

Part (a) of Theorem 5.2 was already proved by EISENHART [3]. Part (b) may be new. It would be of interest to prove or disprove the two implications in (a) of Theorem 5.2 without the assumption that $C(x) \equiv 1$, but still assuming that $A_i := g_i^{-1}$ and C satisfy condition (iii) of Theorem 3.1.

Part (b) of Theorem 5.2 is related to ROBERTSON's [14] result that the equation

$$\Delta u + k^2(E-V(x))u = 0 \tag{5.9}$$

(Δ given by (5.1), $k \neq 0$) separates into second order o.d.e.'s simultaneously for all constants E if and only if (i) the coefficients $A_i := g_i^{-1}$ satisfy condition (iii)" of Theorem 3.1 with $C(x) \equiv 1$, (ii) $\partial \log(g^{\frac{1}{2}} g_i^{-1})/\partial x_i$ only depends on x_i and (iii) V is of the form $V(x) = \sum_i c_i(x)/g_i(x)$. Our theorem shows that, in case of a diagonal Ricci tensor (for instance, on an Einstein manifold), Robertson's conditions already hold if (5.9) separates for one specific value of E.

We conclude this paper with an example of a quite general class of separable p.d.e.'s of the form $\Delta u + \alpha u = 0$.

<u>EXAMPLE 5.3</u>. Consider a formal n-dimensional Riemannian manifold Ω with diagonal fundamental tensor

$$g_i(x) := \frac{1}{f_i(x_i)} \prod_{k\neq i} (x_i - x_k),$$

where $f_i(x_i) \neq 0$ and $x_i \neq x_j$ $(i \neq j)$ for $x \in \Omega$. Then

$$\Delta = \sum_{i=1}^{n} \frac{f_i(x_i)}{\prod_{k\neq i}(x_i-x_k)} \left(\frac{\partial^2}{\partial x_i^2} + \tfrac{1}{2}\frac{f_i'(x)}{f_i(x_i)}\frac{\partial}{\partial x_i}\right).$$

It turns out that the p.d.e.

$$\Delta u + \alpha_1 u = 0 \tag{5.10}$$

separates into the o.d.e.'s

$$(5.11)\qquad X_i''(x_i) + \tfrac{1}{2}\frac{f_i'(x_i)}{f_i(x_i)}X_i'(x_i) + \frac{\sum_{j=1}^{n}\alpha_j x^{n-j}}{f_i(x_i)}X_i(x_i) = 0.$$

If n is small and $f_i(x_i)$ is a polynomial of low degree, the equations (5.11) yield well-known equations of mathematical physics. For instance, if n = 2 and $f_i(x) = x(1-x)$ then (5.11) becomes Mathieu's equation (cf. ERDÉLYI [5, 16.2(3)]) and (5.10) is just the two-variable Helmholtz equation in elliptic coordinates (cf. MILLER [9, p.19]), disguised in algebraic form. The case that n = 3 and f is a fourth degree polynomial arises from a certain R-separable form of the three-variable Laplace equation (cf. MILLER [9, p.209]).

LITERATURE

[1] ATKINSON, F.V., *Multiparameter eigenvalue problems*, Vol. I, Matrices and compact operators, Academic Press, New York, 1972.

[2] DIEUDONNÉ, J., *Éléments d'analyse*, Tome I, Fondements de l'analyse moderne, Gauthier-Villars, Paris, 1972.

[3] EISENHART, L.P., *Separable systems of Stäckel*, Annals of Math. 35 (1934), 284-305.

[4] EISENHART, L.P., *Riemannian geometry*, Princeton University Press, Princeton, second printing, 1949.

[5] ERDÉLYI, A., W. MAGNUS, F. OBERHETTINGER & F.G. TRICOMI, *Higher Transcendental Functions*, Vol. III, McGraw-Hill, New York, 1955.

[6] KÄLLSTRÖM, A. & B.D. SLEEMAN, *Multiparameter spectral theory*, Ark. Mat. 15(1977), 93-99.

[7] KALNINS, E.G. & W. MILLER, Jr., *Killing tensors and variable separation for Hamilton-Jacobi and Helmholtz equations*, preprint, 1979.

[8] KOORNWINDER, T.H., *Review of "Symmetry and separation of variables"*, by W. Miller, Jr., Bull. Amer. Math. Soc., to appear.

[9] MILLER, Jr., W., *Symmetry and separation of variables*, Addison-Wesley, Reading (Mass.), 1977.

[10] MOON, P. & D.E. SPENCER, *Theorems on separability in Riemannian n-space*, Proc. Amer. Math. Soc. 3(1952), 635-642.

[11] MORSE, Ph.M. & H. FESHBACH, *Methods of theoretical physics*, part I, McGraw-Hill, New York, 1953.

[12] NIESSEN, H.-D., *Algebraische Untersuchungen über separierbare Operatoren*, Math. Z. 94(1966), 328-348.

[13] OSBORNE, A. & A.E.G. STUART, *On the separability of the sine-Gordon equation and similar quasilinear partial differential equations*, J. Mathematical Phys. 19(1978), 1573-1579.

[14] ROBERTSON, H.P., *Bemerkung uber separierbare Systeme in der Wellenmechanik*, Math. Ann. 98(1928), 749-752.

[15] SNEDDON, I.N., *Elements of partial differential equations*, McGraw-Hill, New York, 1957.

[16] STÄCKEL, P., *Uber die Integration der Hamilton-Jacobischen Differentialgleichung mittels Separation der Variabelen*, Habilitationsschrift, Halle, 1891.

GENERATION OF LIMIT CYCLES FROM SEPARATRIX POLYGONS IN THE PHASE PLANE

J.W. Reyn

University of Technology
Delft The Netherlands

SUMMARY

Two-dimensional autonomous systems may have solution curves which form separatrix polygons in the phase plane. These are polygons the corner points of which are saddle points and the sides of which are separatrices connecting these saddle points. They are structurally unstable, and in this paper we will study the change in the phase portrait due to arbitrary small changes in the right hand sides of these systems. In particular, attention will be given to the generation of limit cycles from these polygons. The number of limit cycles generated by a separatrix polygon is seen to be related to the eigenvalues of the locally linearized system in the saddle points. For separatrix polygons with two saddle points, criteria, involving these eigenvalues are given when exactly one or exactly two limit cycles can be generated. For separatrix polygons with three or more saddle points similar criteria are given to ensure the generation of at least one, two, or more (till n for a n sided polygon) limit cycles.

1. INTRODUCTION

1. In this paper we consider solutions of the system

$$\frac{dx}{dt} = P(x,y) \ , \ \frac{dy}{dt} = Q(x,y) \tag{1}$$

in a neighbourhood of a separatrix polygon and study the change in phase portrait when the right hand sides of (1) are perturbed by arbitrary small functions. In particular, attention will be given to the generation of limit cycles from these polygons. A separatrix polygon is a polygon the corner points of which are saddle points and the sides are formed by separatrices connecting these saddle points. The simplest type is the saddle to saddle loop, which is a curve formed by two paths: an equilibrium point of the saddle point type and a separatrix connecting the saddle point with itself. This may be done in two ways: in the "small" loop, the region within the loop does not contain the remaining separatrices (fig. 1), whereas for the "large" loop (fig. 2) the opposite statement can be made.

Note that the saddle to saddle loop may be a limit continuum for integral curves near the loop in a region inside the loop for a "small" loop and outside the loop for a "large" loop. In figs. 1 and 2 arrows indicate the direction of increasing t, which is the counterclockwise direction; obviously reversing the signs of P and Q in (1) yields the same phase portraits with reversed t direction.

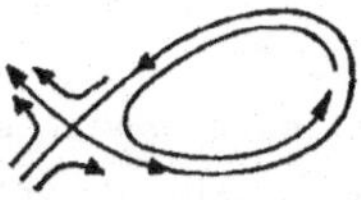

Fig. 1. "Small" saddle to saddle loop.

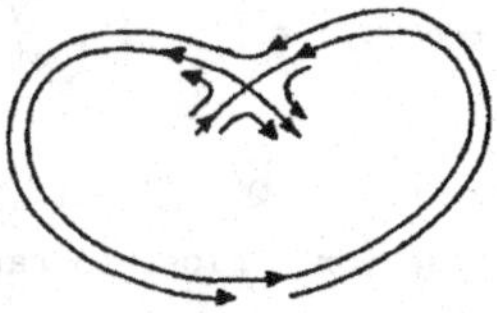

Fig. 2. "Large" saddle to saddle loop.

Separatrix polygons with two saddle points and three saddle points are shown in figs. 3 and 4, respectively. "Small" and "large" polygons may again be distinguished in an obvious way. In both figures the separatrices not forming the polygons are either outside or inside the polygon, so that the polygon may be a limit continuum for integral curves near the polygon in a region inside the polygon for a "small" polygon and outside the polygon for a "large" polygon.

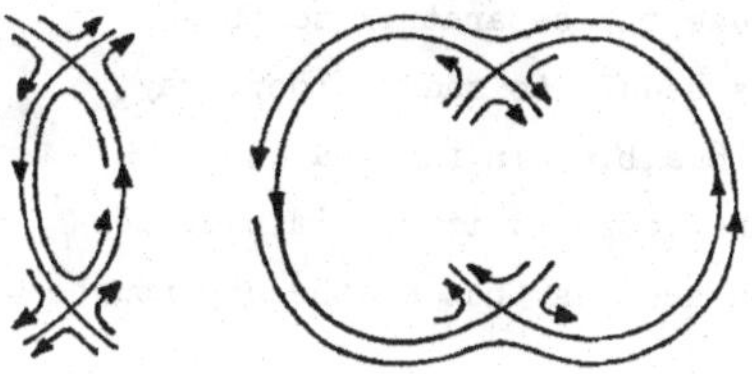

Fig. 3. Separatrix polygons with two saddle points.

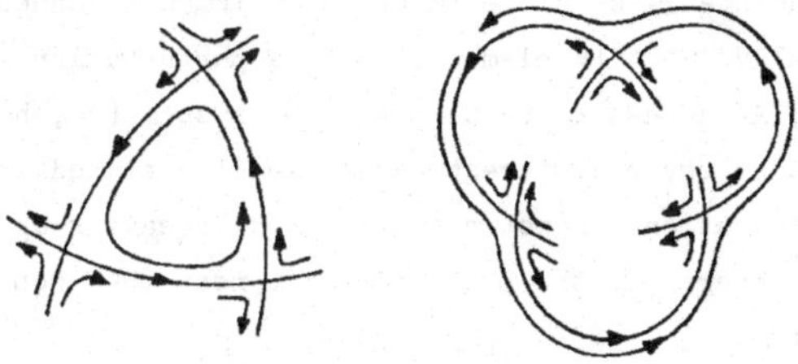

Fig. 4. Separatrix triangles.

Since we are mainly interested in the generation of limit cycles, we will restrict ourselves to such cases and do not consider other separatrix polygons, such as sketched for instance in fig. 5. Also, it will not effect the generality of the results in this paper if we restrict ourselves to "small" polygons with the increasing t direction in the counter clockwise sense.

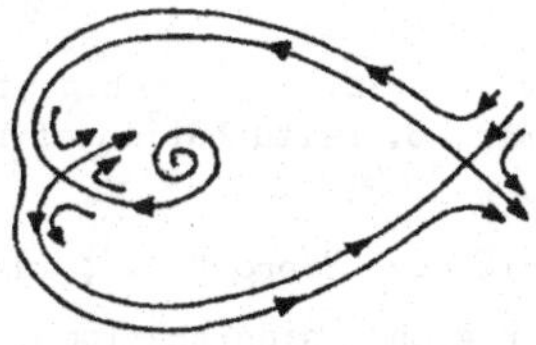

Fig. 5. Separatrix polygon which is not a limit continuum.

2. SADDLE TO SADDLE LOOPS

2. For the saddle to saddle loop we may refer to [1], [2], from which also the following definitions may be briefly collected. Suppose system (1) is perturbed in the sense that the system

$$\frac{dx}{dt} = P(x,y) + p(x,y) = \bar{P}(x,y) \ , \ \frac{dy}{dt} = Q(x,y) + q(x,y) = \bar{Q}(x,y) \qquad (2)$$

is obtained, where $P(x,y)$, $Q(x,y)$, $p(x,y)$, $q(x,y)$ are C_1 functions. We will say that system (2) is δ close (or δ close to rank 1) to system (1) in a region G, if at any point of that region for some $\delta > 0$ there is: $|p(x,y)| < \delta$, $|q(x,y)| < \delta$, $|p_x(x,y)| < \delta$, $|p_y(x,y)| < \delta$, $|q_x(x,y)| < \delta$, $|q_y(x,y)| < \delta$. System (1) is said to be structurally stable in a region G if there exists a $\delta_0 > 0$, such that for all systems (2) which are δ close to system (1) with $0 \le \delta \le \delta_0$, the phase portrait in G is the same as for system (1)*. Otherwise, the system (1) is called structurally unstable in G. Two phase portraits are the same if there exists a homeomorphism which maps G onto itself, such that the integral curves of (1) are mapped onto the integral curves of (2) and vice versa. In a small enough neighbourhood of a saddle point of system (1), the system is structurally stable. The saddle point may then be called a structurally stable element. A separatrix connecting two saddle points is a structurally unstable element, since a perturbation may cause the separatrix to "break up". As a result, perturbation of system (1), having a saddle to saddle loop, may lead to three different cases, as illustrated in fig. 6 a,b,c. In fig. 6a the perturbation has not resulted in a "breaking up" of the loop and if no limit cycles are generated, the phase portrait near the loop is the same as that of the unperturbed system.

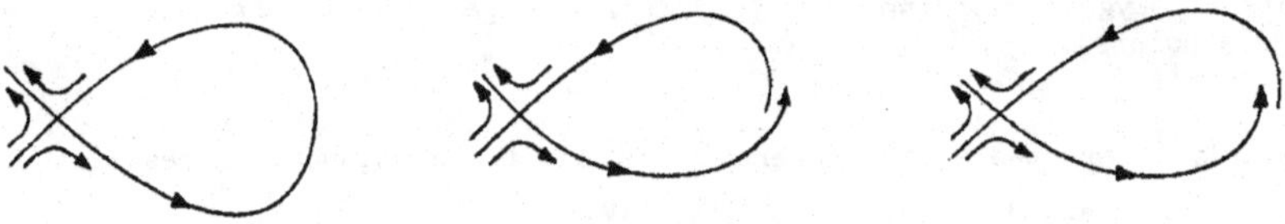

a. zero flow: z. b. outflow: o. c. inflow: i.

Fig. 6. Perturbation of a saddle to saddle loop.

We will name this case "zero flow", indicating that no integral curves are entering or leaving the region inside the loop. In fig. 6b the perturbation causes the loop to "break up" in such a way that "outflow" occurs, whereas in fig. 6c the case of "inflow" is illustrated. Obviously, in the latter two cases the phase portrait is different from the system with the loop, independent of the number of limit cycles which may occur. If the number of limit cycles is the same for figs. 6b and 6c, the phase portrait is the same. We will consider them to be two different cases, however, since the direction of increasing t must be reversed if one case is to be transformed into the other.

---

* A more precise definition has been given in [2], taking into account that the region G has to meet certain requirements.

3. The generation of limit cycles from a saddle to saddle loop may be briefly summarized, using [2], pp. 286-321. If the loop is a limit continuum, arbitrary small perturbations of the right hand sides of eqs. (1) can always be found such that at least one closed path is generated. In fact, if the loop is stable (a ω limit continuum for integral curves near the loop) at least one closed path is generated if inflow takes place and if the loop is unstable (a α limit continuum for integral curves near the loop) at least one closed path is generated in the outflow case. Further statements cannot be made without taking the eigenvalues of the locally linearized system in the saddle point into consideration. They are also indicative for the stability of the loop. In fact, if $\mu < 0$, $\lambda > 0$ are these eigenvalues and $\alpha = \frac{|\mu|}{\lambda}$, if $\alpha > 1$ the loop is stable, if $\alpha < 1$ unstable, whereas for $\alpha = 1$ it may be either stable, unstable of have a neighbourhood containing only closed paths. Then, if $\alpha \neq 1$, it may be demonstrated that no closed path is generated if the loop does not break up. Furthermore, if $\alpha > 1$, exactly one limit cycle is generated, - which is stable -, if inflow takes place and no limit cycle in the outflow case. If $\alpha < 1$, exactly one limit cycle is generated, - which is unstable -, in the case of outflow and no limit cycle in the inflow case. Thus if $\alpha \neq 1$, perturbation leads to the generation of at most one limit cycle.
If $\alpha = 1$, a more refined analysis is needed; work in this direction has been done by Leontovich [6].

4. SEPARATRIX POLYGONS WITH TWO SADDLE POINTS

4. We now consider perturbations of separatrix polygons with two saddle points. If we indicate by z: zero flow, by i: inflow, by o: outflow, then the number of combinations of two letters out of these three letters is equal to 9. Some of these combinations may be transformed into each other by cyclic permutation. Since this means merely a rotation of the phase portrait, we only consider those combinations which cannot be transformed into each other by cyclic permutation. This leaves 6 combinations, which may be written as: zz, zi, zo, ii, io, oo. The combination io gives rise to three cases depending on whether the inflowing separatrix returns to its saddle point i(z)o, flows into the inner region i(i)o, or flows out i(o)o.

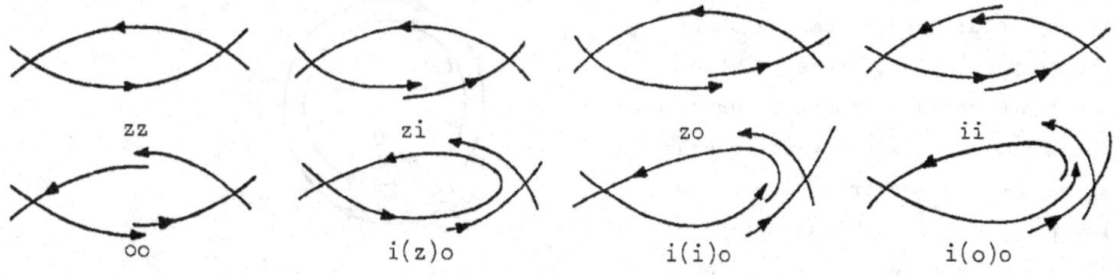

Fig. 7. Perturbation of a separatrix polygon with two saddle points.

The number of cases to be considered is thus equal to 8; they are illustrated in fig. 7. Note that in one case a saddle to saddle loop is generated.

5. We now consider the generation of limit cycles from a separatrix polygon with two saddle points. As in the case of the loop, if the separatrix polygon is a limit continuum, arbitrary small perturbations of the right hand sides of eq. (1) can always be found such that at least one closed path is generated.

Theorem 1. Let system (1) be of class 1 and possess a separatrix polygon with two saddle points being stable (unstable). Then for any $\varepsilon > 0$, there exists a $\delta > 0$ such that if system (2) is δ close (to rank 1) to system (1) and if total inflow occurs (cases zi, ii, i(i)o) [total outflow occurs (cases zo, oo, i(o)o)] or case i(z)o with an unstable (stable) saddle to saddle loop occurs, then in the ε-neighbourhood of the polygon there exists at least one closed path.

Proof. The proof uses the same type of arguments as given in the saddle to saddle loop case, which may be found in [2], p. 309. It is based on a study of the behaviour of the succession function, which is also the basis of all the principal results of this paper.

Let L be a path of (1) or (2) which intersects a transversal l of the vector field at some point M_0 and let the following intersection point of L with l when increasing t be M_1, then the mapping f: $M_0 \to M_1$ will be called the succession function (or Poincaré map).

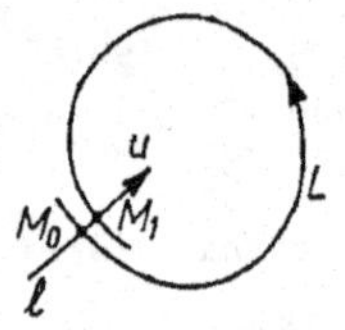

Fig. 8. Succession function.

If the transversal l is given by the parameter equations

$$x = g_1(u) \ , \ y = h_1(u) \tag{3}$$

where g_1, h_1 are C_1 functions, the succession function is also C_1.

Consider the case that the polygon is stable and construct a transversal through some point on the polygon, such that points on l for $u > 0$ ($u < 0$) are inside (outside) the polygon. Then, for a given $\varepsilon > 0$, a $\delta > 0$ can be found such that if (2) is δ close to (1) in a ε-neighbourhood of the polygon there exists a path L_1, which spirals outwards with increasing t, as may be deduced from the stability of the polygon. Also, there exists in this neighbourhood a path L_2, surrounding L_1 which spirals inwards with increasing t, as may be deduced from either the inflow property or the instability of the saddle to saddle loop in the case i(z)o.

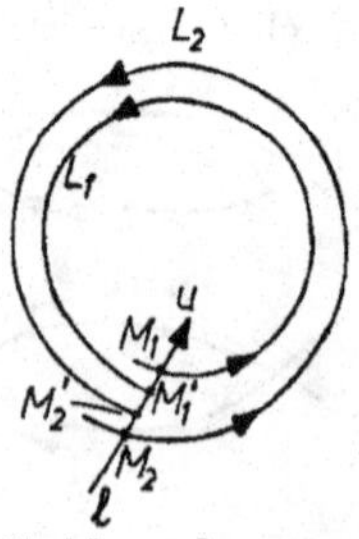

Fig. 9. Stable polygon and inflow.

Thus if u_1 corresponds to M_1, the intersection point of L_1 and l, and $u_2 < u_1$ to M_2 the intersection point of L_2 and l, one obtains $f(u_1) < u_1$ and $f(u_2) > u_2$. Since f(u) is continuous, there is at least one point $u_2 < u_* < u_1$, where $f(u_*) = u_*$. As a result there is at least one closed path in the ε-neighbourhood of the polygon.

The case that the polygon is unstable may be treated similarly.

6. We now take the eigenvalues of the locally linearized system in the saddle points into consideration. As before, let $\mu_i < 0$, $\lambda_i > 0$ be these eigenvalues in the ith saddle point and $|\frac{\mu_i}{\lambda_i}| = \alpha_i$, and choose any numbering of the saddle points. Then the polygon is stable if $\alpha_1\alpha_2 > 1$, unstable if $\alpha_1\alpha_2 < 1$, whereas for $\alpha_1\alpha_2 = 1$ it may be either stable, unstable or have a neighbourhood containing only closed paths. This theorem was shown for analytic systems by Dulac [4] and by Reyn [7] for the case that P(x,y) and Q(x,y) are Lipschitz continuously differentiable functions.

Theorem 2. Let system (1) be Lipschitz continuously differentiable and have a separatrix polygon with two saddle points, which is a limit continuum, and $\alpha_1 \neq 1$, $\alpha_2 \neq 1$, $\alpha_1\alpha_2 \neq 1$. Then for any $\varepsilon > 0$, there exists a $\delta > 0$ such that if system (2) is δ close (to rank 1) to system (1) and (i) $(1 - \alpha_1)(1 - \alpha_2) > 0$, system (2) has at most one limit cycle; (ii) $(1 - \alpha_1)(1 - \alpha_2) < 0$, system (2) has at most two limit cyles in the ε-neighbourhood of the polygon. Perturbations which generate exactly one limit cycle in case (i) and exactly two limit cycles in case (ii) can be given.

Proof. The proof follows by considering all the cases zz, ..., i(o)o, as illustrated in fig. 7, and study the generation of limit cycles in each case. Since in this way more information is obtained then is given in the theorem, we state the results in a number of lemmas.

The proof rests on properties of the succession function, which may be obtained using lemma 1 of ref. [7]. Referring to fig. 10, we may, in addition to the transversal l with parameter u on it [with u > 0 (u < 0) inside (outside) the separatrix polygon] and given by eqs. (3), construct another transversal m through a point on the other separatrix of the polygon given by the parametric equations:

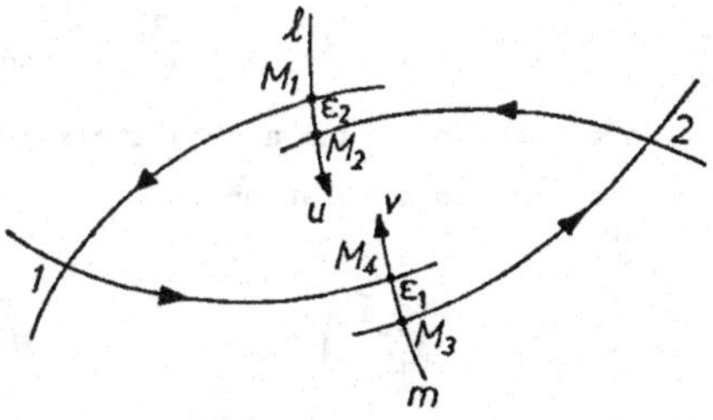

Fig. 10. Notations used for the succession function.

$$x = g_2(v) \ , \ y = h_2(v) \qquad (4)$$

where g_2, h_2 are C_1 functions, and v > 0 (v < 0) inside (outside) the polygon. If u and v are arc lengths along the transversals, [7] shows that the correspondence function v = v(u) may be written as

$$v(u) = v(M_4) + \bar{A}_1\{u - u(M_1)\}^{\bar{\alpha}_1}[1 + \bar{a}_1\{u - u(M_1)\}] \tag{5}$$

where $\bar{A}_1 > 0$ is a constant and $\bar{a}_1 \to 0$ as $u - u(M_1) \to 0$; $\bar{\alpha}_1$ is the α_1 of the perturbed system. Similarly, we obtain for the correspondence function $f\{u(v)\}$

$$f\{u(v)\} = u(M_2) + \bar{A}_2\{v - v(M_3)\}^{\bar{\alpha}_2}[1 + \bar{a}_2\{v - v(M_3)\}] \tag{6}$$

where $\bar{A}_2 > 0$ is a constant and $\bar{a}_2 \to 0$ as $v - v(M_3) \to 0$; $\bar{\alpha}_2$ is the α_2 of the perturbed system. As a result of eqs. (5), (6) the succession function $f = f(u)$ may be written as

$$f(u) = u(M_2) + \bar{A}_2[v(M_4) - v(M_3) + \bar{A}_1\{u - u(M_1)\}^{\bar{\alpha}_1}[1 + \bar{a}_1\{u - u(M_1)\}]]^{\bar{\alpha}_2} \cdot$$

$$[1 + \bar{a}_2\{v(M_4) - v(M_3) + \bar{A}_1\{u - u(M_1)\}^{\bar{\alpha}_1}[1 + \bar{a}_1\{u - u(M_1)\}]\}] \tag{7}$$

Now let $v(M_4) - v(M_3) = \varepsilon_1$, $u(M_2) - u(M_1) = \varepsilon_2$, $u - u(M_1) = \bar{u}$, $f(u) - u(M_1) = F(\bar{u})$, then eq. (7) can be written as:

$$F(\bar{u}) = \varepsilon_2 + \bar{A}_2^*[\varepsilon_1 + \bar{A}_1^*\bar{u}^{\bar{\alpha}_1}]^{\bar{\alpha}_2} \tag{8}$$

where $\bar{A}_1^*(\bar{u};\bar{p},\bar{q}) = \bar{A}_1[1 + \bar{a}_1(\bar{u})]$, and $\lim_{\bar{u},\bar{p},\bar{q}\to 0} \bar{A}_1^*(\bar{u};\bar{p},\bar{q}) = A_1 > 0$, constant.

$\bar{A}_2^*(\bar{u};\bar{p},\bar{q}) = \bar{A}_2[1 + \bar{a}_2\{\varepsilon_1 + \bar{A}_1^*\bar{u}^{\bar{\alpha}_1}\}]$, and $\lim_{\bar{u},\bar{p},\bar{q}\to 0} \bar{A}_2^*(\bar{u};\bar{p},\bar{q}) = A_2 > 0$, constant.

Obviously, a closed path (and a separatrix polygon) is a fixed point of the Poincaré map and corresponds to a solution of

$$u = \varepsilon_2 + \bar{A}_2^*[\varepsilon_1 + \bar{A}_1^*\bar{u}^{\bar{\alpha}_1}]^{\bar{\alpha}_2} \tag{9}$$

If $\bar{u} = 0$ is a solution of (9), it represents a separatrix polygon; if $\bar{u} \neq 0$ is a solution it represents a closed path. We are interested in those solutions of (9), for which $\lim_{\bar{p},\bar{q}\to 0} \bar{u} = 0$, being closed paths generated by the separatrix polygon.

In the following we also need the derivative of the succession function. Its existence can be shown as follows. For the derivative of the correspondence funtion $v = v(u)$ may be written [2, p. 291].

$$v'(u) = \frac{\Delta_1(u)}{\Delta_2(u)} \exp \int_{t(u)}^{t(v(u))} \bar{\sigma}(t)dt \tag{10}$$

where $\bar{\sigma}(t) = \mathrm{div}[\bar{P}\{x(t),y(t)\}\ ,\ \bar{Q}\{x(t),y(t)\}]$, with $x = x(t)$, $y = y(t)$ the solution through point u on l, t(u) and t{v(u)} are the values of t in the points u on l and v(u) on m, respectively and $\Delta_1(u)$, $\Delta_2(u)$ are given by

$$\begin{aligned}\Delta_1(u) &= P\{g_1(u),h_1(u)\}h_1'(u) - Q\{g_1(u),h_1(u)\}g_1'(u)\\ \Delta_2(u) &= P[g_2\{v(u)\},h_2\{v(u)\}]h_2'\{v(u)\} - Q[g_2\{v(u)\},h_2\{v(u)\}]g_2'\{v(u)\}\end{aligned} \tag{11}$$

According to lemma 2 of [7], eq. (10) may also be written as

$$v'(u) = \{u - u(M_1)\}^{\bar{\alpha}_1-1}\ \frac{\Delta_1(u)}{\Delta_2(u)}\ \exp B_1\{u - u(M_1)\} \tag{12}$$

where $B_1\{u - u(M_1)\}$ is a bounded function for $u - u(M_1) \to 0$. Alternatively, differentiation of (5) yields:

$$v'(u) = \bar{\alpha}_1\bar{A}_1\{u-u(M_1)\}^{\bar{\alpha}_1-1}\ [1 + \bar{a}_1\{u-u(M_1)\} + \bar{\alpha}_1^{-1}\{u-u(M_1)\}\bar{a}_1'\{u-u(M_1)\}] \tag{13}$$

Comparison of (12) and (13) gives

$$\lim_{u-u(M_1)\to 0}\ \bar{a}_1\{u - u(M_1)\}\bar{a}_1'\{u - u(M_1)\} = A \tag{14}$$

where A is some constant. Since $\bar{a}_1 \to 0$ as $u - u(M_1) \to 0$ it can be shown that $A = 0$*. As a result (13) may be written as

$$v'(u) = \bar{\alpha}_1\bar{A}_1\{u - u(M_1)\}^{\bar{\alpha}_1-1}\ [1 + \bar{b}_1\{u - u(M_1)\}] \tag{15}$$

where $\bar{b}_1 \to 0$ as $u - u(M_1) \to 0$. Similarly $\frac{df}{dv}$ may be written as

$$\frac{df}{dv} = \bar{\alpha}_2\bar{A}_2\{v - v(M_3)\}^{\bar{\alpha}_2-1}\ [1 + \bar{b}_2\{v - v(M_3)\}] \tag{16}$$

* Given y(x) is continuously differentiable on $0 < x < x_1$ for some x_1, $\lim_{x\to 0} y(x) = 0$, $\lim_{x\to 0} xy'(x) = A$, then $A = 0$. When I made this statement to dr. J.G. Besjes, he could not resist giving the following proof. One may write $y'(x) = \frac{A+\varepsilon(t)}{x}$ on $0 < x < x_1$, where $\varepsilon(x)$ is continuous on $0 < x < x_1$ and $\lim_{x\to 0} \varepsilon(x) = 0$. Then for some $0 < \xi < x_1$, $y(x) - y(\xi) = \int_\xi^x y'(t)dt = \int_\xi^x \frac{A+\varepsilon(t)}{t}\,dt = [A + \varepsilon(t^*)] \int_\xi^x \frac{dt}{t} = [A + \varepsilon(t^*)]\ln \frac{x}{\xi}$ for some $t^*(\xi < t^* < x)$. Thus $A + \varepsilon(t^*) = \frac{y(x)-y(\xi)}{\ln x-\ln\xi}$. Take the limit $\xi \to 0$, then the limit $x \to 0$; then it follows that $A = 0$.

where $\bar{b}_2 \to 0$ as $v - v(M_3) \to 0$. As a result there follows:

$$F'(\bar{u}) = \bar{\alpha}_1\bar{\alpha}_2\bar{A}_1^*\bar{A}_2^*\bar{u}^{\bar{\alpha}_1-1}[\varepsilon_1 + \bar{A}_1^*\bar{u}^{\bar{\alpha}_1}]^{\bar{\alpha}_2-1}[1 + \bar{b}_1(u)][1 + \bar{b}_2\{\varepsilon_1 + \bar{A}_1^*\bar{u}^{\bar{\alpha}_1}\}] \tag{17}$$

Also of interest is the quantity

$$J = \int_{t_0}^{t_0+\tau} \operatorname{div}[\bar{P}\{x(t),y(t)\} , \bar{Q}\{x(t),y(t)\}]dt \tag{18}$$

for a periodic solution $x = x(t)$, $y = y(t)$ with period τ. If $J \neq 0$, this solution represents a simple limit cycle, stable if $J < 0$, unstable if $J > 0$. If $J = 0$, the closed path is a multiple limit cycle of imbedded in a region of closed paths [2]. In order to study the behaviour of J on a closed path, when it approaches the separatrix polygon, we use the relation

$$J = \ln F'(\bar{u}) = (\bar{\alpha}_1 - 1)\ln \bar{u} + (\bar{\alpha}_2 - 1)\ln\{\varepsilon_1 + \bar{A}_1^*\bar{u}^{\bar{\alpha}_1}\} + \ln \bar{B}_2^*(\bar{u}) \tag{19}$$

where $\bar{B}_2^*(\bar{u})$ is a bounded function near $\bar{u} = 0$ and $\bar{u}$ satisfies eq. (9).

<u>Lemma 1. Case zz</u>. If $\alpha_1\alpha_2 \neq 1$ no limit cycle is generated and the separatrix polygon keeps the same stability.

<u>Proof</u>. In this case $\varepsilon_1 = \varepsilon_2 = 0$ and from (8) follows

$$F(\bar{u}) = (\bar{A}_1^*)^{\bar{\alpha}_2}\bar{A}_2^*\bar{u}^{\bar{\alpha}_1\bar{\alpha}_2} \tag{20}$$

and the separatrix polygon (represented by $\bar{u} = 0$) has the same stability as the unperturbed one, since $\bar{\alpha}_1\bar{\alpha}_2 - \alpha_1\alpha_2$ can be made arbitrary small. Eq. (9) becomes

$$\bar{u} = (\bar{A}_1^*)^{\bar{\alpha}_2}\bar{A}_2^*\bar{u}^{\bar{\alpha}_1\bar{\alpha}_2} \tag{21}$$

which in view of the limiting properties of $\bar{A}_1^*$, $\bar{A}_2^*$ admits no solutions for which $\lim_{\bar{p},\bar{q}\to 0} \bar{u} = 0$. As a result no limit cycle is generated.

The proof may also be obtained by noting that if a limit cycle would be generated, eq. (19) shows that the singular behaviour of J is given by

$$J = (1 - \bar{\alpha}_1\bar{\alpha}_2)(- \ln \bar{u}) \tag{22}$$

This shows that $J > 0$ (< 0) if $\bar{\alpha}_1\bar{\alpha}_2 < 1$ (> 1), which means that such a limit cycle is simple and has the same stability as the separatrix polygon enclosing it. There thus exists a region in between the polygon and such a limit cycle containing no other closed paths and no singular points. This is an obvious contradiction.

Lemma 2. Case zi. If $\alpha_1\alpha_2 > 1$ exactly one limit cycle is generated, and this limit cylcle is stable. If $\alpha_1\alpha_2 < 1$ no limit cycle is generated.

Proof. Without loss of generality we may take $\varepsilon_1 > 0$, $\varepsilon_2 = 0$. From eq. (8) follows for the succession function

$$F(\bar{u}) = \bar{A}_2^*[\varepsilon_1 + \bar{A}_1^*\bar{u}^{\bar{\alpha}_1}]^{\bar{\alpha}_2} \tag{23}$$

Thus $F(0) > 0$ and $\lim_{\bar{p},\bar{q}\to 0} F(0) = 0$. Moreover, eqs. (9), (17) show that on a closed path $[F(\bar{u}) = \bar{u}]$ the derivative of the succession function is

$$F'(\bar{u}) = (\bar{A}_2^*)^{\frac{1-\bar{\alpha}_2}{\bar{\alpha}_2}} \bar{B}_2^*(\bar{u})\bar{u}^{\frac{\bar{\alpha}_1\bar{\alpha}_2-1}{\bar{\alpha}_2}} \tag{24}$$

If $\alpha_1\alpha_2 > 1$, the separatrix polygon is stable and Theorem 1 shows that then for any $\varepsilon > 0$, there exists a $\delta > 0$ such that if system (2) is δ close to system (1) at least one closed path exists in the ε-neighbourhood of the polygon. If such a closed path approaches the separatrix polygon, $\bar{u} \to 0$ and eq. (24) shows that $\lim_{\bar{u}\to 0} F'(\bar{u}) = 0$, and $0 < F'(\bar{u}) < 1$ for $\bar{u}$ small enough. Thus such a closed path is a simple limit cycle which is stable. Moreover this limit cycle is unique, since more solutions of $F(\bar{u}) = u$ are only possible if $F'(\bar{u}) \geq 1$ for $\bar{u}$ small enough can occur.
If $\alpha_1\alpha_2 < 1$ and a closed path approaches the separatrix polygon for $\bar{u} \to 0$, eq. (24) shows that $\lim_{\bar{u}\to 0} F'(\bar{u}) \to +\infty$, and for $\bar{u}$ small enough $1 < F'(\bar{u}) < \infty$. Since $F(\bar{u})$, by definition, is univalent, there cannot be a point $F(\bar{u}) = \bar{u}$. As a result no limit cycle will be generated.

The proof may also be obtained by noting that if a limit cycle would be generated, eqs. (9), (19) show that the singular behaviour of J is given by

$$J = \frac{1 - \bar{\alpha}_1\bar{\alpha}_2}{\bar{\alpha}_2} (- \ln \bar{u}) \tag{25}$$

If $\alpha_1\alpha_2 > 1$, then $J < 0$ for $\bar{u}$ small enough, which means that a closed path is a simple limit cycle which is stable. It is impossible to have more than one limit cycle, since otherwise there would be two concentric stable limit cycles, adjacent to each other, which is an obvious contradiction.

If $\alpha_1\alpha_2 < 1$, then $\bar{\alpha}_1\bar{\alpha}_2 < 1$ for δ small enough and eq. (25) shows that $J > 0$ for $\bar{u}$ small enough, which means that such a limit cycle is simple and unstable. However, this is impossible, since such a limit cycle cannot be a limit continuum for the paths entering the region between the outmost limit cycle and the broken polygon.

Lemma 3. Case zo. If $\alpha_1\alpha_2 < 1$ exactly one limit cycle is generated and this limit cycle is unstable. If $\alpha_1\alpha_2 > 1$ no limit cycle is generated.

Proof. The case zo may be brought back to the case zi by replacing P and Q in eq. (1) by -P and -Q. As a result the arrows, indicating the direction of increasing t reverse and inflow becomes outflow. Also, the signs of the eigenvalues in the saddle points change as a result of which α_i changes to its inverse, as does $\alpha_1\alpha_2$, and the lemma readily follows.

Lemma 4. Case ii. If $\alpha_1\alpha_2 > 1$ exactly one limit cycle is generated, and this limit cycle is stable. If $\alpha_1\alpha_2 < 1$ no limit cycle is generated.

Proof. In this case $\varepsilon_1 > 0$, $\varepsilon_2 > 0$. The proof runs essentially the same as for the case zi. From eq. (8) the succession function is obtained to be

$$F(\bar{u}) = \varepsilon_2 + \bar{A}_2^*[\varepsilon_1 + \bar{A}_1^*\bar{u}^{\bar{\alpha}_1}]^{\bar{\alpha}_2} \tag{26}$$

thus $F(0) > 0$ and $\lim_{\bar{p},\bar{q}\to 0} F(0) = 0$. Eqs. (9), (17) show that on a closed path $[F(\bar{u}) = \bar{u}]$ the derivative of the succession function is

$$F'(\bar{u}) = (\bar{A}_2^*)^{\frac{1-\bar{\alpha}_2}{\bar{\alpha}_2}}\,\bar{B}_2^*(\bar{u})\bar{u}^{\bar{\alpha}_1-1}(\bar{u}-\varepsilon_2)^{\frac{\bar{\alpha}_2-1}{\bar{\alpha}_2}} \tag{27}$$

Since for $\bar{u} > 0$, by definition $\varepsilon_2 < \bar{u}$, the limiting behaviour of $F'(\bar{u})$ for $\bar{u} > 0$ is as in eq. (24). Without loss of generality one may choose $\alpha_2 > 1$ ($\alpha_2 < 1$) if $\alpha_1\alpha_2 > 1$ ($\alpha_1\alpha_2 < 1$). The rest of the proof is the same as in case zi.

Lemma 5. Case oo. If $\alpha_1\alpha_2 < 1$ exactly one limit cycle is generated, and this limit cycle is unstable. If $\alpha_1\alpha_2 > 1$ no limit cycle is generated.

Proof. The case oo may be brought back to the case ii by replacing P and Q in (1) by -P and -Q, as is done in lemma 3.

Lemma 6. Case i(z)o. Let the saddle to saddle loop be formed with saddle point 1. Then if $\alpha_1\alpha_2 > 1$ ($\alpha_1\alpha_2 < 1$) and $\alpha_1 < 1$ ($\alpha_1 > 1$), exactly one limit cycle is generated, and the limit cycle is stable (unstable), if $\alpha_1 > 1$ ($\alpha_1 < 1$) no limit cycle is generated.

Proof. In this case $\varepsilon_1 > 0$, $\varepsilon_2 < 0$ and eq. (6) yields

$$\bar{A}_2^*\varepsilon_1^{\bar{\alpha}_2} + \varepsilon_2 = 0 \tag{28}$$

For the succession function then follows with eqs. (8), (28)

$$F(\bar{u}) = \bar{A}_2^*[-\varepsilon_1^{\bar{\alpha}_2} + \{\varepsilon_1 + \bar{A}_1^*\bar{u}^{\bar{\alpha}_1}\}^{\bar{\alpha}_2}] \tag{29}$$

and for the derivative of the succession function with (17), (19)

$$F'(\bar{u}) = \bar{B}_2^*\bar{u}^{\bar{\alpha}_1-1}\{\varepsilon_1 + \bar{A}_1^*\bar{u}^{\bar{\alpha}_1}\}^{\bar{\alpha}_2-1} \tag{30}$$

Consider first the case $\alpha_1\alpha_2 > 1$, $\alpha_1 < 1$, then $\alpha_2 > 1$, thus the separatrix polygon is stable and the saddle to saddle loop is unstable and Theorem 1 shows that there is at least one closed path in the ε-neighbourhood of the separatrix polygon. Obviously $\bar{u}(\varepsilon_1) \equiv 0$ satisfies the relation

$$G(\bar{u},\varepsilon_1) \equiv F(\bar{u}) - \bar{u} = 0 \tag{31}$$

and represents the saddle to saddle loop. Now replace $\bar{u}$ in (31) by

$$\bar{u} = \gamma(\alpha_2A_1A_2)^{\frac{1}{1-\alpha_1}}\varepsilon_1^{\frac{\alpha_2-1}{1-\alpha_1}} \tag{32}$$

then

$$G(\gamma,\varepsilon_1) = (\gamma^{\alpha_1} - \gamma)(\alpha_2A_1A_2)^{\frac{1}{1-\alpha_1}}\varepsilon_1^{\frac{\alpha_2-1}{1-\alpha_1}}[1 + g_1(\gamma,\varepsilon_1)] \tag{33}$$

where $g_1(\gamma,\varepsilon_1) \to 0$ as $\varepsilon_1 \to 0$, uniformly in γ. In eq. (33) $\gamma = 0$ corresponds to the saddle to saddle loop and $\gamma = 1$ to a closed path. If eq. (32) with $\gamma = 1$ is used in eq. (30), for the derivative of the succession function on a closed path follows

$$\lim_{\bar{u},\bar{p},\bar{q}\to 0} F'(\bar{u}) = \alpha_1 < 1 \tag{34}$$

from which follows that such a closed path is a (simple) stable limit cycle. As pointed out previously in similar cases then there cannote be more than one limit cycle.

If $\alpha_1\alpha_2 > 1$, $\alpha_1 > 1$ eq. (29) shows that $F(0) = 0$ for all $\varepsilon_1 \geq 0$ small enough and eq. (30) that $F'(0) = 0$ for all $\varepsilon_1 \geq 0$ small enough. From eq. (30) it may further be concluded that there exists a $\varepsilon^* > 0$ and $\bar{u}^* > 0$, such that for $0 < \varepsilon < \varepsilon^*$ on $0 < \bar{u} < \bar{u}^*$, $|F'(\bar{u})| < \alpha < 1$, as a result of which there is no point $F(\bar{u}) = \bar{u}$ in this interval. This indicates that there is no closed path in a contracting neighbourhood of the separatrix polygon.

The case $\alpha_1\alpha_2 < 1$ may be brought back to the case $\alpha_1\alpha_2 > 1$ by replacing P and Q in (1) by -P and -Q, as is done in lemma 3. □

For the remaining two lemmas we wish to define inner saddle points. Referring to fig. 7, we may regard the cases i(i)o and i(o)o as bifurcations of the case i(z)o through the breaking up of the saddle to saddle loop. The inner saddle point in i(i)o or i(o)o is then that saddle point that corresponds to the saddle point connected with the loop in case i(z)o. Intrinsically defined, it is that saddle point for which its separatrices "shield off" the other (outer) saddle point from the interior region.

Lemma 7. Case i(i)o. Let saddle point 1 be the inner saddle point. Then if $\alpha_1\alpha_2 > 1$, and $\alpha_1 \neq 1$, exactly one limit cycle is generated, and this limit cycle is stable. If $\alpha_1\alpha_2 < 1$ and $\alpha_1 < 1$ no limit cycle is generated; if $\alpha_1 > 1$ there are three possible cases: either no limit cycle, one semistable limit cycle of multiplicity 1, or two simple limit cycles (a stable one enclosing an unstable one) are generated.

Lemma 8. Case i(o)o. Let saddle point 1 be the inner saddle point. Then if $\alpha_1\alpha_2 < 1$, and $\alpha_1 \neq 1$, exactly one limit cycle is generated, and this limit cycle is unstable. If $\alpha_1\alpha_2 > 1$ and $\alpha_1 > 1$ no limit cycle is generated; if $\alpha_1 < 1$ there are three possible cases: either no limit cycle, one semistable limit cycle of multiplicity 1, or two simple limit cycles (an unstable one enclosing a stable one) are generated.

Proof of lemmas 7 and 8. The cases i(i)o and i(o)o can be brought over in each other by replacing P and Q by -P and -Q. We will prove the first part of lemma 7 and the second part of lemma 8; thus always dealing with $\alpha_1\alpha_2 > 1$.

In both cases $\varepsilon_1 > 0$, $\varepsilon_2 < 0$. We will first consider $\alpha_1\alpha_2 > 1$, $\alpha_1 < 1$, thus $\alpha_2 > 1$. With eq. (8) for the succession function a closed path must satisfy the relation

$$G(\bar{u},\varepsilon_1,\varepsilon_2) \equiv F(\bar{u}) - \bar{u} = 0 \tag{35}$$

In this relation, make the substitutions

$$\bar{u} = \gamma_1(\alpha_2 A_1 A_2)^{\frac{1}{1-\alpha_1}} \varepsilon_1^{\frac{\alpha_2-1}{1-\alpha_1}} \tag{36}$$

$$\varepsilon_2 = - A_2\varepsilon_1^{\alpha_2} + \gamma_2(\alpha_2 A_1 A_2)^{\frac{1}{1-\alpha_1}} \varepsilon_1^{\frac{\alpha_2-1}{1-\alpha_1}} \tag{37}$$

then

$$G(\gamma_1,\varepsilon_1,\gamma_2) = (\gamma_2 + \gamma_1^{\alpha_1} - \gamma_1)(\alpha_2 A_1 A_2)^{\frac{1}{1-\alpha_1}} \varepsilon_1^{\frac{\alpha_2-1}{1-\alpha_1}} [1 + g_2(\gamma_1,\varepsilon_1,\gamma_2)] \tag{38}$$

where $g_2(\gamma_1,\varepsilon_1,\gamma_2) \to 0$ as $\varepsilon_1 \to 0$, uniformly in γ_1 and γ_2. Thus, on a closed path we have the relation

$$\gamma_2 + \gamma_1^{\alpha_1} - \gamma_1 = 0 \tag{39}$$

For the derivative of the succession function on a closed path follows with eqs. (17), (36)

$$\lim_{\bar{u},\bar{p},\bar{q}\to 0} F'(\bar{u}) = \alpha_1 \gamma_1^{\alpha_1 - 1} \tag{40}$$

For arbitrary small perturbations, case i(i)o corresponds to $-\gamma_2 < 0$, case i(z)o to $\gamma_2 = 0$ and case i(o)o to $-\gamma_2 > 0$. From (39) follows that $\gamma_1 > 1$ for $-\gamma_2 < 0$ and (40) then shows that such a closed path is a (simple) limit cycle which is stable. Obviously in case i(i)o there will then be exactly one limit cycle. If $\gamma_2 = 0$, eq. (39) shows that either $\gamma_1 = 1$ or $\gamma_1 = 0$, as a result of which, for case i(z)o we find again the results obtained in lemma 6. If $-\gamma_2 > 0$, eq. (39) shows that $0 < \gamma_1 < 1$ and eq. (40) that there are three cases: (i) $0 < \gamma_1 < \alpha_1^{1/1-\alpha_1}$, $F' > 1$, (simple) unstable limit cycle; (ii) $\gamma_1 = \alpha_1^{1/1-\alpha_1}$, $F' = 1$; (iii) $\alpha_1^{1/1-\alpha_1} < \gamma_1 < 1$, $F' < 1$ (simple) stable limit cycle. For arbitrary small perturbations there are thus three cases: (i) $-\gamma_2 > \gamma_2(\alpha_1^{1/1-\alpha_1}) = \alpha_1^{1/1-\alpha_1}(1 - \alpha_1^{\alpha_1})$, there are no closed paths; (ii) $-\gamma_2 = \alpha_1^{1/1-\alpha_1}(1 - \alpha_1^{\alpha_1})$, there is a single limit cycle which is not simple, since $F' = 1$; (iii) $-\gamma_2 < \alpha_1^{1/1-\alpha_1}(1 - \alpha_1^{\alpha_1})$; there are two (simple) limit cycles, an unstable one enclosing a stable one. Since case (ii) may also be understood as a limiting case of case (iii), the limit cycle in case (ii) is semistable of multiplicity 1.

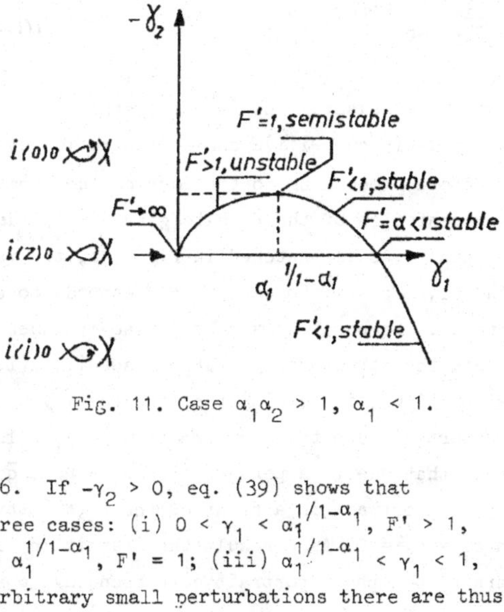

Fig. 11. Case $\alpha_1\alpha_2 > 1$, $\alpha_1 < 1$.

We now consider $\alpha_1\alpha_2 > 1$, $\alpha_1 > 1$. Then we make the substitutions:

$$\bar{u} = \gamma_1 \varepsilon_1 \tag{41}$$

$$\varepsilon_2 = -A_2\varepsilon_1^{\alpha_2} + \gamma_2\varepsilon_1 \tag{42}$$

in eq. (35) with the result that now

$$G(\gamma_1,\varepsilon_1,\gamma_2) = (\gamma_2 - \gamma_1)\varepsilon_1[1 + g_3(\gamma_1,\varepsilon_1,\gamma_2)] \tag{43}$$

where $g_3(\gamma_1,\varepsilon_1,\gamma_2) \to 0$ as $\varepsilon_1 \to 0$, uniformly in γ_1 and γ_2. Thus on a closed path there is the relation

$$\gamma_2 - \gamma_1 = 0 \tag{44}$$

valid. Moreover eqs. (17), (41) lead to

$$\lim_{\bar{u},\bar{p},\bar{q}\to 0} F'(\bar{u}) = 0 \tag{45}$$

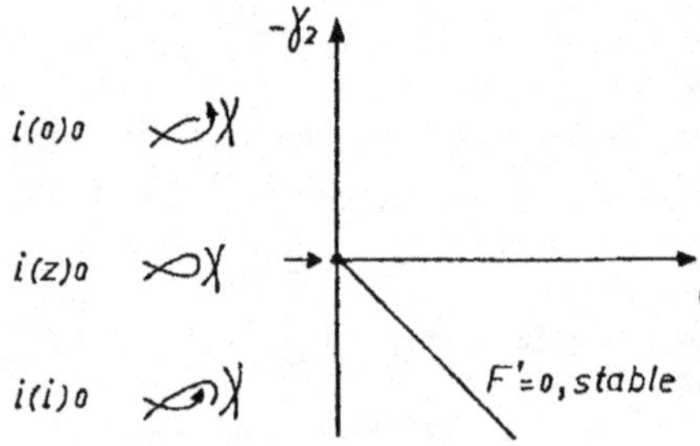

Fig. 12. Case $\alpha_1\alpha_2 > 1$, $\alpha_1 > 1$.

Thus there is exactly one (simple) stable limit cycle in case i(i)o and no generation of limit cycles in case i(o)o. □

7. Before proceeding to the case of the separatrix triangle, we wish to reorganize the presentation of the results given in lemmas 1 - 8 by representing them in the $\varepsilon_1,\varepsilon_2$ plane. As is sketched in fig. 13a, the various cases may be assigned to points in the $\varepsilon_1,\varepsilon_2$ plane. So case zz corresponds to $\varepsilon_1 = \varepsilon_2 = 0$, case ii to $\varepsilon_1 > 0$, $\varepsilon_2 > 0$, and so on. The point $\varepsilon_1 = \varepsilon_2 = 0$ also includes the unperturbed system. Sectors indicate the structurally stable cases ii, i(i)o, i(o)o, oo, separated from each other by lines, indicating the structurally unstable cases zi, i(z)o, zo. Since the phase portrait due to a perturbation $\bar{p}_1,\bar{q}_1$ followed by a perturbation $\bar{p}_2,\bar{q}_2$, is the same as that due to a perturbation $\bar{p}_1 + \bar{p}_2$, $\bar{q}_1 + \bar{q}_2$, the topological character of the phase portrait in a point of the $\varepsilon_1,\varepsilon_2$ plane is independent of the path followed from $\varepsilon_1 = \varepsilon_2 = 0$. As a result the transition from the separatrix polygon phase portrait to a phase portrait corresponding to a given pair $\varepsilon_1,\varepsilon_2$ may also be obtained by changing $\varepsilon_1,\varepsilon_2$ first to some value on a circle around the origin and then moving tangentially along this circle. As may easily be seen from the phase portraits in fig. 7, the change of their topological character, when moving along such a (small) circle is in accordance with the possible bifurcations. This point of view may further be explored in relation to the determination of the number of limit cycles that are generated from the separatrix polygon. This number, as obtained from the lemmas 1 - 8, is indicated in fig. 13b,c. Since the number of limit cycles is a piecewise constant function of ε_1 and ε_2, in order to know this function, it suffices to know its points of discontinuity and the number of limit cycles near such points. Now, when moving along the circumference of a circle around the origin, only the

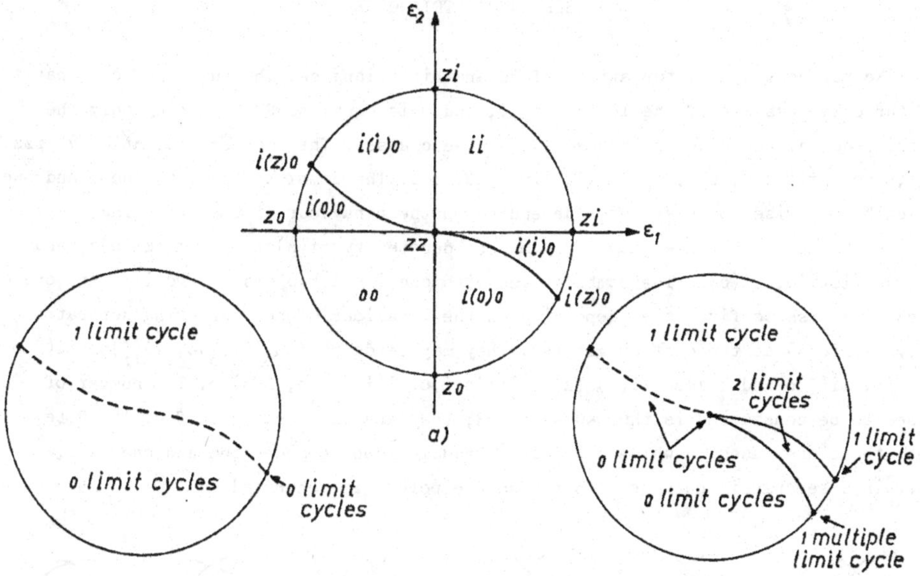

b) $\alpha_1>1, \alpha_2>1, \alpha_1\alpha_2>1$ c) $\alpha_1<1, \alpha_2>1, \alpha_1\alpha_2>1$

Fig. 13. Number of limit cycles in the ε_1, ε_2 plane.

saddle to saddle loop of case i(z)o will change the number of limit cycles as a result of bifurcation from a figure containing saddle points, since it is the only case wherein such a figure is a limit continuum. In addition to lemma 1 (case zz) possibly only lemma 6 (case i(z)o) would then be needed to determine the number of limit cycles as a function of ε_1 and ε_2, the bifurcation from case i(z)o being the known case of bifurcation from a saddle to saddle loop. If $\alpha_1 > 1$, $\alpha_2 > 1$, $\alpha_1\alpha_2 > 1$, fig. 13b shows that this procedure yields the correct result for the number of limit cycles, in particular for the maximum number that can be generated due to a perturbation. If $\alpha_1 < 1$, $\alpha_2 > 1$, $\alpha_1\alpha_2 > 1$ lemma 6 yields the number of limit cycles for case i(z)o and the bifurcation from the saddle to saddle loop gives the correct number of limit cycles near the line of discontinuity, corresponding to case i(z)o. In the entire region above this line the correct number is found by extending the local result near the line of discontinuity i(z)o. Below this line, extending the local result near i(z)o leads to two values of the number of limit cycles: 0 and 2, and it is not possible from bifurcations from the figures containing saddle points to tell whether there is a line of discontinuity with a jump from 0 to 2, and whether there are more of these lines of discontinuity. These lines are apparently related to multiple limit cycles, being the only possibility left to generate limit cycles. Yet, also for $\alpha_1 < 1$, $\alpha_2 > 1$, $\alpha_1\alpha_2 > 1$ the correct maximum number of limit cycles that can be generated is obtained from a local analysis near the line of discontinuity corresponding with the saddle to saddle loop.

5. SEPARATRIX TRIANGLES

8. We now consider perturbations of separatrix triangles. The number of combinations of three letters out of the letters z, i, and o is equal to 27. Counting only the cyclic equivalence classes leaves 11 of these classes. They may be indicated by: zzz, zzi, zzo, zii, zio, zoi, zoo, iii, iio, ioo, ooo. The combinations with one i and one o, each give rise to three cases depending on the behaviour of the inflowing separatrix at the outflow gate. They may be denoted by zi(z)o, zi(i)o, zi(o)o, and iz(z)o, iz(i)o, iz(o)o. The combinations with one i and two o's or two i's and one o each give rise to five cases depending on the behaviour of the inflowing separatrix (separatrices) at the outflow gate(s). They may be denoted by $ii\binom{z}{o}o$, $ii\binom{i}{z}o$, $ii\binom{i}{i}o$, $ii\binom{i}{o}o$, $ii\binom{o}{o}o$ and i(z)oo, i(i)o(z)o, i(i)o(i)o, i(i)o(o)o, i(o)oo. The number of cases to be considered is thus equal to 23; they are illustrated in fig. 14. Note that there are cases in which a saddle to saddle loop is generated and one case, wherein a separatrix polygon with two saddle points is generated.

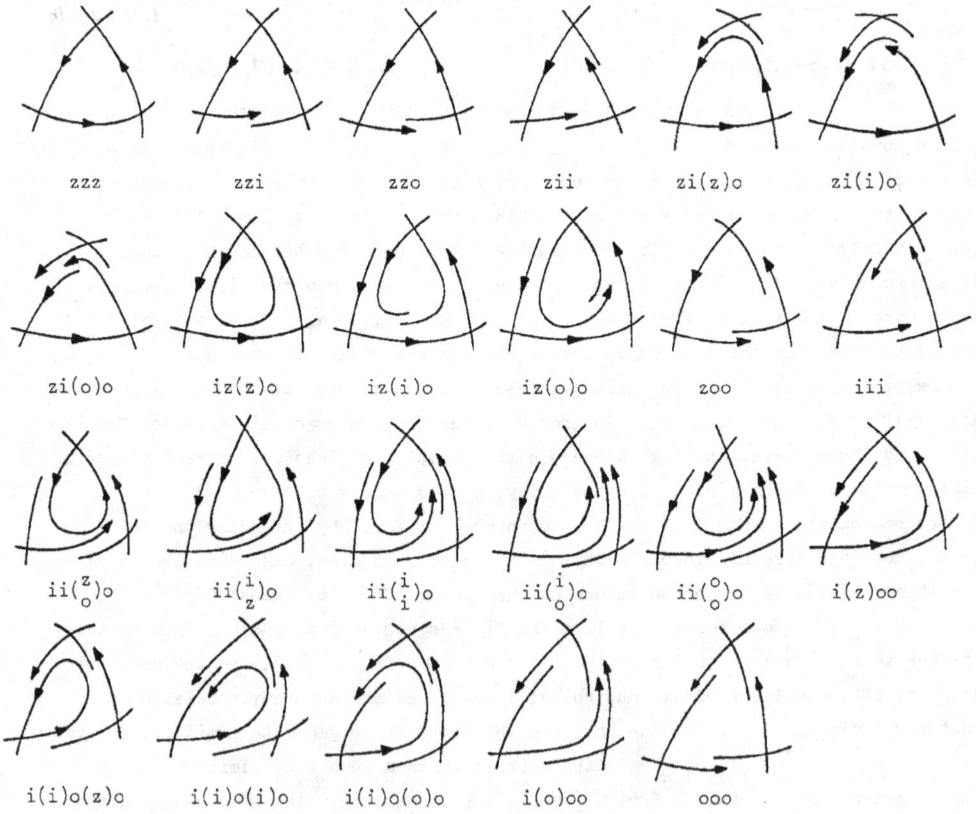

Fig. 14. Perturbation of a separatrix triangle.

9. We now consider the generation of limit cycles from a separatrix triangle. As for the saddle to saddle loop and the separatrix polygon with two saddle points arbitrary small perturbations can always be found such that at least one closed path is generated. More precisely, using the same arguments as in the proof of theorem 1 there may be shown.

Theorem 3. Let system (1) be of class 1 and possess a separatrix triangle being stable (unstable). Then, for any $\varepsilon > 0$, there exists a $\delta > 0$, such that if system (2) is δ close (to rank 1) to system (1) and if total inflow (cases zzi, zii, zi(i)o, iz(i)o, iii, ii($^{i}_{z}$)o, ii($^{i}_{i}$)o, ii($^{i}_{o}$)o, i(i)o(i)o) [total outflow (cases zzo, zi(o)o, iz(o)o, zoo, ii($^{o}_{o}$)o, i(z)oo, i(i)o(o)o, i(o)oo, ooo)] or an unstable (stable) separatrix polygon with two saddle points (case zi(z)o) or an unstable (stable) saddle to saddle loop occurs (cases iz(z)o, ii($^{z}_{o}$)o, i(i)o(z)o), then there exists at least one closed path in the ε-neighbourhood of the separatrix triangle.

10. As before let $\mu_i < 0$, $\lambda_i > 0$ be the eigenvalues of the locally linearized system in the ith saddle point and $|\mu_i|/\lambda_i = \alpha_i$, and choose any numbering of the saddle points. Then if $\alpha_1\alpha_2\alpha_3 > 1$, the triangle is stable, if $\alpha_1\alpha_2\alpha_3 < 1$ unstable, whereas for $\alpha_1\alpha_2\alpha_3 = 1$ it may be either stable, unstable or have a neighbourhood containing only closed paths [4], [7]. As indicated in fig. 15, we may introduce transversals l, m, and n on the separatrices 3 - 1, 1 - 2, and 2 - 3, respectively, and u, v, and w as parameters on them. Let be $u - u(M_1) = \bar{u}$, $f(u) - u(M_1) = F(\bar{u})$, $v(M_4) - v(M_3) = \varepsilon_1$, $w(M_6) - w(M_5) = \varepsilon_2$, $u(M_2) - u(M_1) = \varepsilon_3$, then similar to the case for the separatrix polygon with two saddle points,

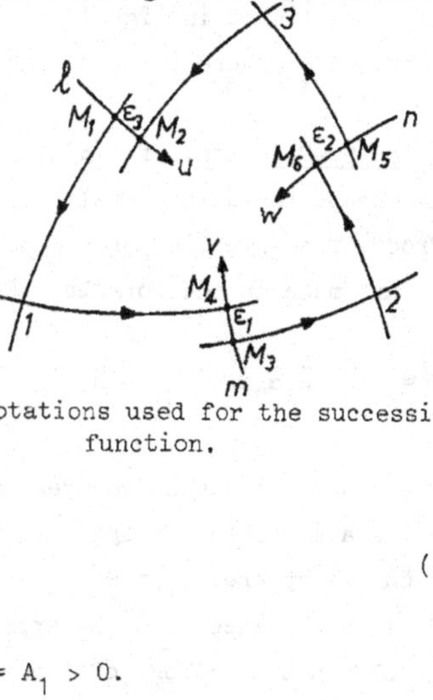

Fig. 15. Notations used for the succession function.

$$F(\bar{u}) = \varepsilon_3 + \bar{A}_3^*[\varepsilon_2 + \bar{A}_2^*[\varepsilon_1 + \bar{A}_1^*\bar{u}^{\bar{\alpha}_1}]^{\bar{\alpha}_2}]^{\bar{\alpha}_3} \tag{46}$$

where $\bar{A}_1^*(\bar{u};\bar{p},\bar{q}) = \bar{A}_1[1 + \bar{a}_1(\bar{u})]$; $\lim_{\bar{u},\bar{p},\bar{q}\to 0} \bar{A}_1^*(\bar{u},\bar{p},\bar{q}) = A_1 > 0.$

$$\bar{A}_2^*(\bar{u};\bar{p},\bar{q}) = \bar{A}_2[1 + \bar{a}_2[\varepsilon_1 + \bar{A}_1^*\bar{u}^{\bar{\alpha}_1}]] \; ; \; \lim_{\bar{u},\bar{p},\bar{q}\to 0} \bar{A}_2^*(\bar{u},\bar{p},\bar{q}) = A_2 > 0.$$

$$\bar{A}_3^*(\bar{u};\bar{p},\bar{q}) = \bar{A}_3[1 + \bar{a}_3[\varepsilon_2 + \bar{A}_2^*[\varepsilon_1 + \bar{A}_1^*\bar{u}^{\bar{\alpha}_1}]]] \; ; \; \lim_{\bar{u},\bar{p},\bar{q}\to 0} \bar{A}_3^*(\bar{u},\bar{p},\bar{q}) = A_3 > 0.$$

and $\bar{a}_1$, $\bar{a}_2$, and $\bar{a}_3$ vanish with their arguments. Similarly for the derivation of the succession function follows

$$F'(\bar{u}) = \bar{B}_3^*(\bar{u})\bar{u}^{\bar{\alpha}_1 - 1}[\varepsilon_1 + \bar{A}_1^*\bar{u}^{\bar{\alpha}_1}]^{\bar{\alpha}_2 - 1}[\varepsilon_2 + \bar{A}_2^*[\varepsilon_1 + \bar{A}_1^*\bar{u}^{\bar{\alpha}_1}]^{\bar{\alpha}_2}]^{\bar{\alpha}_3 - 1} \tag{47}$$

where $\bar{B}_3^{*}(\bar{u}) = \bar{\alpha}_1\bar{\alpha}_2\bar{\alpha}_3\bar{A}_1^{*}\bar{A}_2^{*}\bar{A}_3^{*}[1+\bar{b}_1(\bar{u})][1+\bar{b}_2[\varepsilon_1+\bar{A}_1^{*}\bar{u}^{\bar{\alpha}_1}]][1+\bar{b}_3[\varepsilon_2+\bar{A}_2^{*}[\varepsilon_1+\bar{A}_1^{*}\bar{u}^{\bar{\alpha}_1}]^{\bar{\alpha}_2}]]$,

$$\lim_{\bar{u},\bar{p},\bar{q}\; 0} \bar{B}_3^{*}(\bar{u}) = B_3 = \alpha_1\alpha_2\alpha_3A_1A_2A_3 > 0$$

and $\bar{b}_1$, $\bar{b}_2$, $\bar{b}_3$ vanish with their arguments. On a closed path, moreover

$$J = (\bar{\alpha}_1 - 1)\ln \bar{u} + (\bar{\alpha}_2 - 1)\ln[\varepsilon_{.} + \bar{A}_{.}^{*}\bar{u}^{\bar{\alpha}_1}] + (\bar{\alpha}_3 - 1)\ln[\varepsilon_2 + \bar{A}_2^{*}[\varepsilon_1 + \bar{A}_1^{*}\bar{u}^{\bar{\alpha}_1}]^{\bar{\alpha}_2}] + \\ + \ln \bar{B}_3^{*}(\bar{u}), \qquad (48)$$

where $\ln \bar{B}_3^{*}(\bar{u})$ is a bounded function near $\bar{u} = 0$ and $\bar{u}$ is a fixed point of the Poincaré map satisfying

$$\bar{u} = \varepsilon_3 + \bar{A}_3^{*}[\varepsilon_2 + \bar{A}_2^{*}[\varepsilon_1 + \bar{A}_1^{*}\bar{u}^{\bar{\alpha}_1}]^{\bar{\alpha}_2}]^{\bar{\alpha}_3} \qquad (49)$$

For the separatrix triangle we may, similar to the previous case, represent the various cases listed in fig. 14 in the ε_1, ε_2, ε_3 space, and in fact, if we take case zzz apart, on a sphere in this space around the origin. In analogy with lemma 1, we have:

<u>Lemma 9. Case zzz</u>. If $\alpha_1\alpha_2\alpha_3 \neq 1$ no limit cycle is generated and the separatrix triangle keeps the same stability.

<u>Proof</u>. The same line of arguments may be used as in lemma 1 when the proper changes are made in the formulae. For instance (22) must be replaced by

$$J = (1 - \bar{\alpha}_1\bar{\alpha}_2\bar{\alpha}_3)(-\ln \bar{u}) \qquad (50)$$

□

The remaining cases are presented in fig. 16. The case zi(z)o is the only case containing a separatrix polygon with two saddle points and is represented by three points on the sphere: $\varepsilon_1 = 0$, $\varepsilon_2 > 0$, $\varepsilon_3 < 0$; $\varepsilon_1 < 0$, $\varepsilon_2 = 0$, $\varepsilon_3 > 0$; $\varepsilon_1 > 0$, $\varepsilon_2 < 0$, $\varepsilon_3 = 0$. In accordance with the approach outlined in section 7 we first consider this case to obtain the analogy of lemma 6.

<u>Lemma 10. Case zi(z)o</u>. Let the separatrix polygon with two saddle points be formed with the saddle points 1 and 2. Then if $\alpha_1\alpha_2\alpha_3 > 1$ ($\alpha_1\alpha_2\alpha_3 < 1$) and $\alpha_1\alpha_2 < 1$ ($\alpha_1\alpha_2 > 1$) exactly one limit cycle is generated, and this limit cycle is stable (unstable), if $\alpha_1\alpha_2 > 1$ ($\alpha_1\alpha_2 < 1$) no limit cycle is generated.

<u>Proof</u>. In this case $\varepsilon_1 = 0$, $\varepsilon_2 > 0$, $\varepsilon_3 < 0$ and

$$\bar{A}_3^{*}\varepsilon_2^{\bar{\alpha}_3} + \varepsilon_3 = 0 \qquad (51)$$

For the succession function follows with (46), (51)

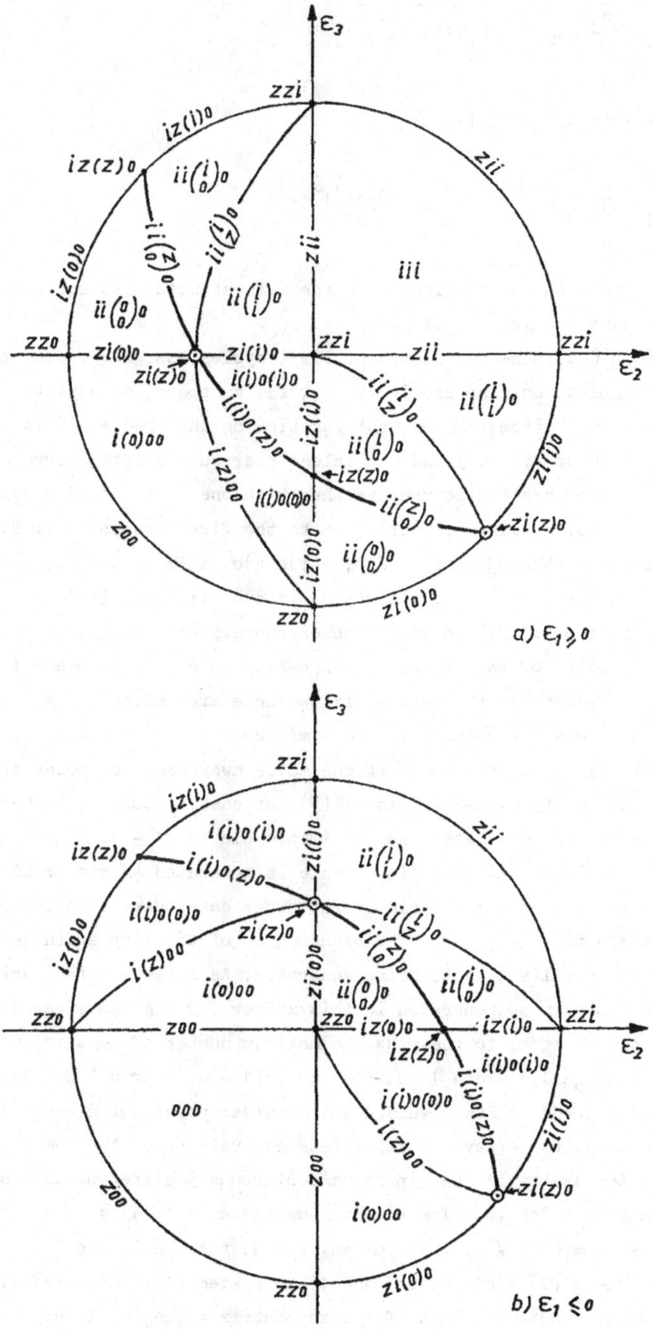

Fig. 16. Areas on a sphere in the ε_1, ε_2, ε_3 space, with cases.

$$F(\bar{u}) = \bar{A}_3^*[-\varepsilon_2^{\bar{\alpha}_3} + [\varepsilon_2 + \bar{A}_1^{*\bar{\alpha}_2}\bar{A}_2^*\bar{u}^{\bar{\alpha}_1\bar{\alpha}_2}]^{\bar{\alpha}_3}] \tag{52}$$

and for its derivative from (47), (51)

$$F'(\bar{u}) = \bar{A}_1^{*\bar{\alpha}_2-1}\bar{B}_3^*(\bar{u})\bar{u}^{\bar{\alpha}_1\bar{\alpha}_2-1}[\varepsilon_2 + \bar{A}_1^{*\bar{\alpha}_2}\bar{A}_2^*\bar{u}^{\bar{\alpha}_1\bar{\alpha}_2}]^{\bar{\alpha}_3-1} \tag{53}$$

It may be noted, that these equations have the same structure as eqs. (29), (30) in lemma 6 and the same analysis applies.

We now try to determine the number of limit cycles assigned to the various areas, lines and points on a sphere in the ε_1, ε_2, ε_3 space, as depicted in fig. 16. Since this number is a piecewise constant function on the sphere, it is sufficient to know the points of discontinuity and the values near such points, corresponding to cases, containing a separatrix polygon (either with one or two saddle points) or to multiple limit cycles. We restrict ourselves to the first possibility. These points form a closed curve on the sphere: zi(z)o, i(i)o(z)o, iz(z)o, ii$\binom{z}{o}$o, zi(z)o, i(i)o(z)o, iz(z)o, ii$\binom{z}{o}$o, zi(z)o, i(i)o(z)o, iz(z)o, ii$\binom{z}{o}$o, zi(z)o, all points, except for case zi(z)o, corresponding to cases containing a saddle to saddle loop. We may start in a point of this line, corresponding to case zi(z)o and lemma 10 tells that if the saddle points in the separatrix twoangle are indicated by i and j, that if $(1 - \alpha_1\alpha_2\alpha_3)(1 - \alpha_i\alpha_j) > 0$ there are no limit cycles, if $(1 - \alpha_1\alpha_2\alpha_3)(1 - \alpha_i\alpha_j) < 0$ exactly one limit cycle is generated. If one moves away from the point zi(z)o over the line of discontinuity either to case ii$\binom{z}{o}$o or case i(i)o(z)o, in both cases a saddle to saddle loop is generated and no limit cycle if $(1 - \alpha_i\alpha_j)(1 - \alpha_i) > 0$. If $(1 - \alpha_i\alpha_j)(1 - \alpha_i) < 0$ exactly one limit cycle is generated in either of the cases ii$\binom{z}{o}$o or i(i)o(z)o and no limit cycle in the other case. Here i indicates the saddle point in the saddle to saddle loop. The breaking up of the loop again leads to no limit cycle or another limit cycle being generated. As a result, the maximum number of limit cycles that can be generated in this manner for a given separatrix triangle (given α_1, α_2, α_3) is equal to one plus the maximum number of negative numbers in the two numbers $(1 - \alpha_1\alpha_2\alpha_3)(1 - \alpha_i\alpha_j)$ and $(1 - \alpha_i\alpha_j)(1 - \alpha_i)$ for all possible choices of i and j out of 1, 2, and 3. Since we did not consider points of discontinuity related to multiple limit cycles we have so far a weak equivalence of theorem 2.

<u>Theorem 4</u>. Let system (1) be Lipschitz continuously differentiable and have a separatrix triangle, which is a limit continuum and $\alpha_1 \neq 1$, $\alpha_2 \neq 1$, $\alpha_3 \neq 1$, $\alpha_1\alpha_2 \neq 1$, $\alpha_1\alpha_3 \neq 1$, $\alpha_2\alpha_3 \neq 1$, $\alpha_1\alpha_2\alpha_3 \neq 1$. Then for any $\varepsilon > 0$, there exists a $\delta > 0$ such that there exists a system (2) δ close (to rank 1) to system (1), which has at least p + 1 limit cycles in the ε-neighbourhood of the separatrix triangle. Here p is the maximum number of changes through 1 in the shrinking sequence $\alpha_1\alpha_2\alpha_3$, $\alpha_i\alpha_j$, α_i, where $i \neq j$ are chosen out of the numbers 1, 2, and 3. A separatrix triangle may thus generate

at least 1, 2, or 3 limit cycles.

Obviously, from the results for the separatrix polygon with two saddle points, it can be strongly expected that theorem 4 can be supplemented with a statement replacing at least by at most. Using the method, used for the separatrix twoangle, however, is elaborating, and we refrain of doing so. If all α_i-s are on the same side of one we have:

Theorem 5. Let system (1) be Lipschitz contiuously differentiable and have a separatrix triangle, which is a limit contiuum, and $\alpha_i < 1$ ($\alpha_i > 1$) ($i = 1, 2, 3$). Then for any $\varepsilon > 0$, there exists a $\delta > 0$ such that if system (2) is δ close (to rank 1) to system (1), system (2) has at most one limit cycle in the ε-neighbourhood of the triangle, and this limit cycle is unstable (stable).

Proof. A limit cycle generated by the separatrix triangle is represented by a solution of (49) for which $\lim_{\varepsilon_1,\varepsilon_2,\varepsilon_3\to 0} \bar{u}(\varepsilon_1,\varepsilon_2,\varepsilon_3) = 0$. Write eq. (49) as

$$G(\bar{u},\varepsilon_1,\varepsilon_2,\varepsilon_3) \equiv F(\bar{u},\varepsilon_1,\varepsilon_2,\varepsilon_3) - \bar{u} = 0 \tag{54}$$

then

$$\frac{\partial G}{\partial \bar{u}}(\bar{u},\varepsilon_1,\varepsilon_2,\varepsilon_3) \equiv F'(\bar{u}) - 1 \tag{55}$$

and for $\alpha_1 > 1$, $\alpha_2 > 1$, $\alpha_3 > 1$, since eq. (47) shows that then $\lim_{\bar{u},\varepsilon_1,\varepsilon_2,\varepsilon\to 0} F'(\bar{u}) \to 0$, eq. (55) yields

$$\frac{\partial G}{\partial \bar{u}}(0,0,0,0) = -1 \neq 0 \tag{56}$$

as a result of which the implicit function theorem shows, that there is a unique function $\bar{u} = \bar{u}(\varepsilon_1,\varepsilon_2,\varepsilon_3)$ in a neighbourhood of the origin (0,0,0,0). If $\bar{u} > 0$, there is a limit cycle and at most one. Eq. (47) shows that on such a limit cycle $F'(\bar{u}) < 1$, thus it is stable.

The case $\alpha_1 < 1$, $\alpha_2 < 1$, $\alpha_3 < 1$ can be dealt with by replacing P, Q by -P, -Q.

6. SEPARATRIX POLYGONS WITH n SADDLE POINTS

11. The number of combinaitons of n letters out of the three letters z, i, and o is equal to 3^n. The determination of the number of cyclic equivalence classes amounts to the so called colouring problem of a roulette, which is discussed for instance in [3]*. The problem was solved by Jablonski [5], [3, p. 263] and the number of ways to paint the n sectors of a roulette into ≤ p colours equals

---

* This reference was pointed out to me by dr. Th.M. Smits.

$$N = \frac{1}{n} \sum_{d/n} \phi(d) p^{\frac{n}{d}} \tag{57}$$

where d/n means "d divides n" (with 1 and n included) and ϕ is the Euler function, defined by [3, p. 193]

$$\phi(n) = n(1 - \frac{1}{p_1})(1 - \frac{1}{p_2}) \ldots (1 - \frac{1}{p_r}) \; ; \; n > 1 \; , \; \phi(1) = 1 \tag{58}$$

where $n = p_1^{d_1} p_2^{d_2} \ldots p_r^{d_r}$. A table of values of $\phi(n)$ yields

n =	1	2	3	4	5	6	7
$\phi(n)$ =	1	1	2	2	4	2	6

and for p = 3 some values of N are

n =	1	2	3	4	5	6	7
N =	3	6	11	24	51	130	315

Remark, that if n is a prime number, then $\phi(n) = n - 1$ and d = 1 and n; then eq. (57) yields with p = 3

$$N = \frac{1}{n}[3^n + 3(n - 1)] \tag{59}$$

which is a lower bound for N if n is not a prime number.

The combinations having both inflowing and outflowing separatrices give rise to extra cases depending on the behaviour of the inflowing separatrix(ces) at the outflow gate(s). We do not attempt to calculate the number of possible cases. It may be noted, however, that there is always one case, wherein a separatrix polygon with n - 1 saddle points is generated, represented by $\underbrace{z \ldots z}_{n-2} i(z)o$.

12. Using the same arguments as in theorems 1 and 3 there may again be shown, that at least one closed path can be generated by perturbing system (1). In fact:

Theorem 6. Let system (1) be of class 1 and possess a separatrix polygon with n saddle points being stable (unstable). Then for any $\varepsilon > 0$, there exists a $\delta > 0$ such that if system (2) is δ close (to rank 1) to system (1) and if total inflow occurs (total outflow occurs) or an unstable (stable) separatrix polygon occurs, there exists at least one closed path in the ε-neighbourhood of the separatrix polygon.

13. As before let $\mu_i < 0$, $\lambda_i > 0$, $\alpha_i = \frac{|\mu_i|}{\lambda_i}$, then if $\alpha_1 \alpha_2 \ldots \alpha_n > 1$ the polygon is stable, if $\alpha_1 \alpha_2 \ldots \alpha_n < 1$ unstable, and $\alpha_1 \alpha_2 \ldots \alpha_n = 1$ it may be either stable, unstable of have a neighbourhood containing only closed paths [4], [7]. As indicated in fig. 17, we may introduce transversals $l_1, l_2, \ldots, l_n$ on the separatrices n - 1, 1 - 2, ..., (n - 1) - n, respectively, and $u_1, u_2, \ldots, u_n$ as parameters on them. Let be $u_1 - u(M_1) = \bar{u}$, $f(u_1) - u_1(M_1) = F(\bar{u})$, $u_2(M_4) - u_2(M_3) = \varepsilon_1$, $u_3(M_6) - u_n(M_5) = \varepsilon_2$, ..., $u_1(M_2) - u_1(M_1) = \varepsilon_n$, then similar to the previous cases the succession function may be represented by

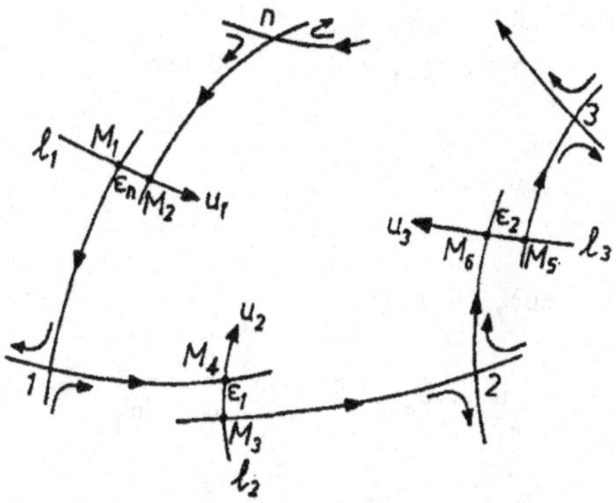

Fig. 17. Notations used for the succession function.

$$F(\bar{u}) = \varepsilon_n + \bar{A}_n^*[\varepsilon_{n-1} + \bar{A}_{n-1}^*[\varepsilon_{n-2} + \ldots + \bar{A}_2^*[\varepsilon_1 + \bar{A}_1^*\bar{u}^{\bar{\alpha}_1}]^{\bar{\alpha}_2}\ldots]^{\bar{\alpha}_n} \qquad (60)$$

with obvious properties of $\bar{A}_i^*$ $(i = 1, 2, \ldots, n)$.

Similarly, for the derivative may be written

$$F'(\bar{u}) = \bar{B}_n^*(\bar{u})\bar{u}^{\bar{\alpha}_1 - 1}[\varepsilon_1 + \bar{A}_1^*\bar{u}^{\bar{\alpha}_1}]^{\bar{\alpha}_2 - 1} \ldots [\varepsilon_{n-1} + \bar{A}_{n-1}^*[\varepsilon_{n-2} + \ldots + \bar{A}_2^* \cdot$$
$$\cdot [\varepsilon_1 + \bar{A}_1^*\bar{u}^{\bar{\alpha}_1}]^{\bar{\alpha}_2}\ldots]^{\bar{\alpha}_{n-1}}]^{\bar{\alpha}_n - 1}$$

with obvious properties of $\bar{B}_n^*(\bar{u})$, whereas

$$J = \ln F'(\bar{u}) \qquad (62)$$

Following the procedure outlined before, there follows similar to lemmas 1, 9:

Lemma 11. Case $\underbrace{zz\ldots z}_{n}$. If $\alpha_1\alpha_2\ldots\alpha_n \neq 1$ no limit cycle is generated and the separatrix polygon keeps the same stability.

Proof. The same line of arguments may be used as in lemma 1, 9 when the proper changes are made in the formulae. For instance (22) must be replaced by

$$J = (1 - \bar{\alpha}_1\bar{\alpha}_2\ldots\bar{\alpha}_n)(-\ln\bar{u}) \qquad (63)$$

The analogy with lemmas 6 and 10 reads:

Lemma 12. Case $\underbrace{zz\ldots z}_{n-2}\,i(z)o$. Let the separatrix polygon with n - 1 saddle points be formed with the saddle points 1, 2, ..., n - 1. Then if $\alpha_1\alpha_2\ldots\alpha_n > 1$ $(\alpha_1\alpha_2\ldots\alpha_n < 1)$ and $\alpha_1\alpha_2\ldots\alpha_{n-1} < 1$ $(\alpha_1\alpha_2\ldots\alpha_{n-1} > 1)$ exactly one limit cycle is generated, and this limit cycle is stable (unstable), if $\alpha_1\alpha_2\ldots\alpha_{n-1} > 1$

$(\alpha_1\alpha_2\ldots\alpha_{n-1} < 1)$ no limit cycle is generated.

Proof. In this case $\varepsilon_1 = \varepsilon_2 = \ldots \varepsilon_{n-2} = 0$, $\varepsilon_{n-1} > 0$, $\varepsilon_n < 0$ and

$$\bar{A}_n^{*}\varepsilon_{n-1}^{\bar{\alpha}_n} + \varepsilon_n = 0 \tag{64}$$

For the succession function follows with (60), (64)

$$F(\bar{u}) = \bar{A}_n^{*}[-\varepsilon_{n-1}^{\bar{\alpha}_n} + [\varepsilon_{n-1} + \bar{A}_{n-1}^{*}\bar{A}_{n-2}^{*\bar{\alpha}_{n-1}}\ldots\bar{A}_1^{*\bar{\alpha}_2\ldots\bar{\alpha}_{n-1}}\bar{u}^{\bar{\alpha}_1\bar{\alpha}_2\ldots\bar{\alpha}_{n-1}}]^{\bar{\alpha}_n}] \tag{65}$$

and for its derivative from (61), (64)

$$F'(\bar{u}) = \bar{A}_1^{*\bar{\alpha}_2\ldots\bar{\alpha}_{n-1}-1}\ldots\bar{A}_{n-2}^{*\bar{\alpha}_{n-1}-1}\bar{B}_n^{*}(\bar{u})\bar{u}^{\bar{\alpha}_1\bar{\alpha}_2\ldots\bar{\alpha}_{n-1}-1}[\varepsilon_{n-1} + \bar{A}_{n-1}^{*}\bar{A}_{n-2}^{*\bar{\alpha}_{n-1}}\ldots\bar{A}_1^{*\bar{\alpha}_2\ldots\bar{\alpha}_{n-1}} \cdot \bar{u}^{\bar{\alpha}_1\ldots\bar{\alpha}_{n-1}}]^{\bar{\alpha}_n-1} \tag{66}$$

It may be noted, that these equations have the same structure as eqs. (29), (30) in lemma 6 and the same analysis applies. □

Similar to theorems 4 and 5 we have theorems 7 and 8.

Theorem 7. Let system (1) be Lipschitz continuously differentiable and have a separatrix polygon with n saddle points, which is a limit continuum and $\alpha_i \neq 1$, $\alpha_i\alpha_j \neq 1$, ..., $\alpha_1\alpha_2\ldots\alpha_n \neq 1$ (i, j, ... = 1, 2, ..., n). Then for any $\varepsilon > 0$, there exists a $\delta > 0$ such that there exists a system (2) δ close (to rank 1) to system (1), which has at least $p+1$ limit cycles in the ε-neighbourhood of the separatrix polygon. Here p is the maximum number of changes through 1 in the shrinking sequence $\alpha_1\alpha_2\ldots\alpha_n$, $\alpha_1\alpha_2\ldots\alpha_n\alpha_i^{-1}$, $\alpha_1\alpha_2\ldots\alpha_n\alpha_i^{-1}\alpha_j^{-1}$, ..., where $i \neq j \neq \ldots$ are to be chosen out of the numbers 1, 2, ..., n. A separatrix polygon may thus generate at least 1, 2, ..., n limit cycles.

Theorem 8. Let system (1) be Lipschitz continuously differentiable and have a separatrix polygon with n saddle points, which is a limit continuum, and $\alpha_i < 1$ $(\alpha_i > 1)$ $(i = 1, 2, \ldots, n)$. Then for any $\varepsilon > 0$, there exists a $\delta > 0$ such that if system (2) is δ close (to rank 1) to system (1), system (2) has at most one limit cycle in the ε-neighbourhood of the polygon, and this limit cycle is unstable (stable).

Theorem 8 was also given in [2, p. 314].

REFERENCES

[1] Andronov, A.A., Gordon, J.J., Leontovich, E.A. and Maier, A.G.; Qualitative Theory of Second-Order Dynamic Systems, Israel Program for Scientific Translation, Jerusalem, 1973.

[2] Andronov, A.A., Gordon, J.J., Leontovich, E.A. and Maier, A.G.; Theory of Bifurcations of Dynamic Systems on a Plane, Israel Program for Scientific Translation, Jerusalem, 1971.

[3] Comtet, L.; Advanced Combinatorics; the art of finite and infinite expansions; revised and enlarged edition, Reidel, Dordrecht, 1974.

[4] Dulac, H.; Sur les cycles limites, Bull. Soc. Math. de France, Vol. 51, pp. 45-188, 1923.

[5] Jablonski; Théorie des permutations et des arrangements complets, Journal de Liouville, 8, pp. 331-349, 1892.

[6] Leontovich, E.A.; On the generation of limit cycles from separatrices, Dokl. Akad. Nank. U.S.S.R., Vol. 78, no. 4, pp. 641-644, 1951.

[7] Reyn, J.W.; A stability criterion for separatrix polygons in the phase plane, Nieuw Archief voor Wiskunde (3), XXVII, pp. 238-254, 1979.

NORMAL SOLVABILITY
OF LINEAR PARTIAL DIFFERENTIAL OPERATORS IN $C^\infty(\Omega)$

Boris Sagraloff

Introduction

Surjectivity statements for LPDO's in Fréchet spaces have been established by Malgrange [5], Hörmander [3], Browder [1] and Tréves [12],[13],[14]. Malgrange proved the surjectivity of LPDO's with constant and C^∞-coefficients in the local Sobolev spaces $H_s^{loc}(\Omega)$ and the Fréchet spaces $C^\infty(\Omega)$ respectively. Hörmander was concerned with the surjectivity of LPDO's with constant coefficients in the local weighted Sobolev spaces $B_{p,k}^{loc}(\Omega)$. The importance of Hörmander's result lies in the fact that he was able to specify the regularity of the solutions more precisely than Malgrange. Browder stated conditions for the surjectivity of LPDO's with variable coefficients in the Fréchet spaces $L_p^{loc}(\Omega)$. Using methods different to those of Malgrange, Trèves reproved the surjectivity of LPDO's with C^∞-coefficients in $C^\infty(\Omega)$.

LPDO's with variable coefficients in Fréchet spaces are not, however,always surjective. Instead of studying the surjectivity of LPDO's with variable coefficients it is thus more appropriate to establish whether the range of these operators is closed or the operators themselves are open or normally solvable. The problem of normal solvability for LPDO's in Banach spaces has been discussed by e.g. Lions, Magenes [4] and Goldberg [2].

Normal solvability of LPDO's in the Fréchet spaces $B_{p,k}^{loc}(\Omega)$ was firstly considered by the author in [10]. In the present paper we study the normal solvability of LPDO's in $C^\infty(\Omega)$.

Section 1 contains the new version of the closed-range theorem in Fréchet spaces given in [9],[10],[7], (see Theorem (1.3) in the present paper). For a closed linear operator T we consider both the global adjoint T' and the semiglobal adjoints $T'^{(p,q)}$ which are closed linear maps between Banach spaces. We state that T is open iff both the $T'^{(p,q)}$ are open and certain linking conditions between T' and the seminormed structure of the Fréchet spaces under study are fulfilled. In this way we are able to transfer the problem of normal solvability in Fréchet spaces to the corresponding problem in Banach spaces.

In [10] we applied Theorem (1.3) to obtain statements about normal solvability of LPDO's in the Fréchet spaces $B_{p,k}^{loc}(\Omega)$. In the present paper we use (1.3) to deduce assertions about the normal solvability of LPDO's in $C^\infty(\Omega)$.

An essential advantage of our methods is the systematical use of Theorem (1.3) and its simple proof. The methods of the above mentioned authors are discussed in brief and compared with our own:

For a LPDO P Malgrange and Browder considered the global adjoint P' for which they showed 1) *a priori estimates* and 2) *convexity conditions*. With the aid of these properties they proved that P' is injective and the range of P' is weakly or strongly closed respectively. Finally they applied Banach's closed-range theorem in Fréchet spaces to obtain the surjectivity of P.

It is a well-known fact that the proof of Banach's closed-range theorem is rather complicated owing to several reductions. In a first step the proof of the assertions has to be reduced to the assertions when the linear operator T is continuous. Secondly, when T is continuous, further reduction is necessary so that T' will become injective. For these reductions an intrinsic study of the weak and strong topology on the duals of sub- and quotient spaces is necessary. (Cf. e.g. [1], [11].)

The closed-range Theorem (1.3) is the abstract functional analytical way of directly proving the openness of P from 1) and 2) which Tréves [12] realized to some extent. An analysis of Malgrange's proof led him to a special case of Theorem (1.3). In addition, he assumed that T is a continuous linear operator defined on the whole of the space under study and that T' is injective.

Trèves' proof is however rather complicated. He established a 'duality theory' which is very unusual. He considered systems of semi-norms which generate the topologies of the spaces under study. He regarded these systems as 'dual spaces' and defined 'adjoint operators' between them. A considerable part of the monograph [12] is concerned with establishing this 'duality theory' but Trèves naturally has to make use of the usual duality theory as well. This leads to a further difficulty because he has to equate statements about adjoint operators in the usual dual spaces with the corresponding ones in 'his dual spaces'.

In the papers [9] and [10] the author proved Theorem (1.3) with geometrical methods by using a result from Banach's closed-range theorem for pairs of subspaces [6]. In [7] Mennicken and the present author gave a very *elementary* proof of Theorem (1.3). They require neither the above mentioned reductions nor the study of various topologies on the dual spaces. They only use the open-mapping theorem and some polar-formulas which are easily deduced from the theorem of Hahn-Banach. It becomes quite clear from [7] that Trèves' calculus in [12] is superfluous when proving (1.3) and related theorems.

In Section 2 we provide the space $C^{\infty}(\Omega)$ with a suitable system of seminorms which generates the topology and we determine the duals of the generating seminormed spaces.

The main result of Section 3 is a new closed-range theorem for LPDO's in $C^{\infty}(\Omega)$ obtained from Theorem (1.3) with the aid of the preparatory work in Section 2. The linking conditions of (1.3) establish an interesting generalization of Trèves' P-convexity in $C^{\infty}(\Omega)$. As a conclusion of these results we obtain a theorem which shows that the normal solvability of P is equivalent to the generalized P-convexity, if the openness of the semiglobal adjoints $P'^{(p,q)}$ is given by a string of a priori estimates.

In [10] we showed that the uniqueness in the Cauchy problem in the complement of a compact subset in Ω implies the P-convexity of Ω. Using this result and the uniqueness theorems of Holmgren and Calderon we deduce normal solvability for elliptic and principally normal LPDO's in $C^{\infty}(\Omega)$. These results generalize the surjectivity statement of Malgrange [5] for elliptic LPDO's and supplement the semiglobal solvability statement for principally normal LPDO's in the monograph [3].

Finally we consider the elliptic LPDO of Plis [8] which is not surjective and show that this operator has a closed range of finite codimension in $C^{\infty}(\Omega)$.

This paper is part of the author's habilitation thesis. I would like to thank Prof. R.Mennicken for his valuable help and suggestions.

1. The closed-range theorem

(X,τ) and (Y,σ) are two locally convex Hausdorff spaces; $X'=(X,\tau)'$ and $Y'=(Y,\sigma)'$ are their duals. The topologies τ and σ are defined by families of continuous seminorms Γ_τ and Γ_σ which we call a basis system for the topologies τ and σ respectively.

For $p\in\Gamma_\tau$ we denote the dual of the seminormed space (X,p) by X'^p. If we define the dual seminorm of p by

(1.1) $\quad p(x') = \inf\{\rho>o : \forall\, x\in X \quad |\langle x,x'\rangle| \leq \rho p(x)\}\ (\leq\infty) \qquad (x'\in X'),$

then it is easy to prove that (X'^p,p) is a Banach space.

In the following, T is a closed linear relation from X to Y with domain D(T) in X. G(T) is the graph, R(T) the range and N(T) the null space of T. Besides the global adjoint T', we define for $p\in\Gamma_\tau$ and $q\in\Gamma_\sigma$ the semiglobal (p,q)-adjoint $T'^{(p,q)}$ of T by

(1.2) $\quad G(T'^{(p,q)}) = G(-T)^{\perp(X\times Y,\, Y'^q\times X'^p)}$.

This is a linear closed relation from $D(T'^{(p,q)})\subset X'^p$ to Y'^q which is also given by

$$D(T'^{(p,q)}) = \{y'\in Y'^q :\ \exists\, x'\in X'^p\ \forall\, x\in D(T)\ \forall\, y\in Tx \quad \langle x,x'\rangle = \langle y,y'\rangle\}$$

$$T'^{(p,q)}y' = T'y'\cap X'^p \qquad (y'\in D(T'^{(p,q)})).$$

According to [7], theorem (2.8) and (4.2) we can state the following *closed-range*

(1.3) <u>Theorem</u>. *Let* X *and* Y *be Fréchet spaces.*
Then the following assertions are equivalent:

a) T *is normally solvable, i.e.*

(1.4) $\quad R(T) = N(T')^{\perp(Y,Y')}$.

b) *For each* $p\in\Gamma_\tau$ *there is a* $q\in\Gamma_\sigma$ *such that*

1) $R(T')\cap X'^p\subset T'(Y'^q)$;

2) $T'^{(p,q)}$ *is* (q,p)-*open.*

c) *For each* $p\in\Gamma_\tau$ *there is a* $q\in\Gamma_\sigma$ *such that*

1) $R(T')\cap X'^p\subset T'(\mathrm{Ker}(q)^{\perp(Y,Y')})$,

2) $N(T')^{\perp(Y,Y')}\subset R(T) + \mathrm{Ker}(q)$.

It is not difficult to prove that $T'^{(p,q)}$ is (q,p)-open iff the following statement is true: there is a constant $C_{(p,q)} > o$ such that the estimates

(1.5) $\quad \forall\ (x',y')\in G(T')\cap(Y'^q\times X'^p) \qquad \mathrm{dist}_q(y',N(T')\cap Y'^q) \leq C_{(p,q)}\, p(x')$

are fulfilled, where $\operatorname{dist}_q(y',N(T')\cap Y'^q)=\inf\{q(y'-z'): z'\in N(T')\cap Y'^q\}$. (Cf.[6], p.133.) Thus the assertion b)2) in Theorem (1.3) is valid iff the estimates (1.5) are satisfied.

2. Seminormed dual spaces of $C^\infty(\Omega)$

$C^\infty(\Omega)$ denotes the space of complex valued functions which are defined and infinitely differentiable in the open set $\Omega\subset\mathbb{R}^d$. $C^\infty(\Omega)$ is provided with the topology of uniform convergence on compact subsets of the functions and of their derivatives. This topology is referred to as τ_{C^∞}. $(C^\infty(\Omega),\tau_{C^\infty})$ is a Fréchet space.

Theorem (1.3) is to be applied to LPDO's which are maps of $C^\infty(\Omega)$ into itself. For this purpose a basis system $\Gamma_{\tau_{C^\infty}}$ of seminorms on $C^\infty(\Omega)$ has to be defined. The notations which follow have been adopted from [3] and [10].

Let us choose a sequence $(\Omega_n)_0^\infty$ of open and bounded subsets of Ω such that

(2.1) $\quad \overline{\Omega}_n\subset\Omega_{n+1} \quad (n\in\mathbb{N}), \quad \Omega=\bigcup_{n\in\mathbb{N}}\Omega_n$.

We set for $n\in\mathbb{N}$

(2.2) $\quad \Phi_n=\{\varphi\in C_0^\infty(\Omega_{n+1}): \exists\ U\subset\mathbb{R}^d \text{ open}, \overline{\Omega}_n\subset U, \varphi_{|U}=1\}$

and define for a number $\nu\in\mathbb{R}$ a seminorm s_ν^n on the local Sobolev space $H_\nu^{loc}(\Omega)$ of order ν by

(2.3) $\quad s_\nu^n(u)=\inf\{|\varphi u|_\nu : \varphi\in\Phi_n\} \qquad (u\in H_\nu^{loc}(\Omega))$.

In [10], Lemma (2.17), we showed

(2.4) $\Gamma_{\tau_\nu}^{loc}=\{s_\nu^n: n\in\mathbb{N}\}$ *is a basis system for the topology* τ_ν^{loc} *on* $H_\nu^{loc}(\Omega)$ *for each* $\nu\in\mathbb{R}$.

Since

(2.5) $\quad i_\nu : (C^\infty(\Omega),\tau_{C^\infty}) \longleftrightarrow (H_\nu^{loc}(\Omega),\tau_\nu^{loc}) \qquad (\nu\in\mathbb{R})$

is a continuous embedding, it follows that $\overline{s}_\nu^n:=s_\nu^n\circ i_\nu$ is a τ_{C^∞}-continuous seminorm on $C^\infty(\Omega)$ for each $\nu\in\mathbb{R}, n\in\mathbb{N}$.

Moreover

(2.6) $\Gamma_{\tau_{C^\infty}}=\{\overline{s}_\nu^n: \nu\in\mathbb{R}, n\in\mathbb{N}\}$ *is a basis system for the topology* τ_{C^∞} *on* $C^\infty(\Omega)$.

The *proof* is clear if we observe (2.4) and that $(C^\infty(\Omega),\tau_{C^\infty})$ is the projective limit of the spaces $(H_\nu^{loc}(\Omega),\tau_\nu^{loc})$ with respect to the embeddings i_ν. (Cf.[3],p.45.)

If $C_o^\infty(\Omega)$ is provided with the Schwartz topology $\tau_{C_o^\infty}$, then

(2.7) $\quad i : \quad (C_o^\infty(\Omega),\tau_{C_o^\infty}) \longleftrightarrow (C^\infty(\Omega),\tau_{C^\infty})$

is a continuous embedding. The dual map

(2.8) $\quad i' : \quad C^\infty(\Omega)' \longrightarrow D'(\Omega)$ *is an isomorphism onto* $\mathcal{E}'(\Omega) = \{u\in D'(\Omega): \operatorname{supp} u \subset \Omega, \text{ compact}\}$. (Cf.e.g.[3], p.12.)

We note the following trivial consequence of (2.8): if we define

(2.9) $\quad v(u) := \langle u, i'^{-1} v\rangle \qquad (u\in C^\infty(\Omega), v\in \mathcal{E}'(\Omega))$,

then not only $\langle C^\infty(\Omega), C^\infty(\Omega)'\rangle$ but also $(C^\infty(\Omega), \mathcal{E}'(\Omega))$ is a dual pair. It therefore follows immediately from (2.9) that for each $\overline{s}_\nu^n \in \Gamma_{\tau_{C^\infty}}$

(2.10) $\quad i'(\operatorname{Ker}(\overline{s}_\nu^n)^{\perp\langle C^\infty(\Omega), C^\infty(\Omega)'\rangle}) = \operatorname{Ker}(\overline{s}_\nu^n)^{\perp(C^\infty(\Omega), \mathcal{E}'(\Omega))}$.

(2.11) *For each* $\overline{s}_\nu^n \in \Gamma_{\tau_{C^\infty}}$ *the inclusions*

$$\mathcal{E}'(\Omega_n) \subset \operatorname{Ker}(\overline{s}_\nu^n)^{\perp(C^\infty(\Omega), \mathcal{E}'(\Omega))} \subset \mathcal{E}'(\overline{\Omega}_n)$$

are valid.

We first *prove* the second inclusion. The definition of $\overline{s}_\nu^n$ leads to

(2.12) $\quad C_o^\infty(\Omega\setminus\overline{\Omega}_n) \subset \operatorname{Ker}(\overline{s}_\nu^n)$.

Thus we obtain for each $v\in\operatorname{Ker}(\overline{s}_\nu^n)$

$$v(\varphi) = 0 \qquad (\varphi\in C_o^\infty(\Omega\setminus\overline{\Omega}_n)),$$

and therefore $\operatorname{supp} v \subset \overline{\Omega}_n$.

For the proof of the first inclusion it suffices to show

(2.13) $\quad \operatorname{Ker}(\overline{s}_\nu^n) \subset \{u\in C^\infty(\Omega): u_{|\Omega_n} = 0\}$.

We observe that for each $u\in C^\infty(\Omega)$

$$\int_{\Omega_n} u(x)\,\psi(x)\,dx = \int_{\mathbb{R}^d} (\varphi u)(x)\,\psi(x)\,dx$$

$$= (2\pi)^{-n}\int_{\mathbb{R}^d} (\widehat{\varphi u})(\xi)\,\hat{\psi}(\xi)\,d\xi \qquad (\varphi\in\Phi_n, \psi\in C_o^\infty(\Omega_n)).$$

Using Schwartz's inequality this leads to

$$\left|\int_{\Omega_n} u(x)\,\psi(x)\,dx\right| \le |\psi|_{-\nu}\,|\varphi(i_\nu u)|_\nu \qquad (\varphi\in\Phi_n, \psi\in C_o^\infty(\Omega_n)).$$

Hence, if we take the infimum over $\varphi\in\Phi_n$, we obtain

$$\left|\int_{\Omega_n} u(x)\,\psi(x)\,dx\right| \le |\psi|_{-\nu}\,\overline{s}_\nu^n(u) \qquad (\psi\in C_o^\infty(\Omega_n)),$$

and therefore (2.13).

In the following we determine the duals of the seminormed spaces $(C^\infty(\Omega), \overline{s}_\nu^n)$. Let $i'^{(\nu,n)}$ denote the restriction of i' to the dual space

(2.14) $\quad C^\infty(\Omega)'^{(\nu,n)} = (C^\infty(\Omega),\bar{s}^n_\nu)' \quad (\subset C^\infty(\Omega)') \qquad (\nu\in\mathbb{R}, n\in\mathbb{N})$

of the seminormed space $(C^\infty(\Omega),\bar{s}^n_\nu)$. We can assert that

(2.15) $i'^{(\nu,n)}$ *is a topological isomorphism from* $(C^\infty(\Omega)'^{(\nu,n)},\bar{s}^n_\nu)$ *onto* $H^C_{-\nu}(\bar{\Omega}_n),||_{-\nu})$, *where* $H^C_{-\nu}(\bar{\Omega}_n) = H_{-\nu} \cap \mathcal{E}'(\bar{\Omega}_n)$.

Proof. We consider the sequence

$$H^{loc}_\nu(\Omega)' \xrightarrow{i'_\nu} C^\infty(\Omega)' \xrightarrow{i'} \mathcal{E}'(\Omega)$$

and the composed map $J_\nu = i' \circ i'_\nu$. Let J^n_ν denote the restriction of J_ν to the space

$$H^{loc}_\nu(\Omega)'^n = (H^{loc}_\nu(\Omega),s^n_\nu)' \quad (\subset H^{loc}_\nu(\Omega)').$$

According to [10], (2.24)

$$J^n_\nu : (H^{loc}_\nu(\Omega)'^n, s^n_\nu) \longrightarrow \mathcal{E}'(\Omega)$$

is a topological isomorphism onto $(H^C_{-\nu}(\bar{\Omega}_n),||_{-\nu})$.

Thus it suffices to prove that the restriction

$$i'^n_\nu : (H^{loc}_\nu(\Omega)'^n, s^n_\nu) \longrightarrow C^\infty(\Omega)'$$

of i'_ν is a topological isomorphism onto $(C^\infty(\Omega)'^{(\nu,n)},\bar{s}^n_\nu)$.

For proof we observe that

$$i_\nu : (C^\infty(\Omega),\bar{s}^n_\nu) \longleftrightarrow (H^{loc}_\nu(\Omega),s^n_\nu)$$

is injective, continuous and has a s^n_ν-dense range in $H^{loc}_\nu(\Omega)$. Thus it is easy to show that i'^n_ν is a $(s^n_\nu,\bar{s}^n_\nu)$-continuous isomorphism onto $C^\infty(\Omega)'^{(\nu,n)}$. Finally the $(s^n_\nu,\bar{s}^n_\nu)$-openness follows from the open-mapping theorem in Banach spaces.

3. Normal solvability and C^∞-P_n- convexity

Let P be a LPDO with infinitely differentiable coefficients, which is considered as a map from $D'(\Omega)$ into itself. The formal adjoint ${}^tP$ is also regarded as a map from $D'(\Omega)$ into itself, i.e. ${}^tP$ is the dual of the restriction of P to $C^\infty_o(\Omega)$ with respect to the dual pair $(C^\infty_o(\Omega),D'(\Omega))$.

P is obviously τ_{C^∞}-continuous as a map from $C^\infty(\Omega)$ into itself. This map is referred to as P^∞. The dual operator of P^∞ with respect to $<C^\infty(\Omega),C^\infty(\Omega)'>$ satisfies the relation

(3.1) $\quad i'P^{\infty\prime}i'^{-1} = {}^tP_{\mathcal{E}'}$,

where the restriction of ${}^tP$ to $\mathcal{E}'(\Omega)$ is denoted by ${}^tP_{\mathcal{E}'}$.

For *proof* we consider the following commutative diagram :

$$\begin{array}{ccc} C_0^\infty(\Omega) & \xrightarrow{\ i\ } & C^\infty(\Omega) \\ P\big\downarrow & & \big\downarrow P^\infty \\ C_0^\infty(\Omega) & \xrightarrow[\ i\ ]{} & C^\infty(\Omega) \end{array} .$$

This leads us to the following commutative dual diagram:

$$\begin{array}{ccc} D'(\Omega) & \xleftarrow{\ i'\ } & C^\infty(\Omega)' \\ {}^tP\big\uparrow & & \big\uparrow P^{\infty\prime} \\ D'(\Omega) & \xleftarrow[\ i'\ ]{} & C^\infty(\Omega)' \end{array} ,$$

i.e. $i'P^{\infty\prime} = {}^tPi'$. Owing to $R(i') = \mathcal{E}'(\Omega)$ we obtain (3.1).

We consider the 'linking conditions' b)1) and c)1) of Theorem (1.3) in the following case under study :

(3.2) $\quad (X,\tau) = (Y,\sigma) = (C^\infty(\Omega), \tau_{C^\infty})$,

(3.3) $\quad p = \bar{s}^n_\nu \in \Gamma_{\tau_{C^\infty}}\ ,\ q = \bar{s}^k_\mu \in \Gamma_{\tau_{C^\infty}}$.

Because of (3.1) and (2.15), it is clear that (1.3),b)1) is equivalent to

(3.4) $\quad R({}^tP_{\mathcal{E}'}) \cap H^c_{-\nu}(\bar{\Omega}_n) \subset {}^tP(H^c_{-\mu}(\bar{\Omega}_k))$.

Owing to (3.1),(2.15),(2.10) and the second inclusion in (2.11) we obtain from (1.3),c)1)

(3.5) $\quad R({}^tP_{\mathcal{E}'}) \cap H^c_{-\nu}(\bar{\Omega}_n) \subset {}^tP(\mathcal{E}'(\bar{\Omega}_k))$.

Using the first inclusion in (2.11) it is seen that (3.5) implies (1.3),c)1) for $p = \bar{s}^n_\nu$ and $q = \bar{s}^{k+1}_\nu$.

Theorem (1.3) and formula (3.5) motivates the following

(3.6) <u>Definition</u>. Ω *is called* C^∞-P_n-convex, *if*

(3.7) $\quad \forall\ \nu \in \mathbb{R},\ K \subset \Omega$ compact $\exists\ K' \subset \Omega$ compact $\quad R({}^tP_{\mathcal{E}'}) \cap H^c_\nu(K) \subset {}^tP(\mathcal{E}'(K'))$.

In particular, we call Ω C^∞- P -convex, *if*

(3.8) $\quad \forall\ \nu \in \mathbb{R},\ K \subset \Omega$ compact $\exists\ K' \subset \Omega$ compact $\quad ({}^tP_{\mathcal{E}'})^{-1}(H^c_\nu(K)) \subset \mathcal{E}'(K')$.

It should be noted that there is a relation between the present definition of C^∞-P_n-convexity and the P_n-convexity of Ω defined in [10], (3.10). If we set

(3.9) $\quad G_\nu = \{ v \in \mathcal{E}'(\Omega) : {}^tPv \in H_\nu \} \qquad (\nu \in \mathbb{R})$,

it is easy to see that Ω is C^∞-P_n-convex, iff Ω is P_n-convex with respect to G_ν for each $\nu \in \mathbb{R}$.

Furthermore our definition of C^∞-P-convexity agrees with that of

Trêves in [12], p.62. However, Trèves only used the term if ${}^{t}P_{\mathcal{E}'}$ is injective.

It is obvious that Ω is C^∞-P_n-convex if Ω is C^∞-P-convex. The inverse conclusion is also valid iff there is a compact subset K of Ω such that

(3.10) $\quad N({}^{t}P_{\mathcal{E}'}) \subset \mathcal{E}'(K)$.

It is clear that this inclusion is fulfilled for a suitable compact set $K\subset\Omega$, if the null space of ${}^{t}P_{\mathcal{E}'}$ is of finite dimension. It should however be noted that there are LPDO's P and open sets $\Omega\subset\mathbb{R}^d$, such that Ω is C^∞-P_n-convex and not C^∞-P-convex. (Cf.[10], Example (4.15),2).)

The P-convexity of Ω with respect to the space $G=C_o^\infty(\Omega) \subset \mathcal{E}'(\Omega)$ was first intoduced for LPDO's with constant coefficients by Hörmander [3], p.80. For these operators the P-convexity of Ω with respect to $C_o^\infty(\Omega)$ implies the P-convexity of Ω with respect to $G = \mathcal{E}'(\Omega)$. (Cf.[14], p.394.) Furthermore, if we observe that here ${}^{t}P_{\mathcal{E}'}$ is injective (because of the existence of a fundamental solution in $D'(\Omega)$), then Ω is clearly C^∞-P_n [P]-convex, iff Ω is P-convex with respect to $C_o^\infty(\Omega)$.

The 'seminormed normal solvability' statement c)2) of Theorem (1.3) gives rise to another

(3.11) <u>Definition</u>. *We call* P^∞ *semiglobally normally solvable, if for each open set* $\Omega'\subset\subset\Omega$ *we have*

(3.12) $\quad \forall\ f\in N({}^{t}P_{\mathcal{E}'})^\perp\ \exists\ u\in C^\infty(\Omega) \qquad (f-P^\infty u)|_{\Omega'} = o$.

If instead of (3.12) we have

(3.13) $\quad \forall\ f\in C^\infty(\Omega)\ \exists\ u\in C^\infty(\Omega) \qquad (f-P^\infty u)|_{\Omega'} = o$,

then P^∞ *is called semiglobally solvable.*

The term 'semiglobal solvability' was originally introduced by Trèves in [12] in connection with statements on surjectivity. It is clear that if P^∞ is semiglobally solvable the range of P^∞ is τ_{C^∞}-dense in $C^\infty(\Omega)$, i.e. ${}^{t}P_{\mathcal{E}'}$ is injective. The new and more general definition 'semiglobal normal solvability' of P^∞ is based on the fact that P^∞ can have a closed range without being surjective. In this case P^∞ is then semiglobally normally solvable but not semiglobally solvable. (See Example (3.38).)

Having accomplished this preparatory work, we are now able to state the following main

(3.14) <u>Theorem</u>. *The following assertions are equivalent:*

a) $P^\infty : C^\infty(\Omega) \longrightarrow C^\infty(\Omega)$ *is normally solvable, i.e.*

(3.15) $\quad R(P^\infty) = N({}^{t}P_{\mathcal{E}'})^\perp$.

b) *For each compact set* $K\subset\Omega$ *and each number* $\nu\in\mathbb{R}$ *there is a compact set* $K'\subset\Omega$ *and a number* $\mu\in\mathbb{R}$ *such that*

1) $R({}^tP_{\mathcal{E}'})\cap H^c_\nu(K) \subset {}^tP(H^c_\mu(K'))$;

2) *there is a constant* $C>0$, *such that*

$$\forall\ v\in H^c_\mu(K') \quad ({}^tPv\in H^c_\nu(K) \tag{3.16}$$
$$\Rightarrow \operatorname{dist}_{|\,|_\mu}(v, N({}^tP_{\mathcal{E}'})\cap H^c_\mu(K')) \le C|{}^tPv|_\nu).$$

c) 1) Ω *is* C^∞-P_n-convex *and* 2) P^∞ *is semiglobally normally solvable.*

Proof. In the case (3.2) under study, the assumptions of Theorem (1.3) are fulfilled. From (3.1) and (2.9) it is easy to obtain

$$N({}^tP_{\mathcal{E}'})^{\perp(\mathcal{E}'(\Omega),C^\infty(\Omega))} = N(P^\infty{}')^{\perp\langle C^\infty(\Omega)',C^\infty(\Omega)\rangle}. \tag{3.17}$$

Therefore the statement a) of Theorem (1.3) is equivalent to (3.15).

There is obviously no loss of generality if one considers the statements b) and c) with respect to the sequence $(\Omega_n)_0^\infty$ of open and relatively compact subsets of Ω.

We now prove that the assertion a) is equivalent to b). For this purpose it suffices to show that b) is equivalent to the statement (1.3), b). We have already proved that (3.4) is equivalent to the linking condition (1.3), b)1), if the seminorms p and q are defined by (3.3). Furthermore for these p and q the estimates (1.11) and also (1.3), b)2) are true iff (3.16) is valid; this equivalence follows from (2.15) and the formulas (3.1) and (3.17).

It remains to be proved that a) $\Leftrightarrow$ c). The assertion a) implies b)1) and therefore also the weaker statement c)1). Furthermore P^∞ is semiglobally normally solvable, if P^∞ is normally solvable. For the inverse implication c) $\Rightarrow$ a) it suffices to show that c) implies the assertion (1.3), c). To this end assume that $p = \bar{s}^n_\nu \in \Gamma_{\tau_{C^\infty}}$ is given. Then it follows from c) that for the given numbers $\nu\in\mathbb{R}$ and $n\in\mathbb{N}$ there is a number $k\in\mathbb{N}$ such that (3.5) and

$$\forall\ f\in N({}^tP_{\mathcal{E}'})^\perp\ \forall\ u\in C^\infty(\Omega) \qquad (f-P^\infty u)|_{\Omega_{k+2}} = 0 \tag{3.18}$$

are true. If we set $q = \bar{s}^{k+1}_\nu$, then we obtain from (3.5) the assertion (1.3),c)1) and also (1.3),c)2) because of (3.18) and the definition of q. This completes the proof.

The surjectivity statements which Trèves studied in [12], Theorem 19.1,p.60 and Theorem 20.1,p.63 are a special case of Theorem (3.14). In [12],a),b) and c) of Theorem (3.14) are only considered for $N({}^tP_{\mathcal{E}'}) = \{0\}$; we denote the corresponding properties by a'),b') and c').

For example, a') means that P^∞ is surjective and c') 2) that P^∞ is semiglobally solvable. Each of these properties a'), b') or c') implies the injectivity of ${}^tP_{\mathcal{E}'}$. Thus, according to Theorem (3.14) the properties a'), b') and c') are also equivalent.

An immediate consequence of Theorem (3.14) is the following surjectivity statement for a LPDO P with *constant* coefficients :
P^∞ is surjective iff Ω is P-convex with respect to $C_o^\infty(\Omega)$ (C^∞-P-convex respectively). (C.f. [3], Cor.3.5.2 and [14], p.394.)

For *proof* we observe that P^∞ is semiglobally solvable and ${}^tP_{\mathcal{E}'}$ is injective because of the existence of a fundamental solution in $D'(\Omega)$ for LPDO's with constant coefficients. The equivalence a) $\Leftrightarrow$ c) of Theorem (3.14) completes the proof.

A further consequence of Theorem (3.14) which is important for LPDO's with *variable* coefficients is the following

(3.19) Theorem. *Let* $P^\infty : C^\infty(\Omega) \longrightarrow C^\infty(\Omega)$ *be a LPDO with coefficients in* $C^\infty(\Omega)$.

Assume that for each compact set $K\subset\Omega$ *and for each number* $\nu\in\mathbb{R}$ *there exists a number* $\mu\in\mathbb{R}$ *such that*

(3.20) $\quad \forall\, v\in \mathcal{E}'(\Omega) \qquad ({}^tPv\in H_\nu^c(K) \Rightarrow v\in H_\mu).$

Suppose in addition that there is a number $\mu'<\mu$ *such that the a priori estimates*

(3.21) $$\forall\, K'\subset\Omega,\text{compact}\ \exists\, C_{K'}>o\ \forall\, v\in H_\mu^c(K') \quad ({}^tPv\in H_\nu \Rightarrow |v|_\mu \le C_{K'}(|v|_{\mu'} + |{}^tPv|_\nu)$$

are valid.

Then the following statements are equivalent:

a) Ω *is* C^∞-P_n-convex.

b) P^∞ *is normally solvable.*

Furthermore, if Ω *is* C^∞-P-convex, *then the range of* P^∞ *has a finite codimension in* $C^\infty(\Omega)$.

Proof. Because of (3.14), a) $\Rightarrow$ c) 1), the implication b) $\Rightarrow$ a) is obvious. Inversely, if a) is fulfilled, then there exists for each compact set $K\subset\Omega$ and $\nu\in\mathbb{R}$ a compact set $K'\subset\Omega$ such that the inclusion in (3.7) is valid. Therefore it follows from (3.20) that there is a number $\mu\in\mathbb{R}$ such that (3.14), b) 1) can be asserted. The estimates (3.16) must now be deduced. For the given numbers $\nu,\mu\in\mathbb{R}$ and the compact set $K'\subset\Omega$ we consider the restriction ${}^tP_{\mathcal{E}',K'}$ of ${}^tP_{\mathcal{E}'}$ to the subspace

$$D({}^tP_{\mathcal{E}',K'}) = \{v\in H_\mu^c(K') : {}^tPv\in H_\nu^c(K')\}$$

of $H^c_\mu(K')$. ${}^tP_{\mathcal{E}',K'}$ is a closed linear map from $(H^c_\mu(K'),||_\mu)$ into $(H^c_\nu(K'),||_\nu)$ with the domain $D({}^tP_{\mathcal{E}',K'})$. According to Sobolev's lemma, the embedding

$$(H^c_\mu(K'),||_\mu) \longleftrightarrow (H^c_{\mu'}(K'),||_{\mu'}) \qquad (\mu'<\mu)$$

is a compact map. (Cf. e.g. [3], p.38.) Therefore it follows from the a priori estimates (3.21) that ${}^tP_{\mathcal{E}',K'}$ is $(||_\mu,||_\nu)$-open and that

(3.22) $\quad \dim N({}^tP_{\mathcal{E}',K'}) = \dim (N({}^tP_{\mathcal{E}'}) \cap H^c_\mu(K')) < \infty$.

(C.f. e.g. [2], p.184.) Thus, because of the openness of ${}^tP_{\mathcal{E}',K'}$ there is a constant $C_{K'}>0$ such that in particular (3.16) is true.

Furthermore, if Ω is C^∞-P-convex, then from the assertion (3.8) and from (3.20) we can assert $N({}^tP_{\mathcal{E}'}) \subset H^c_\mu(K')$ for a compact set $K'\subset\Omega$ and for $\mu\in\mathbb{R}$ mentioned above. Therefore it follows from (3.22) and the proved normal solvability of P^∞

$$\operatorname{codim} R(P^\infty) = \dim N({}^tP_{\mathcal{E}'}) = \dim N({}^tP_{\mathcal{E}',K'}) < \infty .$$

We now give some examples of applications of the last Theorem:

(3.23) <u>Examples</u>. 1) Let P be an elliptic differential operator of order m with coefficients in $C^\infty(\Omega)$. Then, according to [15], p.352 for each compact set $K\subset\Omega$ and $\nu\in\mathbb{R}$, (3.20) as well as (3.21) are fulfilled for $\mu=\nu-m$ and $\mu'=\mu-1$.

2) Let P be a principally normal differential operator of order m with coefficients in $C^\infty(\Omega)$. Assume furthermore that there exists a function $\psi\in C^2(\Omega)$ with $\psi'\neq 0$ in Ω and pseudo-convex level surfaces throughout Ω. Then, according to [3], p.207 for each compact set $K\subset\Omega$ and $\nu\in\mathbb{R}$, (3.20) and (3.21) are fulfilled for $\mu=\nu-m+1$ and $\mu'=\mu-1$.

For the next Theorem we require the following statement which is only a slight alteration of the Definition (3.11).

(3.24) P^∞ *is semiglobally normally solvable, if for each open set* $\Omega'\subset\subset\Omega$ *the assertion (3.12) is fulfilled for* $u\in C^\infty(\Omega')$ *instead of for* $u\in C^\infty(\Omega)$.

Proof. Let Ω'' be an open set with $\Omega''\subset\subset\Omega$. There is an open set Ω' such that $\Omega''\subset\subset\Omega'\subset\subset\Omega$. Thus there exists a $\varphi\in C^\infty_0(\Omega'')$ with the property $\varphi_{|\Omega''}=1$. Because of the above assumption, for each $f\in N({}^tP_{\mathcal{E}'})^\perp$ there is a $u\in C^\infty(\Omega')$ such that $(f-P^\infty u)_{|\Omega'}=0$. It remains to be noted that $v=\varphi u\in C^\infty(\Omega)$ satisfies $(f-P^\infty v)_{|\Omega''}=0$.

For *hypoelliptic* LPDO's (see [14], p.535) only a *single* a priori estimate has to be proved for each compact set $K\subset\Omega$. This is shown by the following

(3.25) <u>Theorem</u>. *Let* P *be hypoelliptic with coefficients in* $C^\infty(\Omega)$. *Assume that for each compact set* $K \subset \Omega$ *there are numbers* $\nu, \mu, \mu' \in \mathbb{R}$, $\mu' < \mu$ *such that*

$$(3.26) \qquad \forall\, \varphi \in C_0^\infty(K) \qquad |\varphi|_\mu \leq C_K(|\varphi|_{\mu'} + |{}^tP\varphi|_\nu)$$

is valid for a suitable constant $C_K > 0$.

Then P^∞ *is semiglobally normally solvable. Furthermore* P^∞ *is normally solvable, if* Ω *is* C^∞-P_n-*convex*.

Proof. Because of Theorem (3.14) we only have to prove that P^∞ is semiglobally solvable.

Let Ω' be an open and bounded set in Ω with $K := \overline{\Omega'} \subset \Omega$. According to the above assumption, there are numbers ν, μ, μ', $\mu' < \mu$, such that (3.26) is fulfilled. If we regard ${}^tP$ as a map from $C_0^\infty(K)$ $(\subset H_\mu)$ into H_ν, then its $(|\,|_\mu, |\,|_\nu)$-closure ${}^tP_{(\mu,\nu)}$ is a linear operator from H_μ into H_ν with the domain

$$(3.27) \qquad D({}^tP_{(\mu,\nu)}) \subset H_\mu^c(K) \qquad (\subset H_\mu).$$

It is obvious that (3.26) is also valid for $u \in D({}^tP_{(\mu,\nu)})$. Therefore, since the embedding

$$(H_\mu^c(K), |\,|_\mu) \longleftrightarrow (H_{\mu'}, |\,|_{\mu'}) \qquad (\mu' < \mu)$$

is compact, it follows in the usual way that ${}^tP_{(\mu,\nu)}$ is a $(|\,|_\mu, |\,|_\nu)$-open map.

Theorem (1.3) is applied to the case

$$(X,\tau) = (H_\mu, |\,|_\mu), (Y,\sigma) = (H_\nu, |\,|_\nu),\ \Gamma_\tau = \{|\,|_\mu\}, \Gamma_\sigma = \{|\,|_\nu\},\ T = {}^tP_{(\mu,\nu)}.$$

Because the assumptions of (1.3) are clearly fulfilled, we can state from the implication b) $\Rightarrow$ a) of (1.3) that the linear *relation* ${}^tP'_{(\mu,\nu)} \subset H_{-\nu} \times H_{-\mu}$ is normally solvable, i.e.

$$(3.28) \qquad R({}^tP'_{(\mu,\nu)}) = N({}^tP_{(\mu,\nu)})^{\perp\langle H_\mu, H_{-\mu}\rangle},$$

where $H'_\mu [H'_\nu]$ is identified with $H_{-\mu} [H_{-\nu}]$.

Suppose that f belongs to $N({}^tP_{\mathcal{E}'})^{\perp(\mathcal{E}'(\Omega), C^\infty(\Omega))}$, then we have

$$(3.29) \qquad z(f) = 0 \qquad (z \in N({}^tP_{(\mu,\nu)})).$$

because of $N({}^tP_{(\mu,\nu)}) \subset N({}^tP_{\mathcal{E}'})$. For the given compact set $K \subset \Omega$, there is a function $\psi = 1$ in a neighbouhood of K. If we observe that the support of each $z \in N({}^tP_{(\mu,\nu)})$ is a subset of K, then (3.29) leads to

$$(\psi z)(f) = z(\psi f)$$

Therefore we obtain $\psi f \in N({}^tP_{(\mu,\nu)})^{\perp\langle H_\mu, H_{-\mu}\rangle}$, if we regard $\psi f \in C_0^\infty(\Omega)$ as belonging to $H_{-\mu}$. From (3.28) it then follows that there exists a $u \in D({}^tP'_{(\mu,\nu)})$, which satisfies

$$(3.30) \qquad v \in D({}^tP_{(\mu,\nu)}) \quad , \langle u, {}^tPv\rangle = \langle \psi f, v\rangle .$$

Since $C_o^\infty(\Omega') \subset D({}^tP_{(\mu,\nu)})$, this shows that $P(u|_{\Omega'}) = f|_{\Omega'}$. Because P is hypoelliptic we have in addition $u|_{\Omega'} \in C^\infty(\Omega')$. Thus, in view of (3.24), the proof is complete.

Theorem (3.25) should be supplemented by the following

(3.31) Remarks. 1) In (3.25) P^∞ is even semiglobally solvable, if we require

$$(3.32) \qquad \forall\ \varphi \in C_o^\infty(K) \qquad |\varphi|_\mu \leq C_K\, |{}^tP\varphi|_\nu \quad ,$$

instead of the weaker assumption (3.26).

This assertion can also be found in [12], p.66. The proof is evident, as may be seen from the proof of Theorem (3.25). We only have to observe that the operators ${}^tP_{(\mu,\nu)}$ are injective because of (3.32).

In the monograph [12] Treves proved (3.31),1) by another method. He used a lemma on the existence of right inverses in Hilbert spaces. (Cf. [12], p.107-108.) Instead of this lemma we apply the closed-range theorem for linear relations in Hilbert spaces.

2) If ${}^tP$ is hypoelliptic, it can be shown that for each compact set $K \subset \Omega$ there exist numbers $\nu, \mu, \mu' \in \mathbb{R}$, $\mu' < \mu$, such that (3.26) is valid. (Cf. [14], (52.1), p.538.) Owing to Theorem (3.25) we can state

(3.33) *If* P *and* ${}^tP$ *are hypoelliptic with coefficients in* $C^\infty(\Omega)$, *then* $P^\infty : C^\infty(\Omega) \longrightarrow C^\infty(\Omega)$ *is semiglobally normally solvable.*

In [10] we noted explicit conditions for normal solvability and in particularly for surjectivity of elliptic and principally normal differential operators which operate between local Sobolev spaces. Here we state similar conditions for LPDO's which map $C^\infty(\Omega)$ into itself.

(3.34) Theorem. *Let* P *be an elliptic PDO of order* m *with coefficients in* $C^\infty(\Omega)$. *Furthermore assume that there exists an open set* $\omega \subset\subset \Omega$ *such that the coefficients of* P *are analytic functions in* $\Omega\setminus\overline{\omega}$. *We state:*

$P^\infty : C^\infty(\Omega) \longrightarrow C^\infty(\Omega)$ *is normally solvable and the codimension of the range of* P^∞ *in* $C^\infty(\Omega)$ *is finite. Moreover, if the coefficients of* P *are analytic functions in* Ω, *then* P^∞ *is surjective.*

Proof. According to Theorem (3.19) and (3.23),1) we have to show that Ω is C^∞-P-convex. It follows from *Holmgren's uniqueness theorem* (cf.[3], p.125) that ${}^tP$ satisfies the uniqueness in the Cauchy problem in $\Omega\setminus\overline{\omega}$ with respect to $D'(\Omega\setminus\overline{\omega})$. (Cf. [10], Definition (4.1).) Because of Theorem (4.5) in [10] Ω is therefore P-convex with respect to $\mathcal{E}'(\Omega)$ and thus also C^∞-P-convex.

If the coefficients of P are analytic functions in Ω, then ${}^tP$ fulfills the uniqueness in the Cauchy problem in the whole of Ω with respect to $D'(\Omega)$. Hence, according to (4.3) in [10], ${}^tP_{\mathcal{E}'}$ is injective. Thus it is clear that P^∞ is surjective, for we have already shown that P^∞ is normally solvable.

(3.35) <u>Remark</u>. Let P_m denote the principal part of P. Instead of the uniqueness condition of Holmgren in $\Omega\setminus\bar\omega$ $[\Omega]$ we suppose in (3.34) that the equation

(3.36) $\quad P_m(x,\xi+\tau N) = o$

does not have any complex double zero τ for any $\xi, N \in \mathbb{R}^d$ with $\xi+\tau N \neq o$ in $x\in\Omega\setminus\bar\omega$ $[x\in\Omega]$. Then the statements in (3.34) remain valid.

For *proof* we have to observe that ${}^tP$ satisfies *Calderon's uniqueness condition*. From Remark 3 in [3], p.203-204 and Corollary 8.9.1, p.225 it therefore follows that ${}^tP$ once more fulfills the uniqueness in the Cauchy problem in $\Omega\setminus\bar\omega$ $[\Omega]$ with respect to $D'(\Omega\setminus\bar\omega)$ $[D'(\Omega)]$. Hence the proof can be completed as in Theorem (3.34).

Statements concerning semiglobal normal solvability of principally normal differential operators have been considered by Hörmander [3], Theorem 8.7.6, p.214. Owing to (3.19),(3.23) 2) and the reasons of the proof of Remark (3.35) we obtain the following global

(3.37) <u>Theorem</u>. *Let* P *be a principally normal differential operator of order* m *with coefficients in* $C^\infty(\Omega)$. *Assume that there exists a function* $\psi\in C^2(\Omega)$ *with* $\psi'\neq o$ *in* Ω *and pseudo-convex level surfaces throughout* Ω *with respect to* P. *Furthermore, let* ω *be an open set with* $\omega\subset\subset\Omega$ *such that* P *satisfies Calderon's uniqueness condition in* $\Omega\setminus\bar\omega$ *(noted in Remark (3.35)).*

Then $P^\infty : C^\infty(\Omega) \longrightarrow C^\infty(\Omega)$ *is normally solvable and its range is of finite codimension in* $C^\infty(\Omega)$. *If Calderon's uniqueness condition is fulfilled in the whole of* Ω, *then* P^∞ *is surjective.*

Finally we give an example of a LPDO which is normally solvable but not surjective.

(3.38) <u>Example</u>. Plis [8], p.610 constructed an elliptic PDO with C^∞-coefficients in $\mathbb{R}^3$ of the form

(3.39) $\quad Pl = (D_1^2 + D_2^2 + D_3^2) + P_o$,

where P_o is a PDO with coefficients identical to zero for $|x|\geq\frac{5}{6}$. He showed that Pl is not injective in $C_o^\infty(\bar K_{5/6}(o))$. We state :

$({}^tPl)^\infty : C^\infty(\mathbb{R}^3) \longrightarrow C^\infty(\mathbb{R}^3)$ is normally solvable, codim $R(({}^tPl)^\infty) < \infty$, but $({}^tPl)^\infty$ is not surjective.

Proof. We set $\omega = K_{5/6}(o), \Omega = \mathbb{R}^3$ and $P = {}^tPl$. Since the assumptions of Theorem (3.34) are fulfilled, $({}^tPl)^\infty$ is normally solvable. But $({}^tPl)^\infty$ is not surjective because codim $R(({}^tPl)^\infty) = \dim N(({}^tPl)_{\mathcal{E}'}) \neq o$.

Bibliographic References

[1] *F.E.Browder,* Functional analysis and partial differential equations II. Math.Ann. 145 (1962), 81-226.

[2] *S.Goldberg,* Unbounded linear operators: theory and applications. New York-London 1966.

[3] *L.Hörmander,* Linear partial differential operators, Grundlehren der math. Wiss. 116, Berlin-Heidelberg-New York 1976.

[4] *J.L.Lions,E.Magenes,* Non-homogeneous boundary value problems and applications, Grundlehren der math. Wiss. 181, Berlin-Heidelberg-New York 1972.

[5] *B.Malgrange,* Existence et approximation des solutions des équations aux derivées partielles et des équations de convolution, Ann. Inst. Fourier Grenoble 6 (1955-1956), 271-355.

[6] *R.Mennicken,B.Sagraloff,* Eine Verallgemeinerung des Satzes vom abgeschlossenen Wertebereich in lokalkonvexen Räumen, manus. math. 18, (1976), 109-146.

[7] *R.Mennicken,B.Sagraloff,* Characterizations of nearly-openness, J. reine angew. Math. (in print.)

[8] *A. Plis,* A smooth linear elliptic differential equation without any solution in a sphere, Com. pure and appl. math. 14 (1961), 599-617.

[9] *B.Sagraloff,* Eine Ergänzung zum Satz vom abgeschlossenen Wertebereich in lokalkonvexen Räumen, manus. math. 22 (1977), 213-224.

[10] *B.Sagraloff,* Normale Auflösbarkeit bei linearen partiellen Differentialoperatoren in lokalen gewichteten Sobolevräumen, J. reine angew. Math. 310 (1979), 131-150.

[11] *H.H.Schaefer,* Topological vector spaces, Graduate texts in mathematics 3, Berlin 1971.

[12] *F.Trèves,* Locally convex spaces and linear partial differential equations, Grundlehren der math. Wiss. 146, Berlin-Heidelberg-New York 1967.

[13] *F.Trèves,* Linear partial differential equations, New York-London-Paris 1970.

[14] *F.Trèves,* Topological vector spaces, distributions and kernels, New York 1967.

[15] *F.Trèves,* Basic linear differential equations, New York 1975.

B.Sagraloff
Fachbereich Mathematik
der Universität Regensburg
Universitätsstr. 31
D-8400 Regensburg

CONNECTION PROBLEMS FOR LINEAR ORDINARY DIFFERENTIAL EQUATIONS IN THE COMPLEX DOMAIN

R. Schäfke and D. Schmidt

1. In two papers [6], [7] the authors considered the following system of linear differential equations

$$(1)\qquad y'(z) = \left(\frac{1}{z}A_0 + \frac{1}{z-1}A_1 + G(z)\right)y(z)$$

where A_0 and A_1 are n by n matrices and $G(z)$ is a matrix-valued function holomorphic in a disk $|z| < r$ with $r > 1$.

(1) has two singular points at 0 and 1 . They are singular points of the first kind and they are the only singular points of (1) within the circle $|z| < r$.

(1) is not a very special differential equation, but an almost general one: All linear differential equations with two singular points z_0 and z_1 of the first kind such that z_1 is neighbouring to z_0 can be transformed into (1). Here "z_1 is neighbouring to z_0" means that the distance between z_1 and z_0 is the smallest among all distances between z_0 and other singular points.

Now it is known that (1) has fundamental solutions at 0 and 1 of the form

$$(2)\qquad \begin{aligned} Y_0(z) &= \sum_{k=0}^{\infty} D_k^0\, z^k\, z^{C_0} \\ Y_1(z) &= \sum_{k=0}^{\infty} D_k^1\, (1-z)^k\, (1-z)^{C_1} \end{aligned}$$

where the n by n matrices C_j, D_k^j can be computed from the data A_0, A_1 and G of (1); see e.g. [1], p. 120 or [2], vol. II, p. 162 ff. They are supposed to be known.

Our final aim is to derive a formula for the connection matrix C between the fundamental solutions (2).

Determine the connection matrix C *in*

$$(3)\qquad Y_0(z) = Y_1(z)C \qquad (z \in \,]0,1[;\ |z-1| < r-1) .$$

The study of such connection formulas is one field in the theory of the global behaviour of solutions of singular linear differential equations. For example (3) tells us, how the solution $Y_o(z)$ which is known near 0 behaves when z approaches 1.

2. But before we come to the general problem (3) we want to discuss a special case which is easier and more interesting in applications. The additional assumption here is that C_1 is diagonalizable or without loss of generality diagonal. Thus we can state the simplified problem as follows.

Assume $C_1 = \operatorname{diag}(\alpha_1,\ldots,\alpha_n)$ *and that*

$$y_j(z) = (1-z)^{\alpha_j} \sum_{k=0}^{\infty} (1-z)^k d_k^j$$

are the columns of $Y_1(z)$. *[We only need that* $y_1(z),\ldots,y_n(z)$ *are linearly independent.] Further let*

(4)
$$y(z) = z^{\alpha} \sum_{k=0}^{\infty} d_k z^k$$

be a vector solution of (1) at 0. *Determine the complex constants* $\gamma_1,\ldots,\gamma_n$ *in*

$$y(z) = \sum_{j=1}^{n} \gamma_j y_j(z) \qquad (z \in]0,1[;\ |z-1| < r-1) .$$

We give a solution of the problem (4) by a limit formula for the $\gamma_1,\ldots,\gamma_n$:

(5) <u>Theorem</u>: *Consider the connection problem (4). Suppose the fundamental set* $y_1,\ldots,y_n$ *to be normalized such that*

$$d_o^1,\ldots,d_o^n \text{ are linearly independent.}$$

Further, let $m \in \mathbb{N}_o$ *be chosen such that* $m > \hat{\gamma} - 1$ *where*

$$\hat{\gamma} := \max_j \operatorname{Re} \alpha_j - \min_j \operatorname{Re} \alpha_j .$$

If c *denotes the vector with components* $\frac{\gamma_j}{\Gamma(-\alpha_j)}$ $(j=1,\ldots,n)$ *and* D_k *the n by n matrix with columns*

$$\sum_{\ell=0}^{m} \prod_{\sigma=1}^{\ell} \left(\frac{-\alpha_j-\sigma}{k+\alpha-\alpha_j-\sigma}\right) d_\ell^j \qquad (j=1,\ldots,n) \quad ,$$

then

$$c = \lim_{k\to\infty} \operatorname{diag}\left(\frac{\Gamma(k+\alpha+1)}{\Gamma(k+\alpha-\alpha_j)}\right) D_k^{-1} d_k \quad ,$$

the convergence being $O(k^{\hat{\gamma}-1-m})$.

Since α, the α_j, C_k and d_ℓ^j can be computed recursively from the differential equation, our formula indeed expresses the vector c (containing the γ_j) as limit of known quantities. The convergence will be the better the larger m is chosen. But then the computation of D_k becomes more extensive.

We can get some γ_j itself out of the limit formula only if the corresponding α_j is <u>not</u> 0 or a positive integer. If α_j is a nonnegative integer, theorem 1 gives γ_j when a preliminary transformation

$$y(z) = (1-z)^{\nu}\, \tilde{y}(z)$$

with suitable $\nu \in \mathbb{C}$ has been performed.

There is a very special case where the formula becomes very simple and which should be mentioned:

(6) <u>Corollary</u>: *Additionaly to the hypothesis of theorem 1 suppose that* $\hat{\gamma} < 1$ *i.e.* $\max_j \operatorname{Re}\alpha_j - \min_j \operatorname{Re}\alpha_j < 1$. *If the* $\delta_k^j \in \mathbb{C}$ *are defined by the linear representation*

$$d_k = \sum_{j=1}^{n} \delta_k^j d_o^j \qquad (k \in \mathbb{N})$$

then we have

$$\frac{\gamma_j}{\Gamma(-\alpha_j)} = \lim_{k\to\infty} k^{\alpha_j+1} \delta_k^j \qquad (j=1,\ldots,n) \quad .$$

3. Before we give an idea of the proof, we include an application in the theory of special functions, which yields new results.

We consider the generalized Heun equation

$$y''(z) + \left(\frac{1-\mu_o}{z} + \frac{1-\mu_1}{z-1} + \frac{1-\mu_2}{z-a}\right) y'(z) + \frac{\beta_o+\beta_1 z+\beta_2 z^2}{z(z-1)(z-a)}\, y(z) = 0 \tag{7}$$

$$[a \in \mathbb{C} \setminus \{0,1\};\ \ \mu := (\mu_o,\mu_1,\mu_2) \in \mathbb{C}^3;\ \ \beta := (\beta_o,\beta_1,\beta_2) \in \mathbb{C}^3] .$$

(7) includes many differential equations of Mathematical Physics as special cases; e.g. the ellipsoidal wave equation or Ince's equation. (7) has in general singular points of the first kind at 0, 1 and a and of the second kind at ∞. The roots of the indicial equations corresponding to 0, 1 and a are 0 and μ_o; 0 and μ_1; 0 and μ_2 respectively.

It turns out, that there is essentially only one "generalized Heun function" (which depends on the parameters). Similar to the hypergeometric equation this is shown by substitutions of the type $y(z) = z^{\gamma_o}(1-z)^{\gamma_1}\,\tilde{y}(\alpha_o + \alpha_1 z)$ with appropriate α's and γ's. That function mentioned above is discribed in

(8) <u>Proposition</u>: *There exists a unique function* η, *holomorphic with respect to* (z,μ,β,a), *where* $|z| < \min(1,|a|)$, *such that* $\eta(\cdot,\mu,\beta,a)$ *is for each* (μ,β,a) *a solution of (7) satisfying*

$$\eta(0,\mu,\beta,a) = \frac{1}{\Gamma}(1-\mu_o) .$$

η *can be expanded into a power series*

$$\eta(z,\mu,\beta,a) = \sum_{k=0}^{\infty} \frac{\tau_k(\mu,\beta,a)}{\Gamma(1-\mu_o+k)\,k!}\, z^k$$

where the τ_k *are determined by a four term recurrence relation*

$$\tau_k = \sum_{\kappa=1}^{3} \varphi_\kappa(k-\kappa)\,\tau_{k-\kappa} \qquad (k \in \mathbb{N})$$

$$\tau_{-2} = \tau_{-1} \equiv 0, \quad \tau_o \equiv 1$$

Here the φ_κ can be obtained from (7). They are polynomials in μ,β and $\frac{1}{a}$.

In terms of η we have the following solutions of (7) near 1

$$\eta_1(z) = \eta(1-z,\ \mu_1,\ \mu_o,\ \mu_2,\ \tilde{\beta},\ 1-a)$$

$$\eta_2(z) = \eta(1-z,\ -\mu_1,\ \mu_o,\ \mu_2,\ \tilde{\beta},\ 1-a)(1-z)^{\mu_1}$$

where $\tilde{\beta} \in \mathbb{C}^3$ is obtained from (μ,β,a) by inserting $y(z) = \tilde{y}(1-z)$

into (7) and reordering in the same form.

Now one wants to determine the α_1 and α_2 in the connection relation between $\eta_o(z) = \eta(z,\mu,\beta,a)$ and η_1, η_2

$$\eta_o(z) = \alpha_1 \eta_1(z) + \alpha_2 \eta_2(z) \qquad z \in]0,1[\text{ near } 1 \quad .$$

Using theorem 1 and some additional considerations α_1 and α_2 and their dependence on the parameters can be determined.

(9) Proposition: *There exists a unique function* q *holomorphic with respect to* (μ,β,a) *where* $|a| > 1$ *, such that*

$$\begin{aligned}\frac{\sin(\pi\mu_1)}{\pi}\,\eta(z,\mu,\beta,a) &= q(\mu_o,-\mu_1,\mu_2,\beta,a)\;\eta(1-z,\mu_1,\mu_o,\mu_2,\tilde{\beta},1-a)\\ &\quad - q(\mu,\beta,a)(1-z)^{\mu_1}\;\eta(1-z,-\mu_1,\mu_o,\mu_2,\tilde{\beta},1-a)\end{aligned}$$

is valid for $z \in \mathbb{C}$ *such that* $|z| < 1$, $|z-1| < \min(1,|a-1|)$ *and* $\arg(1-z) \in]-\pi,\pi[$ *and for all choices of parameters where* $|a| > 1$.

(10) Theorem: *The function* q *of proposition (9) is given by*

$$q(\mu,\beta,a) = \lim_{k\to\infty} \frac{\tau_k(\mu,\beta,a)}{\Gamma(k+1-\mu_o)\,\Gamma(k-\mu_1)} \times \underline{\qquad\qquad}$$

$$\underline{\qquad\qquad} = \left[1 + \sum_{\ell=1}^{m} \frac{\tau_\ell(-\mu_1,\mu_o,\mu_2,\beta,1-a)}{\ell!} \prod_{\sigma=1}^{\ell} (\sigma+\mu_1-k)^{-1}\right]^{-1}$$

the convergence being $O(k^{-m-1})$ *. Here* $\mu,\beta \in \mathbb{C}^3$ *and* $|a| > 1$ *and* m *is an arbitrary nonnegative integer.*

Since $m = 0$ may be chosen in (10) we have especially

$$q(\mu,\beta,a) = \lim_{k\to\infty} \frac{\tau_k(\mu,\beta,a)}{\Gamma(k+1-\mu_o)\,\Gamma(k-\mu_1)} \quad .$$

Hence the values of the function q of (9) which is the essential connection coefficient for (7) can be computed from the four term recurrence relation in (8) for the τ_k .

(9) is only one connection relation between the solutions of (7). The same substitutions of (7), which permit to express all characteristic solutions in terms of η , yield now the full set of connection relations between them in terms of q .

4. Now let me give you an idea of the proof of theorem (5).

The first step is to derive the following asymptotic formula for the coefficients d_k (see (4)).

(11) Lemma: *Consider the connection problem (4). Then*

$$d_k = \sum_{j=1}^{n} \gamma_j \sum_{\ell=0}^{m} \frac{\Gamma(k+\alpha-\ell-\alpha_j)}{\Gamma(k+\alpha+1)} \frac{1}{\Gamma(-\ell-\alpha_j)} d_\ell^j + O\left(k^{-\alpha_- -m-2}\right)$$

as $k\to\infty$ *, where* $\alpha_- = \min_j \operatorname{Re}\alpha_j$ *and* m *is an arbitrary non-negative integer.*

(11) is in fact some kind of asymptotic formula for the d_k , since the quotients of Γ-functions have an asymptotic series in terms of powers of k . (11) looks somewhat complicated, but can be proved in a very simple way:

First, by the Cauchy formula, we have:

$$d_k = \frac{1}{2\pi i} \int_{|z|=\varepsilon} z^{-\alpha-k-1}\, y(z)\,dz$$

Here integration is taken (in positive sense) over the circle $|z|=\varepsilon$. This integration curve can be deformed into $\mathcal{C}_0 - \mathcal{C}_1$, where $\mathcal{C}_0$ and $\mathcal{C}_1$ are shown in figure 1 ($1<\rho<r$ here).

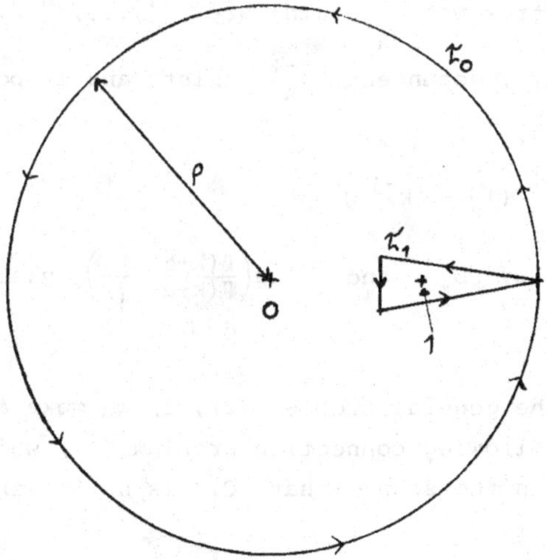

figure 1

Hence we have

$$d_k = \left(\frac{1}{2\pi i}\int_{\mathfrak{k}_0} - \frac{1}{2\pi i}\int_{\mathfrak{k}_1}\right) z^{-\alpha-k-1}\, y(z)\,dz$$

The first integral is $O(\rho^{-k})$ as $k\to\infty$ and can be neglected in an asymptotic formula with a remainder $O(k^{\mu})$.

In the second integral we can use the connection formula (4) and get

$$d_k = \sum_{j=1}^{n} \gamma_j \left(-\frac{1}{2\pi i}\int_{\mathfrak{k}_1} z^{-\alpha-k-1}\, y_j(z)\,dz\right) + O(\rho^{-k})$$

Now by inserting the series for $y_j(z)$ and integrating the first m terms up to $O(\rho^{-k})$, but estimating the remainding terms one can prove the formula (11).

Once (11) has been established the second step is easy. The functional equation for the Γ-function gives

$$d_k = \sum_{j=1}^{n}\left(\sum_{\ell=0}^{m} d_\ell^j \prod_{s=1}^{\ell}\left[\frac{-s-\alpha_j}{k+\alpha-\alpha_j-s}\right]\right)\frac{\Gamma(k+\alpha-\alpha_j)}{\Gamma(k+\alpha+1)}\,\frac{1}{\Gamma(-\alpha_j)}\,\gamma_j + O\left(k^{-\alpha_- -m-2}\right)$$

or in the terms of the theorem

$$(12)\qquad d_k = D_k\, \mathrm{diag}\left(\frac{\Gamma(k+\alpha-\alpha_j)}{\Gamma(k+\alpha+1)}\right) c + O\left(k^{-\alpha_- -m-2}\right) .$$

Now $\lim_{k\to\infty} D_k$ is the matrix whose columns are $d_o^1,\ldots,d_o^n$.

Since they are linearly independent D_k^{-1} exists and is bounded as $k\to\infty$. Using

$$\frac{\Gamma(k+\alpha+1)}{\Gamma(k+\alpha-\alpha_j)} = k^{\alpha_j+1}\,(1+O(k^{-1}))$$

multiplication of (12) by D_k^{-1} and $\mathrm{diag}\left(\frac{\Gamma(k+\alpha+1)}{\Gamma(k+\alpha-\alpha_j)}\right)$ yields the desired result.

5. Before we come to the general problem (2),(3) we make a first approach and study the following connection problem (13) which is a generalization of (4) in the sense, that C_1 is no longer assumed to be diagonalizable .

Let a fundamental solution of (1) at 1 *be given by*

$$Y_1(z) = \sum_{k=0}^{\infty} (1-z)^k D_k^1 (1-z)^{C_1} \qquad |z-1| < r-1$$

and let a single solution of (1) at 0 *be given by*

(13)
$$y_0(z) = z^{\alpha} \sum_{k=0}^{\infty} d_k z^k \qquad |z| < 1 \quad .$$

Determine the connection vector c *in*

$$y_0(z) = Y_1(z)c \qquad z \in]0,1[\quad \textit{near} \quad 1 \quad .$$

For (13) we have obtained a generalization of the limit formula (5) which is somewhat more complicated but is proved in quite the some way

(14) <u>Theorem</u>: *Consider the connection problem (13). Suppose the* n *by* n *matrix* C_1 *satisfies*

$$\sigma(C_1) = \{\lambda \in \sigma(A_1) \mid \lambda - n \notin \sigma(A_1) \quad \textit{for} \quad n = 1,2,\ldots\} \quad .$$

Define

$$\hat{\gamma} := \max\{\mathrm{Re}(\gamma_1 - \gamma_2) \mid \gamma_j \in \sigma(C_1)\}$$
$$d := \max\{\ell \in \mathbb{Z} \mid \ell = \alpha_1 - \alpha_2 \quad \textit{where} \quad \alpha_j \in \sigma(A_1)\} \quad ,$$

Choose an integer $m = d$ *such that* $m > \hat{\gamma} + d - 1$ *and define*

$$D_k := D_0^1 + \sum_{\ell=1}^{m} D^1 \prod_{s=1}^{\ell} \left[(-s-C_1)(k+\alpha-s-C_1)^{-1}\right] \quad .$$

If further we assume that $\sigma + C_1$ *is nonsingular for* $\sigma = 1,\ldots,d$ *we have*

$$\frac{1}{\Gamma}(-C_1)c = \lim_{k\to\infty} \Gamma(k+\alpha+1)\frac{1}{\Gamma}(k+\alpha-C_1)D_k^{-1} d_k$$

the convergence being $O(k^{\hat{\gamma}+d-1-m+\delta})$, $\delta > 0$ *arbitrary.*

The reciprocal Γ-function of a matrix B is here defined by insertion B into the power series of the entire function $\frac{1}{\Gamma}$. It can be computed from the scalar Γ-function and its derivatives and has all properties of the scalar $\frac{1}{\Gamma}$ as a complex function (functional equations, integral representations, asymptotic representations).

The $\delta > 0$ occures in (14), because C_1 need not be diagonal and hence z^{C_1} cannot be estimated as sharp as for a diagonal C_1 .

The first condition on C_1 does not restrict any applications because

a) by almost all common methods of constructing a fundamental solution Y_1 at 1 the C_1 has this property,

b) if a fundamental solution with a C_1 not satisfying that condition or a fundamental set of logarithmic solutions is given we can achieve the form required by the theorem making finitely many manipulations.

There is again a special case where (14) becomes very simple.

(15) <u>Corollary</u>: *Consider the connection problem (13) and suppose that*

$$\max\{\mathrm{Re}(\alpha_1-\alpha_2) \mid \alpha_1 \in \sigma(A_1)\} < 1$$

and hence $C_1 = A_1$ *and* $D_0^1 = I$ *may be chosen in (13). Then*

$$\frac{1}{\Gamma}(-A_1)c = \lim_{k\to\infty} k^{A_1+1} d_k \quad .$$

The corollary follows from (14) if $m = 0$ is chosen, because the condition on A_1 implies $\hat{\gamma} < 1$, $d = 0$ there.

6. Now finally we come to the most general problem (2),(3) i.e. to find the connection matrix between two general fundamental solutions of (1) at 0 and 1 . I shall use a trick and get the general result from the special case (14).

We transform the matrix differential equation corresponding to (1) by

$$Y(z) = X(z)\, z^{C_0}$$

and get

$$X(z) = \frac{1}{z}\hat{A}_0\, X(z) + \left(\frac{1}{z-1} A_1 + G(z)\right) X(z) \tag{16}$$

where $\hat{A}_0$ is a linear operator on the space of n by n matrices defined by

$$\hat{A}_0 Z := A_0 Z - Z C_0 \quad \text{for n by n matrices } Z \ . \tag{17}$$

On the right side of (17) and in the last term of (16) we have the common multiplication of matrices.

From (2) we get that (16) has a "single solution" at 0

(18) $$X_0(z) = \sum_{k=0}^{\infty} z^k D_k^0 \qquad |z| < 1$$

and a "fundamental solution" at 1 (which is now a function whose values are invertible linear transformations on the space of n by n matrices)

$$\hat{X}_1(z)Z = Y_1(z) Z z^{-C_0} \qquad |z-1| < r-1 \quad .$$

The connection problem (3) becomes

Determine the "connection factor" C *in*

(19) $$X_0(z) = \hat{X}_1(z)C \qquad z \in]0,1[\quad \textit{near} \quad 1 \quad .$$

If we show that $\hat{X}_1(z)$ has the form required in (14), we can apply (14). Now $\hat{X}_1$ can be written

$$\hat{X}_1(z) = \hat{H}_1(z)(1-z)^{\hat{C}_1}$$

where $\hat{C}_1 z = C_1 z$ and $\hat{H}_1$ defined by

$$\hat{H}_1(z)Z = \sum_{k=0}^{\infty} (1-z)^k D_k^1 Z z^{-C_0} \qquad \text{for n by n matrices } Z$$

is <u>holomorphic near 1</u> and can be developped into a power series

(20) $$\hat{H}_1(z)Z = \sum_{k=0}^{\infty} (1-z)^k \hat{D}_k^1$$
$$\hat{D}_k^1 Z = \sum_{\mu=0}^{k} D_{k-\mu}^1 Z \begin{pmatrix} C_0+\mu-1 \\ \mu \end{pmatrix} \qquad \text{for matrices } Z \; .$$

Now (14) can be applied to (19) with (18) and (20) and translating the result back to the original problem (2),(3) we get

(21) <u>Theorem</u>: *Consider the connection problem (2),(3). Suppose the n by n matrix* C_1 *satisfies*

$$\sigma(C_1) = \{\lambda \in \sigma(A_1) \mid \lambda - n \notin \sigma(A_1) \quad \textit{for} \quad n = 1,2,\ldots\} \; .$$

Define

$$\hat{\gamma} = \max\{\mathrm{Re}(\gamma_1-\gamma_2) \mid \gamma_j \in \sigma(C_1)\}$$

$$d = \max\{\ell \in \mathbb{Z} \mid \ell = \alpha_1 - \alpha_2 \quad \text{where} \quad \alpha_j \in \sigma(A_1)\} .$$

Choose an integer $m \geqq d$ *such that* $m > \hat{\gamma} + d - 1$ *and define linear operators* $\hat{D}_k$ *on the space of matrices by*

$$\hat{D}_k Z = D_o^1 Z + \sum_{\ell=1}^{m} \sum_{\nu=0}^{\ell} D_{\ell-\nu}^1 \prod_{s=1}^{\ell} \left[(-s-C_1)(k-s-C_1)^{-1} \right] Z \binom{C_o+\nu-1}{\nu} .$$

If finally we assume that $\sigma + C_1$ *is nonsingular for* $\sigma = 1,\ldots,d$ *then*

$$\frac{1}{\Gamma}(-C_1)C = \lim_{k\to\infty} k!\ \frac{1}{\Gamma}(k-C_1)\hat{D}_k^{-1}\ D_k^o$$

the convergence being $O(k^{\hat{\gamma}+d-1-m+\delta})$, $\delta > 0$ *arbitrary.*

The remarks below (14) are valid here again and need not to be repeated. A special case, where (21) becomes very simple should be noted here, too.

(22) Corollary: *Consider the connection problem (2),(3) and suppose that*

$$\max\{\mathrm{Re}(\alpha_1-\alpha_2) \mid \alpha_j \in \sigma(A_1)\} < 1$$

and hence $C_1 = A_1$ *and* $D_o = 1$ *may be chosen in (2). Then*

$$\frac{1}{\Gamma}(-A_1)C = \lim_{k\to\infty} k^{A_1+1}\ D_k^o \quad .$$

References:

[1] E.A.Coddington and N.Levinson, Theory of ordinary differential equations, chapter 4, McGraw-Hill, New York (1955)

[2] F.R.Gantmacher, The theory of matrices, Chelsea Publishing Comp., New York (1971)

[3] K.Heun, Zur Theorie der Riemann'schen Funktionen zweiter Ordnung mit vier Verzweigungspunkten, Math.Ann. 33 (1889), 161-179

[4] E.Hille, Lectures on ordinary differential equations, Addison-Wesley, Reading Mass. (1969)

[5] F.W.J.Olver, Asymptotics and special functions, Academic Press, New York (1974)

[6] R.Schäfke and D.Schmidt, The connection problem for general linear ordinary differential equations at two regular singular points with applications in the theory of special functions, preprint (1978)

[7] R.Schäfke, The connection problem for two neighbouring regular singular points of general linear complex ordinary differential equations, preprint (1978)

[8] D.Schmidt, Spektraleigenschaften und kanonische Fundamentallösungen linearer Differentialgleichungen bei einfachen Singularitäten, Arch.Math. 31 (1978), 302-309

PERIODIC SOLUTIONS OF CONTINUOUS SELF-GRAVITATING SYSTEMS

F. Verhulst
Mathematisch Instituut
Rijksuniversiteit Utrecht
3508 TA UTRECHT
The Netherlands

SUMMARY

A collection of self-gravitating particles can be described by the nonlinear system consisting of the collision-less Boltzmann equation and the appropriate Poisson-equation. Such a system can be studied by associating it with dynamical systems in a finite-dimensional phase-space. The finite-dimensional problems are treated in the frame-work of KAM-theory by Birkhoff normalization and averaging techniques. This leads to a classification of possible two-parameter families of periodic solutions in these dynamical systems.
The asymptotic approximations of the solutions in two degrees of freedom problems with a discrete symmetric potential produce ring-type solutions of the original continuous system.

1. INTRODUCTION

The results which will be described in this paper are applications and, as far as the quantitative theory is concerned, extensions of KAM-theory (derived from Kolmogorov, Arnold and Moser). This theory describes qualitative and quantitative aspects of nonlinear Hamiltonian mechanics with more than one degree of freedom.
The quantitative extensions of the theory which are used here are based on the theory of asymptotic approximations of solutions of nonlinear differential equations, in particular Hamiltonian systems, as has been developed by van der Burgh, Sanders and Verhulst.

Here we shall be concerned with applications to certain continuous systems. In section 2 we formulate the problem of the collision-less Boltzmann equation supplemented by the Poisson-equation. The fundamental approach is to associate this problem with the study of a finite-dimensional phase-space in the frame-work of the theory of characteristics. In astrophysics, strangely enough, most authors avoid the fundamental approach which is possible by using KAM-theory and take recourse to a direct treatment of the partial differential equations by formal and often mathematically not very clear methods; for an interesting survey of part of the astrophysical literature see Toomre (1977).

In section 3 we reduce the number of degrees of freedom to two by either assuming infinitesimal flatness or axi-symmetry. In the subsequent sections we study axi-symmetric rotating systems.

In sections 4 and 5 we reformulate as a resonance (small denominator) problem and we discuss normalization techniques by Birkhoff transformation and averaging.

In sections 6 and 7 we return to the continuous problem in a rather unorthodox way. We first study certain explicit examples at the resonances 1:2, 2:1 and 1:1 where we base ourselves on explicit expressions, approximations of periodic solutions, derived in earlier work. In these examples one obtains ring structures which exist in an axi-symmetric system at the resonances.

In section 7 we argue that the particular examples studied in section 6 represent a generic picture of what happens in an axi-symmetric system. Moreover we mention some open problems.

2. CONTINUOUS AND DISCRETE FORMULATIONS

Several relations can be established between discrete nonlinear Hamiltonian mechanics and the mechanics of continuous systems.

A famous example is the one-dimensional lattice studied by Fermi, Pasta and Ulam; their paper started the development of the theory of nonlinear lattice dynamics and its relations with soliton theory and continuum mechanics. For a survey of this field and appropriate references the reader is referred to the paper by Jackson (1978).

Here we shall be concerned with another relation between discrete and continuous systems which has been furnished by the theory of Lagrangian characteristics for first-order partial differential equations. Consider a continuous system in which the dynamical behaviour is governed by a collective force field U; for instance in the case of galaxies, U is the gravitational field.

If H is the Hamiltonian determining the motion of each fluid element, $q_i (i=1,2,3)$ are three spatial coordinates and p_i are the corresponding three momenta, the distribution function $f(t,p_i,q_i)$ is determined by the collision-less Boltzmann equation

$$\frac{\partial f}{\partial t} + \frac{\partial H}{\partial p_i}\frac{\partial f}{\partial q_i} - \frac{\partial H}{\partial q_i}\frac{\partial f}{\partial p_i} = 0 \tag{1}$$

For self-gravitating systems this equation is supplemented by the Poisson-equation

$$\Delta U = \rho . \tag{2}$$

The density $\rho(t,q_i)$ is obtained by integrating the distribution function over velocity space.

The relation between system (1-2) and studies of a finite-dimensional phase-space was pointed out by Jeans (1916) who observed that the distribution function f is a function of the independent integrals of the Lagrangian equations for the characteristics. We are interested in systems rotating around an axis so we introduce cylindrical coordinates r,θ,z and assume rotation around the z-axis.
The equations of motion for a fluid element become

$$\left.\begin{aligned} \ddot{r} &= r\dot{\theta}^2 - \frac{\partial U}{\partial r} \\ r\ddot{\theta} &= -2\dot{r}\dot{\theta} - \frac{1}{r}\frac{\partial U}{\partial \theta} \\ \ddot{z} &= -\frac{\partial U}{\partial z} \end{aligned}\right\} \tag{3}$$

To solve equation (1) we must solve system (3) for all possible potentials U. One integral of motion is of course the energy; if two more independent integrals can be found the system is integrable and the distribtion function f is a function of these three independent integrals. This happens for instance in the case that U is spherically symmetric. Integrability however constitues the non-generic case (for two degrees of freedom systems see Moser, 1955) and we are left with the problem of describing a in general non-integrable phase-flow in 6-space.

To perform this formidable task we proceed as follows. First we reduce system (3) to a two degrees of freedom problem by an additional assumption; two reductions of this type will be demonstrated in section 3. It should be remarked however that studies of the phase-flow of three degrees of freedom systems have been started (cf. E. van der Aa and J.A. Sanders, 1979) so that this simplification will not be necessary in the near future.

Secondly it can be demonstrated that in certain parts of phase-space, corresponding with regions of small energy in a sense to be defined later, the phase-flow of two degrees of freedom systems can be approximated by an integrable one. The main features of this integrable phase-flow are the short-periodic solutions which will be shown to play a special part in the solutions of the continuous system.

3. REDUCTION TO TWO DEGREES OF FREEDOM PROBLEMS

Here and in the following we shall consider stationary solutions of equations (1-2) only, i.e. f, U and ρ are not explicitly dependent on time t. There are two cases.

A. The system is infinitesimally flat and no motions in the z-direction are starting; we express this by

$$f(r,\theta,z,\dot{r},\dot{\theta},\dot{z}) = 0 \qquad z \neq 0$$

System (3) reduces to

$$\ddot{r} = r\dot{\theta}^2 - \frac{\partial U}{\partial r} \tag{4a}$$

$$r\ddot{\theta} = -2\dot{r}\dot{\theta} - \frac{1}{r}\frac{\partial U}{\partial \theta} \tag{4b}$$

An alternative formulation is found on introducing

$$J = r^2\dot{\theta} \tag{5}$$

Equation (4b) can be replaced by equation (5) and

$$\frac{dJ}{dt} = -\frac{\partial U}{\partial \theta} \tag{6}$$

Note that if the potential is axi-symmetric i.e. $\partial U/\partial\theta = 0$, J is a constant of motion, the angular momentum integral. So this is a useful formulation for perturbation theory as for small deviations from axis-symmetry, the righthandside of equation (6) will be small. The dimension of the problem can be reduced again by the use of a method borrowed from celestial mechanics (cf. Verhulst, 1976). Introduce θ as a time-like variable instead of t and the transformation according to Laplace

$$\frac{1}{r} = x$$

The fourth-order system (4 a-b) becomes with these transformations

$$\frac{d^2x}{d\theta^2} + x = -\frac{1}{J^2}\frac{\partial U}{\partial x} + \frac{1}{J^2x^2}\frac{\partial U}{\partial \theta}\frac{dx}{d\theta} \tag{7a}$$

$$\frac{dJ}{d\theta} = -\frac{1}{Jx^2}\frac{\partial U}{\partial \theta} \tag{7b}$$

Supplemented by the equation to be integrated separately

$$\frac{d\theta}{dt} = Jx^2 \tag{7c}$$

Again, equations (7 a-b) can be useful in a perturbation approach for small deviations from axial symmetry.

B. The system is axi-symmetric which we express by

$$\frac{\partial U}{\partial \theta} = 0$$

Note that the angular momentum J (cf. equations 5 and 6) is now a conserved quantity. According to Ollongren (1962) we introduce a reduced potential ϕ of the form

$$\phi = U + \tfrac{1}{2}\frac{J^2}{r^2} \tag{8}$$

System (3) reduces to

$$\ddot{r} = -\frac{\partial \phi}{\partial r}, \qquad \ddot{z} = -\frac{\partial \phi}{\partial z} \tag{9}$$

From now on we shall study model B and we shall specify our results for potential problems which are discrete-symmetric in z, i.e. $\phi(r,z) = \phi(r,-z)$. This is an appropriate choice for these models. Note however that our methods apply to general Hamiltonian systems; also that this natural choice introduces some interesting degenerations which do not occur in more general Hamiltonian systems.

System (9) admits a continuous set of circular orbits ($r = r_0, z=0$) which correspond successively with the origin of phase-space in a shifted coordinate system $r-r_0=x$, $z=z$ (for details see Ollongren, 1962 or Verhulst, 1979). If ϕ is supposed to be analytic with respect to r and z we have locally for the Hamiltonian

$$H= \tfrac{1}{2}(\dot{x}^2+\dot{z}^2+\omega_1^2\, x^2+\omega_2^2\, z^2) + \phi_3(x,\, z^2) + \phi_4(x,\, z^2) + \ldots \qquad (10)$$

in which $\phi_3, \phi_4, \ldots$ are homogeneous polynomials in x and z.

It is important to realise that we have to study the phase-flow corresponding with H for a continuous set of frequencies ω_1 and ω_2. For instance, in modelling spiral galaxies typical numerical values would be given by

$$\tfrac{1}{2} \leqslant \frac{\omega_2}{\omega_1} \leqslant 3$$

Such numerical values however, depend on the mass-distribution of the model (see Martinet and Mayer, 1975).

The following trick has been used often to perform a local analysis of the phase-flow. Rescale $x=\varepsilon\bar{x}$, $z=\varepsilon\bar{z}$ where ε is a small positive parameter. The equations of motion induced by Hamiltonian (10) become after writing out ϕ_3 and ϕ_4 and dropping the bars

$$\ddot{x}+\omega_1^2\, x= \varepsilon(a_1\, x^2+a_2\, z^2) + \varepsilon^2(b_1\, x^3+b_2\, xz^2) + O(\varepsilon^3) \qquad (11a)$$

$$\ddot{z}+\omega_2^2\, z= \varepsilon 2a_2xz+\varepsilon^2(b_2x^2z+b_3z^3) + O(\varepsilon^3) \qquad (11b)$$

It is clear that ε^2 is a measure for the energy of the system (translated again for galaxies: a measure for the energy of the motion around the circular orbits in the rotating system).

System (11 a-b) is in general not integrable. At small values of the energy however invariant manifolds exist which foliate the energy manifold into tori around the stable periodic solutions. In this case (ε small) the phase flow is dominated by these invariant manifolds in an asymptotic sense (the phase-flow between the tori has measure $O(\varepsilon)$) and we can solve the collision-less Boltzmann equation (1).

4. FORMULATION AS A RESONANCE PROBLEM

The ratio of the basic frequencies ω_1 and ω_2 being rational or irrational plays an important part in the theory of normal forms for Hamiltonian systems. This is directly connected with the problems of small denominators which arise in series defined by certain successive canonical transformations; for a summary of the discussion see Moser, 1973.

In the context in which we study these problems for continuous systems we cannot avoid rational or irrational frequency ratios. We shall incorporate the essentially resonant structure of the system corresponding with Hamiltonian (10) by starting with the rationals and then admitting the irrationals by small, detuning perturbations. We put

$$\frac{\omega_2^2}{\omega_1^2} = \frac{n^2}{m^2}\,[1+\delta(\varepsilon)] \tag{12}$$

Here m and n are relative prime natural numbers $\delta(\varepsilon)$ is a continuous function of the small parameter ε, $\delta(\varepsilon)= o(1)$. In some of the examples it will be convenient to replace $1+\delta$ by its inverse in equation (12).

5. NORMALIZATION TECHNIQUES

Several methods have been developed to treat dynamical systems with resonances like system (11 a-b). The main techniques are Birkhoff normalization and averaging procedures. Both involve nonlinear transformations of the system which lead to asymptotic approximations of the phase-flow valid on a time-scale $O(\varepsilon^{-p})$. It turns out that p depends on the order of resonance, i.e. on m and n. This is not always transparant in the literature on Birkhoff normalization (see Siegel and Moser, 1971 or Arnold, 1974) but it has been one of the results obtained in extending averaging techniques; see van der Burgh (1974), Sanders (1978), Sanders and Verhulst (1979).

For these methods and technical details the reader is referred to the literature cited here. We restrict ourselves to three remarks:
a. Birkhoff normalization involves canonical transformations and this

is not necessarily the case in averaging techniques. It turns out however that the averaging techniques which have been employed conserve the Hamiltonian character of the system, at least to the order of approximation considered up till now.

b. It is not clear from the outset that both techniques will produce identical qualitative *and* quantitative results. For two degrees of freedom problems however the results are to a certain significant order of approximation identical; see Verhulst (1979), sections 5 and 8, Sanders and Verhulst (1979), section 6.

c. A fundamental question is the following. The two techniques produce for two degrees of freedom problems as an approximation the solutions of an integrable system (the integrability of the approximating system does not carry through in general for three or more degrees of freedom). The approximations are valid on a certain time-scale, but what can one deduce about the qualitative problem of the existence of these phenomena in the original problem i.e. the periodic solutions and the invariant tori. The answer is that the qualitative picture found for the integrable system persists in the original non-integrable problem but that in the original system there exist an infinite number of additional phenomena of a smaller size or on a longer time-scale so that we failed to find these in the approximations which were introduced. For a discussion of these problems see Sanders and Verhulst (1979), section 7 and Cushman (1979).

6. CONTINUOUS HAMILTONIAN MECHANICS

In this section we shall use the analysis of equations (11 a-b) to construct solutions of the collision-less Boltzmann equation in the time-independent axi-symmetric case. At the heart of the analysis lies the use of stable periodic solutions which have been found for the system (11 a-b).

The following representation of the solutions is useful

$$x(t) = A(t)\cos[mt+\phi(t)],\quad z(t) = B(t)\cos[nt+\Psi(t)]$$

together with the phase-difference

$$\chi = n\phi - m\Psi$$

We shall call periodic solutions with $\chi(t)\equiv 0, 2\pi$ in-phase solutions, periodic solutions with $\chi(t)\equiv\frac{\pi}{2}$, $3\frac{\pi}{2}$ out-phase solutions.

Periodic solutions with either $A(t)\equiv 0$ or $B(t)\equiv 0$ will be called normal modes

In the analysis two parameter-spaces determine the qualitative and quantitative behaviour of the solutions.

At first the frequency numbers m and n determine the k-jet of the Hamiltonian which has to be studied; examples are given below. The parameters of the associated parameter-space are a_1, a_2, b_1, b_2, b_3,...

Secondly, for a given potential $\phi(r,z)$ the location in the first parameter-space is fixed and we have a two-parameter space generated by the detuning δ and the energy E_0 of the system. In a number of resonance cases and for a given potential we shall show how to use these two-parameter families of periodic solutions to construct continuous solutions. In all examples δ has been supposed to vary as $r-r_0$.

a. *The first-order resonance* m=2, n=1.

We use the results of Verhulst (1979), sections 6 and 7. In this case k=3, i.e. the cubic terms of the potential ϕ determine the main features of the topology of phase-space. We take as an example for the Hamiltonian (10)

$$H=\tfrac{1}{2}\dot{x}^2+\tfrac{1}{2}\dot{z}^2+2(1+\delta)x^2+\tfrac{1}{2}z^2-\tfrac{1}{3}x^3-\tfrac{3}{2}xz^2 \tag{13}$$

From the theory of approximations we have $\delta(\varepsilon)=O(\varepsilon)$.

Apart form the normal mode $z\equiv 0$, two stable families of periodic solutions may exist; the existence depends on the detuning δ and the energy E_0.

For the in-phase solutions we have the parametrization

$$x=\frac{\sqrt{E_0}}{3}\ (-d+\sqrt{d^2+3/2})\ \cos 2\tau \tag{14}$$

$$z=\frac{2}{3}\sqrt{E_0}(3-2d^2+2d\ \sqrt{d^2+3/2})^{\frac{1}{2}}\ \cos\tau$$

in which $d=2\delta/3\sqrt{E_0}$.

The solutions exist and are stable if $d>-\sqrt{2}/2$.

Note however that according to the nature of these approximations we assumed $E_0=O(\varepsilon^2)$ so we cannot increase the energy without bounds. A critical points analysis of the energy surface moreover shows that the energy manifold bifurcates at the critical value

$$H_{cr} = \frac{17}{81} + \frac{2}{9}\delta$$

If $E_0 \geqslant H_{cr}$, the energy surface is not compact.

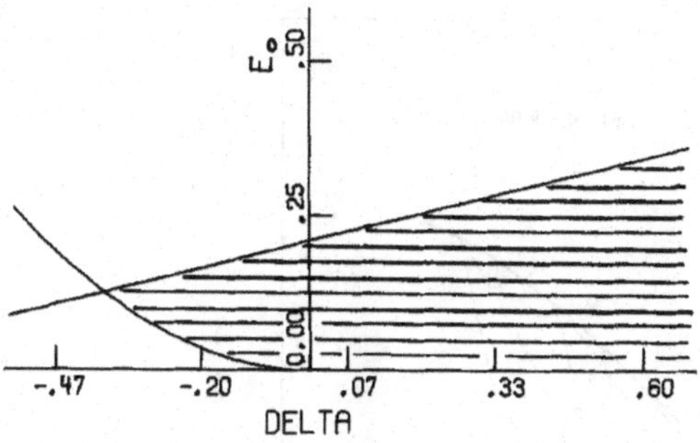

Figure 1

The E_0, δ-parameter space for in-phase periodic solutions of Hamiltonian (13); m:n= 2:1. Each point in the horizontally shaded area corresponds with a stable periodic solution.

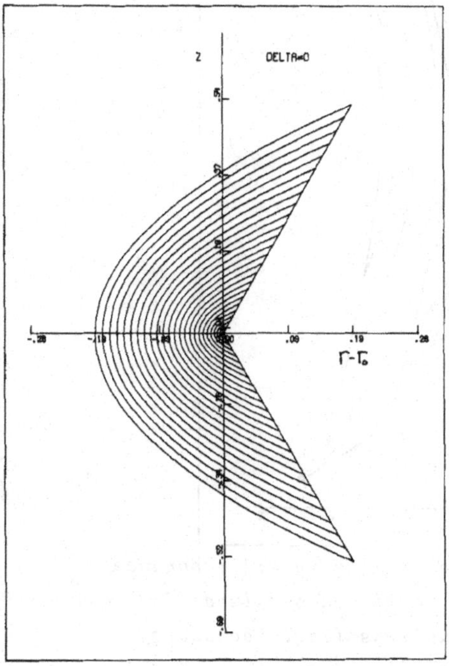

Figure 2

Cross-section of a ring consisting of particles in exact 2:1 resonance with Hamiltonian (13). The orbits were produced with equations (14) where d=0. This cross-section is imbedded in the full solution shown in figure 3.

The existence and stability condition and the compactness condition $0 \leqslant E_0 < H_{cr}$ produce a set in E_0, δ-parameter space which is unbounded (fig. 1). This is unnatural and we have here an example

of the main open problem of this work; cf. section 7.
In figure 2 we show the continous solution in r, z-space at exact resonance $\delta=0$. The actual solution for a galaxy is obtained by rotating this cross-section around the z-axis at r=0.

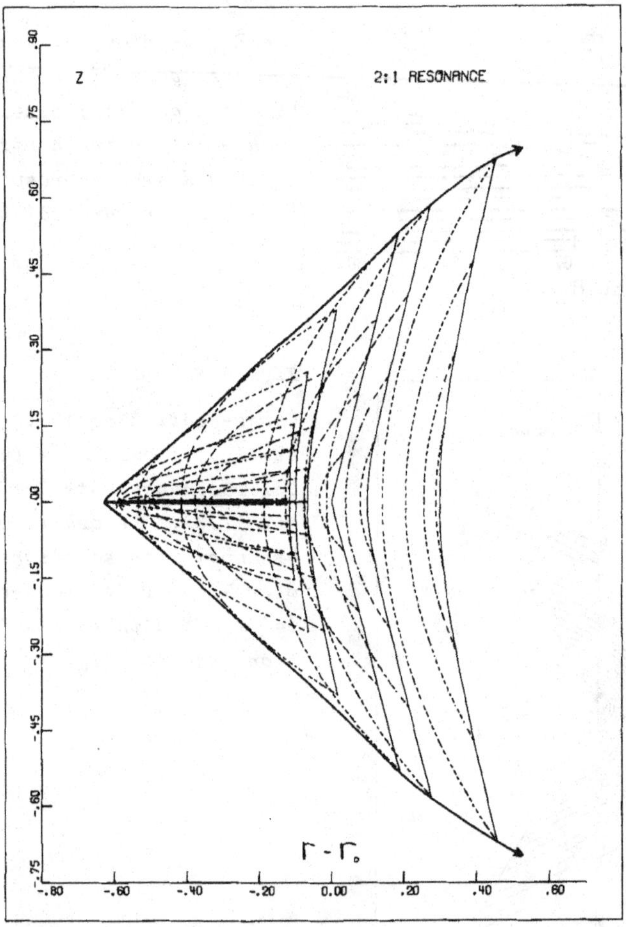

Figure 3

Cross-section of a ring consisting of particles in 2:1 resonance with Hamiltonian (13). For a number of permitted δ-values (figure 1) orbits have been depicted, based on equations (14). Because of the (unrealistic) unboundedness of the E_0, *δ-set the envelope of the orbits should be continued following the arrows. Some envelopes for fixed values of δ have been indicated.*

If we assume that the detuning behaves like $x=r-r_0$, equations (14)

produce together with the existence and stability set in E_0, δ-space figure 3.
The out-phase solutions are parametrized by

$$x = -\frac{\sqrt{E_0}}{3}\left(d+\sqrt{d^2+3/2}\right)\cos 2\tau$$

$$z = \frac{2}{3}\sqrt{E_0}\left(3-2d^2-2d\sqrt{d^2+3/2}\right)^{\frac{1}{2}}\cos\tau \qquad (15)$$

The solutions exist and are stable if $d<\sqrt{2}/2$.

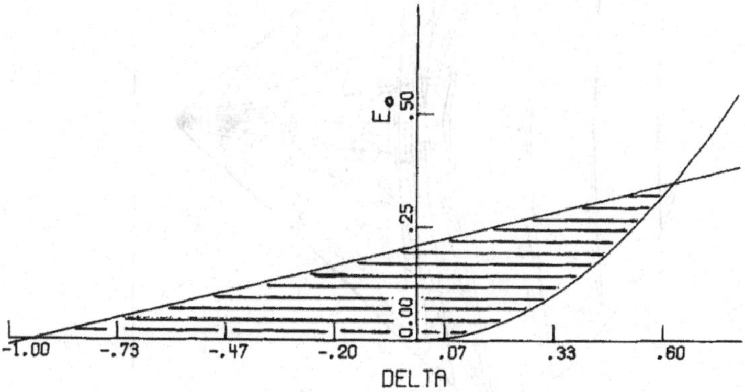

Figure 4

The E_0, δ-parameter space for out-phase periodic solutions of Hamiltonian (13); m:n=2:1.
Each point in the horizontally shaded area corresponds with a stable periodic solution.

This condition and the compactness condition $0\leq E_0<H_{cr}$ provide us with a set in E_0, δ-parameter space corresponding with periodic out-phase solutions; this set is bounded (fig. 4).
Again we use these results to produce a continuous solution in r, z-space (fig. 5).

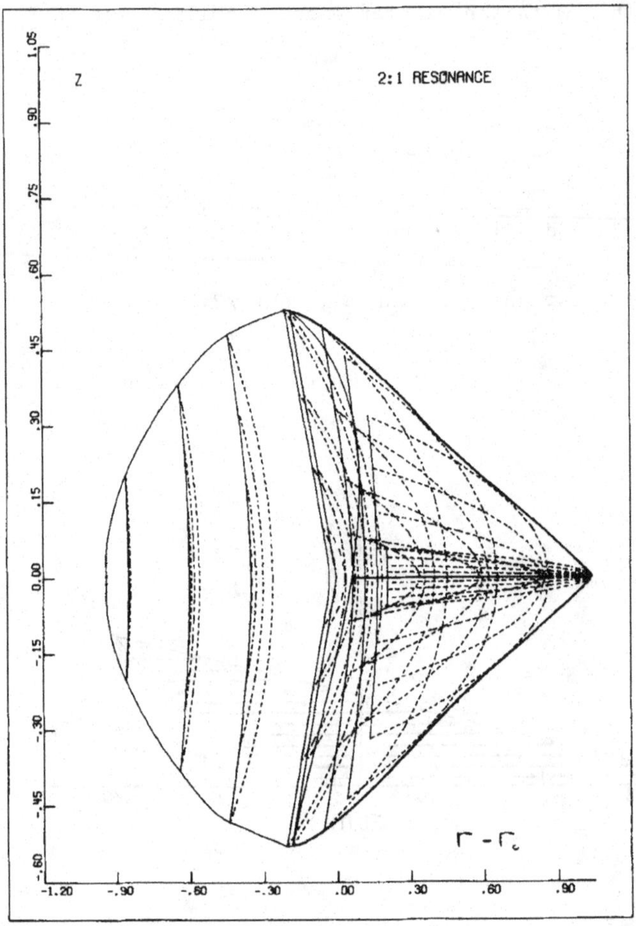

Figure 5

Cross-section of a ring constisting of particles in 2:1 resonance with Hamiltonian (13). For a number of permitted E_0, δ*-values (figure 4) orbits have been depicted based on equations (15). Some envelopes for fixed values of* δ *have been indicated.*

b. *The resonance case m=1, n=2*

In general Hamiltonian mechanics the treatment of this case yields exactly the same qualitative results as the case m=2, n=1 treated above. However, the discrete symmetry in z of the potential ϕ triggers off an interesting degeneration which changes the qualitative and quantitative picture completely.

For reasons of comparison we use a Hamiltonian with the same ϕ_3 as in the preceding case:

$$H= \tfrac{1}{2}\dot{x}^2+\tfrac{1}{2}\dot{z}^2+\tfrac{1}{2}x^2+2(1+\delta)z^2-\tfrac{1}{3}x^3-\tfrac{3}{2}xz^2 \qquad (16)$$

The results obtained here use as a starting point the work of Verhulst (1979), section 11 and Sanders and Verhulst (1979).
In this case k=4 so we should use a 4-jet with non-trivial quartic terms but still the Hamiltonian (16) remains an interesting example. From the theory of approximations we have $\delta(\varepsilon)= O(\varepsilon^2)$.
Because of the discrete symmetry in z the Hamiltonian (16) can be brought in higher order Birkhoff normal form than was expected and we have a so-called higher order resonance.
These types of resonances generally occur only if $m+n\geq 5$. Calculating the differential equation for χ we find

$$\frac{d\chi}{dt} = -\varepsilon^2(\frac{23}{60} A^2+\frac{201}{160} B^2) - \delta(\varepsilon)+O(\varepsilon^3) \qquad (17)$$

The asymptotic approximation of the energy integral is given by

$$2E_0= A^2+4B^2+O(\varepsilon) \qquad (18)$$

So that equation (17) becomes

$$\frac{d\chi}{dt} = -\varepsilon^2 \frac{23}{30} E_0+\varepsilon^2 \frac{133}{480} B^2 - \delta(\varepsilon)+O(\varepsilon^3) \qquad (19)$$

The critical points of this equation will provide us with the existence conditions and the location of the resonance manifold and the periodic solutions.
We find

$$\varepsilon^2B^2= \frac{368}{133} E_0\varepsilon^2+\frac{480}{133} \delta(\varepsilon)+O(\varepsilon^3)$$

With the energy integral we have the condition

$$0\leq B^2\leq \tfrac{1}{2} E_0$$

so that the resonance manifold and its associated periodic solutions exist if

$$-\frac{23}{30} E_0 \leq \delta(\varepsilon) \leq -\frac{201}{320} E_0 \qquad (20)$$

In equation (20) we returned to the original energy of the system E_0 prior to scaling (section 3).
One stable family of in-phase periodic solutions exists in the resonance manifold; the family of out-phase solutions is unstable. The in-phase solutions can be parametrized by

$$x = \left(-\frac{1206}{133} E_0 - \frac{1920}{133} \delta\right)^{\frac{1}{2}} \cos \tau$$

$$z = \left(\frac{368}{133} E_0 + \frac{480}{133} \delta\right)^{\frac{1}{2}} \cos 2\tau \qquad (21)$$

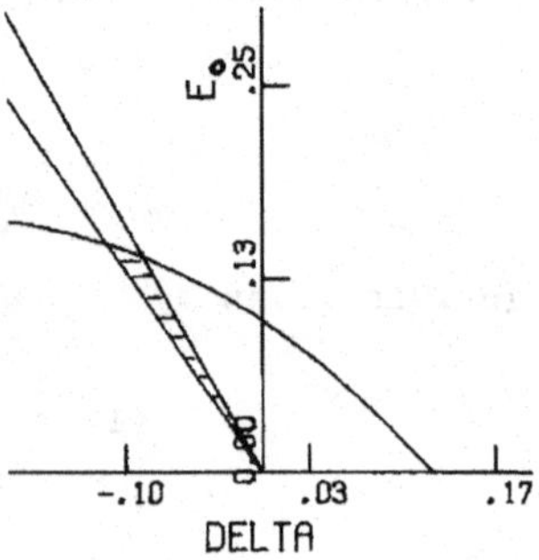

Figure 6

The E_0, δ-parameter space for in-phase periodic solutions of Hamiltonian (16); m:n=1:2.
Each point in the horizontally shaded area corresponds with a stable periodic solution.

Apart from the existence condition (20) we have the compactness condition for the energy manifold. Calculation of the critical points of the Hamiltonian (16) leads to

$$H_{cr} = \frac{8}{81} (1+\delta)^2 (1-8\delta)$$

The compactness condition is $0 \leqslant E_0 < H_{cr}$.
The existence and the compactness condition produce the permitted set of parameter values in E_0, δ-space (figure 6). This set of parameter values generates a continuous solution in r,z-space; see figure 7.

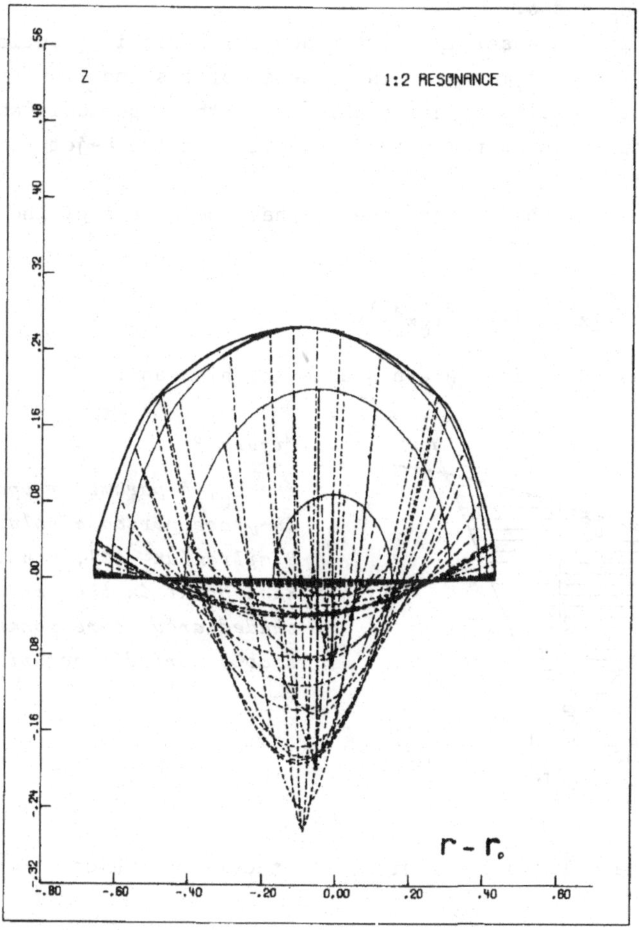

Figure 7

Cross-section of a ring consisting of particles in 1:2 resonance with Hamiltonian (16). For a number of permitted E_0, *δ-values (figure 6) orbits have been depicted based on equations (21). For fixed values of δ some envelopes have been indicated; a change in δ corresponds with a shift of the envelope in the r-coordinate. Note the asymmetry of the ring in z while the potential is symmetric in z.*

c. *The resonance case m=1,n=1*

This is the most complicated case and the toplogy of phase-space associated with Hamiltonian (10) is interesting. A fairly complete analysis can be found in Verhulst (1979), sections 8-10; see also the survey paper by Churchill, Pecelli and Rod (1979).
In this resonance case k=4 so that five parameters play a part

apart from E_0 and δ.
For each point in a set in E_0, δ-space six periodic solutions may exist: 2 normal modes, 2 in-phase solutions and 2 out-phase solutions. The existence and stability of these periodic solutions depend on the 5 parameters $a_1, \ldots, b_3$ in the 4-jet of the Hamiltonian.
As an example we shall study the in-phase solutions of the Hamiltonian

$$H = \tfrac{1}{2}\dot{x}^2 + \tfrac{1}{2}\dot{z}^2 + \tfrac{1}{2}x^2 + \tfrac{1}{2}(1+\delta)z^2 + \tfrac{1}{6}x^3 - \tfrac{1}{3}xz^2 \qquad (22)$$

The out-phase solutions are unstable in this case.

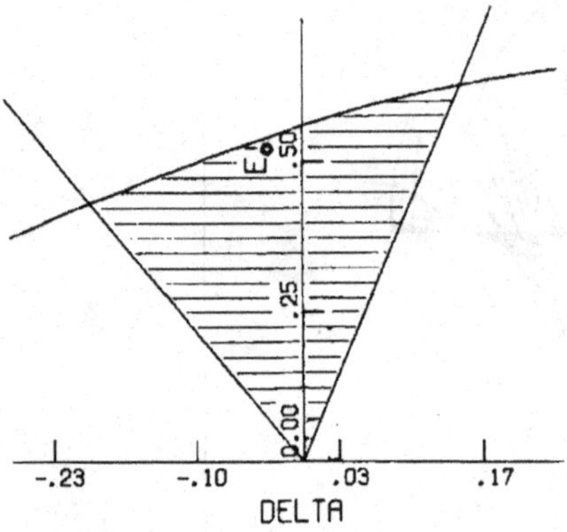

Figure 8

The E_0, δ-parameter space for in-phase periodic solutions of Hamiltonian (22); m:n=1:1. Each point in the horizontally shaded area corresponds with 2 stable periodic solutions.

The in-phase solutions are stable if they exist which is the case if

$$E_0 > \frac{108}{25}\,\delta \text{ and } E_0 > -\frac{54}{25}\,\delta \qquad (23)$$

The compactness condition is obtained again by studying the critical points of the Hamiltonian. From (22) we find $0 \leq E_0 < H_{cr}$ with

$$H_{cr} = \frac{9}{16}(1+\delta)^2(1-\delta)$$

The existence and the compactness condition produce the permitted set of parameter values in E_0, δ-space (figure 8).

For the two in-phase solutions we find the parametrization

$$x = \left(\frac{4}{3} E_0 + \frac{72}{25} \delta\right)^{\frac{1}{2}} \cos \tau \qquad (24)$$

$$z = \pm\left(\frac{2}{3} E_0 - \frac{72}{25} \delta\right)^{\frac{1}{2}} \cos \tau$$

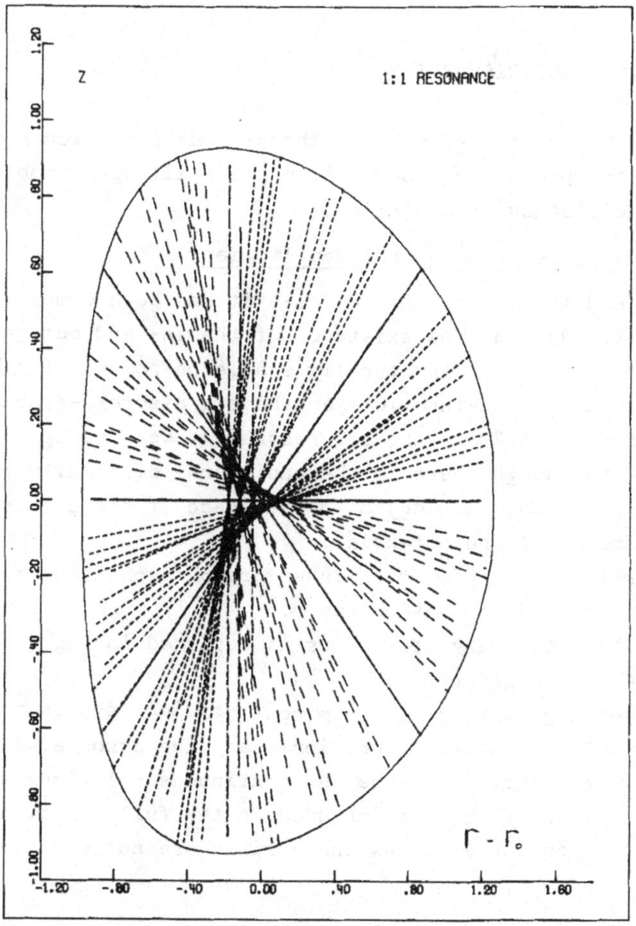

Figure 9

Cross-section of a ring consisting of particles in 1:1 resonance. The two families of orbits at exact resonance δ=0 are indicated by a full line. The two-parameter family of orbits with a plus sign in eqs. (24) has been indicated by ----, the orbits correspon- with a minus sign by — — . For a fixed value of δ≠0 the two families of periodic solutions have an envelope described by 4 triangles.

The projections of the orbits in x,z- or r,z-space are straight lines; for a number of cases these orbits have been depicted in figure 9 together with their envelope. Figure 9 shows a typical cross-section of a ring existing in an axi-symmetric system at the 1:1 resonance.

7. GENERALIZATIONS AND DISCUSSION

The particular examples studied in the preceding section represent generally existing phenomena though there are still open problems. We summarize the discussion as follows:

a. The ring structures in a general Hamiltonian

For the 2:1 and the 1:1 resonance cases it has been demonstrated by Verhulst (1979) that the existence of the in- and out-phase periodic solutions is a structurally stable phenomenon for Hamiltonians which are Morse-functions and have a discrete-symmetric potential. For the 1:2 and the 1:3 resonances we have again structural stability though there is the additional difficulty of a drastic change of the topology of phase-space if one perturbs the discrete symmetry of the potential.

The structural stability of the higher-order resonances, $m+n \geq 5$, will be demonstrated in another paper.

So we claim that the ring structures as depicted in figures 3, 5, 7 and 9 do exist in general.

Whether these rings are actual prominent physical phenomena depends on the relevant distribution function f and the associated existence set in E_0, δ-space. Of course the continuous solutions which we found in our examples, are imbedded in the full solutions of system (1-2) as in our examples these rings are not self-gravitating.

b. An open problem

A fundamental open problem can be posed if the existence set in E_0, δ-space includes values of E_0 and δ which have not been considered in our asymptotic theory, which assumes $E_0 = O(\varepsilon^2)$ and $\delta = O(\varepsilon)$

or $O(\varepsilon^2)$. In our examples this occurs in figures 1, 4 and 8; in figure 6 the parameter set is bounded by small values. This is a difficult problem wich we ignored while using the bifurcation values H_{cr} of the energy manifold. It seems that in general one may expect that on increasing E_0 the stability character of the periodic solutions changes type. The full analysis has been given only in a few isolated examples; see Churchill, Pecelli and Rod , 1979.

The development of a general theory for these questions is highly desirable. This will lead finally, to a morphology of possible envelopes of ring structures.

Ring galaxies

More or less axi-symmetric rings exist in certain galaxies. To model these phenomena a number of authors performed computer experiments describing the penetration of a disk-like galaxy by another concentrated massive galaxy to trigger of rings in a disk; see Theys and Spiegel (1977), Lynds and Toomre (1976). Though the dynamics of these experiments is not compatible with stationary models it is interesting to compare the results because such a massive axially symmetric perturbation can be associated with resonances. The spherical symmetric potential for instance triggers of the 1:1 resonance. So if there is a temporarily large axially symmetric perturbation in a disk-like galaxy the occurrence of transient ring waves should not surprise us.

Acknowlegments

Prof. P.O. Vandervoort (University of Chicago) drew my attention to the astrophysical literature in section 7-c.
The figures in this paper were plotted by Ingrid Birkhoff.

REFERENCES

Arnold, V.I., 1974, The mathematical methods of classical mechanics (in Russian), Moskou; French edition 1976 by ed. Mir , Moskou; English edition 1978 by Springer Verlag, Heidelberg.

Churchill, R.C., Pecelli, G. and Rod, D.L., 1979, A survey of the Hénon-Heiles Hamiltonian with Applications to Related Examples, Volta Memorial Conference, Como 1977, Lecture Notes in Physics, 93, Springer Verlag, Heidelberg.

Cushman, R., 1979, Morse Theory and the Method of Averaging, preprint 119, Mathematisch Instituut, Rijksuniversiteit Utrecht.

Jackson, E.A., Nonlinearity and Irreversibility in Lattice Dynamics, Rocky Mountain J. Math. 8, 127

Jeans, J.H., 1916, Mon. Not. Roy. astr. Soc. 76, 70.

Lynds, R. and Toomre, A., 1976, On the interpretation of Ring Galaxies: the binary ring system II Hz4, Ap.J. 209, 382.

Martinet, L. and Mayer, F., 1975, Galactic orbits and integrals of motion for stars of old galactic populations, Astron. and Astroph. 44, 45.

Moser, J, 1955, Nonexistence of Integrals for Canonical Systems of Differential Equations, Comm. Pure and Appl. Math. 8, 409.

Moser, J., 1973, Stable and random motions in dynamical systems, Princeton Univ. Press., Ann. Math. Studies 77.

Ollongren, A., 1962, Three-dimensional galactic stellar orbits, Bull. Astr. Inst. Neth. 16, 241.

Sanders, J.A., 1978, Are higher order resonances really interesting? Celes. Mech. 16, 421.

Sanders, J.A. and Verhulst, F., 1979, Approximations of higher order resonances with an application to Contopoulos'model problem, in "Asymptotic Analysis, from theory to application", ed. F. Verhulst, Lecture Notes in Math. 711, Springer Verlag, Heidelberg.

Siegel, C.L. and Moser, J.K., 1971, Lectures in Celestial Mechanics, Springer Verlag.

Theys, J.C. and Spiegel, E.A., 1977, Ring galaxies II, Ap. J. 212, 616.

Toomre, A., 1977, Theories of spiral structure, Annual Rev. Astron. and Astroph. 15, 437.

Van der Aa, E. and Sanders, J., 1979, The 1:2:1:-resonance, its periodic orbits and integrals, in "Asymptotic Analysis, from theory to application", ed. F. Verhulst, Lecture Notes in Math. 711, Springer Verlag, Heidelberg.

Van der Burgh, A.H.P., 1974, Studies in the asymptotic theory of nonlinear resonances, thesis Techn. Univ. Delft.

Verhulst, F., 1976, On the theory of averaging, in Long-time predictions in dynamics, eds. V.S. Szebehely and B.D. Tapley, Reidel Publ. Co.

Verhulst, F., 1979, Discrete-symmetric dynamical systems at the main resonances with applications to axi-symmetric galaxies, Phil. Trans. Roy. Soc. London A, 290, 435.